Agricultural Biochemistry

Agricultural Biochemistry

Edited by **Elizabeth Lamb**

SYRAWOOD
PUBLISHING HOUSE

New York

Published by Syrawood Publishing House,
750 Third Avenue, 9th Floor,
New York, NY 10017, USA
www.syrawoodpublishinghouse.com

Agricultural Biochemistry
Edited by Elizabeth Lamb

© 2016 Syrawood Publishing House

International Standard Book Number: 978-1-68286-071-7 (Hardback)

Printed in the United States of America.

Contents

Preface

The purpose of the book is to provide a glimpse into the dynamics and to present opinions and studies of some of the scientists engaged in the development of new ideas in the field from very different standpoints. This book will prove useful to students and researchers owing to its high content quality.

Agricultural biochemistry integrates chemistry and biochemistry and seeks to apply the concepts into agricultural practice. This innovative and comprehensive book combines the well-developed theory and practical applications of agricultural biochemistry through lucid elaborations of selected topics of vital importance such as enzymology, plant biochemistry and genetics, plant physiology, etc. With state-of-the-art inputs by acclaimed experts of this field, this book targets students and professionals alike. Research scholars will also find this book a useful resource material filled with significant topics which can be taken up for research and further study.

At the end, I would like to appreciate all the efforts made by the authors in completing their chapters professionally. I express my deepest gratitude to all of them for contributing to this book by sharing their valuable works. A special thanks to my family and friends for their constant support in this journey.

Editor

Isolation and preliminary characterization of conophor nut (*Tetracarpidium conophorum*) lipase

Victor N. Enujiugha

Department of Food Science and Technology, Federal University of Technology. P.M.B. 704, Akure, Nigeria.
E-mail: venujiugha@yahoo.com.

Lipase in oilseeds helps to hydrolyze the ester bonds of storage triacylglycerols. The crude lipase from the conophor nut (*Tetracarpidium conophorum*) was isolated and assayed via quantification of the free fatty acids liberated by the hydrolysis of the oilseed triacylglycerols. Optimum pH and temperature for the enzyme activity of the conophor nut lipase was pH 8.0 and at 30°C with substantial lipolysis at 80°C, underscoring the thermostability of the enzyme. The effects of different ions on the activity of the isolated lipase were examined. NaCl and EDTA inhibited activity by various degrees, while Ca^{2+} and Hg^+ enhanced the enzyme activity. The results of the present study show that the lipase from conophor nut can favourably be exploited to complement existing lipase sources.

Key words: Conophor nut, lipase activity, temperature, pH.

INTRODUCTION

Lipases (triacylglycerol acyl hydrolases EC 3.1.1.3) are water-soluble enzymes that catalyze the hydrolysis of tri-acylglycerols and a large variety of esters. Oilseed lipases have been discovered to have great commercial prospects as industrial enzymes (Enujiugha et al., 2004). They represent cheap sources of industrial lipases with remarkable potentials for biocatalytic hydrolysis. However, lipase activity is generally absent in ungerminated seeds, and progressively increases during germination. According to Hills et al. (1990) during the growth of oilseed plants, lipases are produced in large amounts to hydrolyze triacylglycerol to fatty acids and glycerol, which supports for the growth of young plants. The natural substrates of lipases are triacylglycerols, having very low solubility in water. Under natural conditions, they catalyze the hydrolysis of ester bonds at the interface between an insoluble substrate phase and the aqueous phase in which the enzyme is dissolved.

The Conophor plant (*T. conophorum* Mull. (Arg) Euphorbiaceae), commonly called the African walnut, is a perennial climbing shrub found in the moist forest zones of Sub-Saharan Africa. It is cultivated principally for the nuts which are cooked and consumed as snacks, along with boiled corn (Enujiugha, 2003). Conophor nut contains 49.2% dry wt. of oil (Enujiugha and Ayodele-Oni, 2003), which is liquid and golden yellow in color, with taste and odor resembling those of linseed oil. The residue after oil expression contains over 50% protein. Gas chromatographic analysis of the seed oil has shown a high level (>66% dry wt.) of the sn-3 fatty acid, linolenic acid (Ogunsua and Adebona, 1983), which is considered essential to the well being, growth and development of children (Innis, 1991). The purpose of this work was to isolate and partially characterize crude lipase from conophor nut and determine its substrate specificity. This is expected to give insight into its potentials for industrial utilization.

MATERIALS AND METHODS

Sources and preparation of materials

The raw conophor nut used in the study was obtained from local farmers at Akure, Ondo state, Nigeria. The nuts were carefully cracked, sliced to about 1-2 cm thickness with a kitchen knife and milled into powder using Waring blender and kept at 4°C prior to analysis. The four oil substrates used for the analysis namely, soybean, cottonseed, groundnut and palm kernel were obtained from the King's market at Akure. All the chemicals and reagents used in the study are of analytical grade and procured from E. Merck AG (Germany).

Crude enzyme preparation

The acetone powder of the conophor nut lipase was prepared according to the method of Hassanien and Mukherjee (1986) with some modifications. 30 g of the seed cotyledon was ground with 30 ml of cold acetone using a Waring blender. The acetone extract

Table 1. Effect of different ions on the activity of lipase from conophor nut on different substrates.

Salt/ion	Soybean oil Activity (%FFA)[b]	Groundnut oil Activity (%FFA)[b]	Palm kernel oil Activity (%FFA)[b]	Cottonseed oil Activity (%FFA)[b]
NaCl	0.07±0.002	0.08±0.008	0.05±0.001	0.05±0.042
CaCl₂	0.33±0.020	0.28±0.035	0.27±0.003	0.26±0.013
HgCl₂	0.27±0.004	0.36±0.020	0.26±0.005	0.24±0.004
EDTA	0.25±0.001	0.23±0.010	0.13±0.001	0.20±0.001
Control	0.27±0.005	0.25±0.003	0.22±0.010	0.20±0.004

[a]Reaction conditions: enzyme-substrate ratio 1:5; 1 ml of 0.01M of salt solutions; phosphate buffer (pH 7) used as control. [b]Values represent means of triplicate determinations (mean ± SME).

was filtered through a cheese cloth (Enujiugha et al., 2004) and washed four times with 20 ml each of cold acetone. The residue was air-dried at room temperature (26°±1°C) to yield the acetone powder which was kept at 4°C.

Assay of enzyme activity

The assay of the lipase activity was carried out using the titrimetric method of Khor et al. (1986) as modified by Enujiugha et al. (2004). The assay mixture contained 5 g of substrate, 2.5 ml of hexane to solubilize the oil, and 1 g of the crude enzyme. The mixture was incubated at 30°C for a period of 1 h with continuous stirring using a magnetic stirrer. Each incubation process was terminated with the addition of 25 ml acetone-ethanol (1:1; v/v) to facilitate the extraction of the free fatty acids liberated. The liberated free fatty acids were quantified by direct titration with 0.01 M NaOH using phenolphthalein as indicator. The lipase activity was expressed as the percentage of free fatty acids liberated after 1 h incubation at 30°C (Wetter, 1957).

Effect of substrate and ions on lipase activity

Four different substrates (soybean oil, cottonseed oil, groundnut oil, and palm kernel oil) were substitutively used at 5 g in the assay mixture with subsequent incubation at room temperature for 1 h at 60% relative humidity (RH). The oils were dried at 50°C for 4 h in an air oven before being used in the assay. Lipase activity was measured for each substrate so as to determine the fatty acids specificity of the lipase from conophor nut.

Approximately 0.01 M solutions of sodium chloride (NaCl), calcium chloride (CaCl2), mercury chloride (HgCl2) and ethylene diamine tetra-acetic acid (EDTA) were prepared. The control solution was prepared using 0.01 M phosphate buffer at pH 7. Using the method of Mukundan et al. (1985), 1 ml of each of the solutions was added to separate assay mixtures and then incubated for 1 h with continuous stirring. After which the lipase activity was quantified as described earlier.

Effect of pH and temperature on enzyme activity

Phosphate buffer (5 ml) at different pH (4-9) was added to 5 g of substrate, 1 g of enzyme preparation and 2.5 ml hexane. The mixture was incubated at 30°C for 1 h with continuous stirring, and the activity determined for each pH. A thermostatic, water-jacketed reaction chamber with shaker was employed to determine the temperature dependence of lipase activity. The assay mixtures were incubated at different temperatures (30 to 80°C) for 1 h and the activities measured.

RESULTS AND DISCUSSION

Rate of lipolysis of different substrates

The present study has highlighted the potential for exploitation of industrial lipase from a common and inexpensive plant source. The results of the lipolysis of soybean, cottonseed, groundnut and palm kernel oils by the lipase in conophor nut are presented in Table 1. In systems without added ions, lipolysis was more pronounced in soybean oil, and groundnut oil (comprising <C20 fatty acids). In a previous work, Enujiugha et al. (2004) reported a pronounced lipolysis in palm kernel and coconut oil with short–chain FFA. It has also been observed by Huang and Moreau (1978) that oilseed lipases are more active on triacylglycerols containing short-chain fatty acids. The above results are in conformity with the observation of Mukundan et al. (1985) that the short–chain fatty acids, owing to their comparatively higher water solubility, will have a smaller inhibitory effect in the lipid phase of the triglyceride emulsion, while the long chain fatty acids, due to their lipophilic nature may cause more inhibition in the lipid phase. The amount of oils available at the interface determines the activity of the lipases. Hexane was added to the assay mixture according to the method of Khor et al. (1986) so as to increase the interfacial area of the activity of the enzyme. Mukherjee (1990) observed that the actual site of lipolysis is at the interface. Also, Enujiugha et al. (2004) reported that the enzymatic activity of a lipase is related to the interfacial area of the water-insoluble substrate.

Effects of ions on lipolysis

The effects of ions on the activities of lipase from conophor nut were found to be variable (Table 1). Sodium chloride was observed to significantly reduce the activity in all the substrates which it was exposed to, although Mukundan et al. (1985) observed that the chloride ion did not cause any inhibitory effect but the metal ions did. This means that the Na$^+$ could therefore be main causative effect for the decline in activity. The Ca^{2+} and Hg^{2+} enhanced the activity considerably. The observed calcium effect is in agreement with the observation of

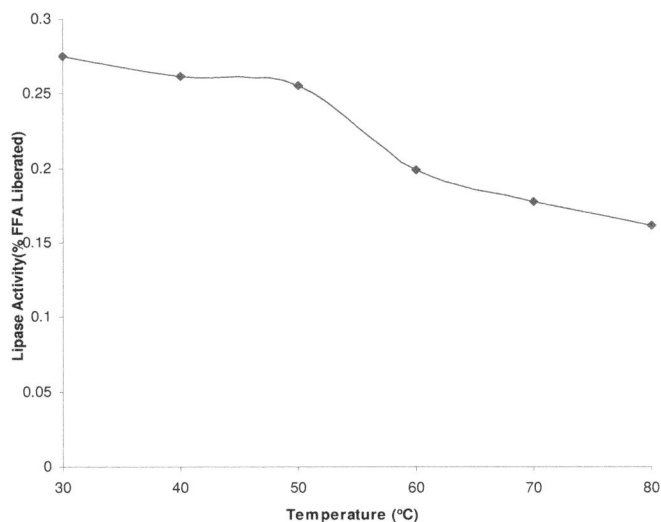

Figure 1. Effects of temperature on conophor nut lipase activity using soybean oil in hexane (enzyme-substrate ratio1:5; pH 8.0; 1 h incubation period).

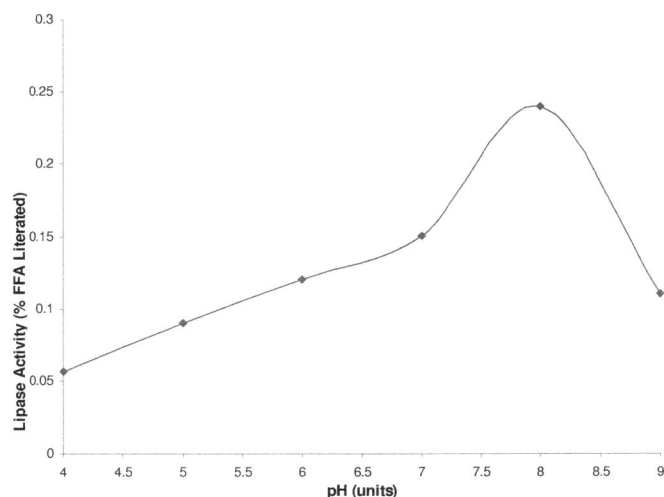

Figure 2. Effect of pH on the activity of conophor nut lipase at 30°C using soybean oil in hexane (enzyme-substrate ratio1:5; 1 h incubation period).

Enujiugha et al. (2004) that calcium ion in the reaction mixture brought about 64% increases in activity. Abigor et al. (2002) and Haas et al. (1992) also observed an increase in lipase activity by calcium ion inclusion. The inhibition of activity by EDTA was more pronounced in the palm kernel oil. The inhibition of the activity could be attributed to its chelating process of the system and thereby disrupting the formation of the enzyme substrate complex. This invariably affects the formation of the end product (Enujiugha et al., 2004). A slight difference was observed in the effect of Hg^{2+} on the enzyme activity. The Hg^{2+} slightly favored lipolysis in the oils, which is in contrast with the observation of Sanders and Pattee (1975) that 10^{-4} M solution of Hg^{2+} brought about complete inhibition of lipase activity in peanut. The inhibition of lipase activity by mercury chloride is an indication of the presence of sulfhydryl group in the enzyme molecule (Mukundan et al., 1985). Although the Hg^{2+} did not inhibit lipolysis in this study, there was indication that it could inhibit lipolysis in cottonseed oil as there was a gradual drop in activity relative to other substrates.

Effects of temperature and pH on lipase activity

Figures 1 and 2 show the respective effects of temperature and pH on the activity of the lipase from conophor nut (*T. conophorum*). Abigor et al. (1985) observed that purified oil palm lipase has been shown to have optimal activity at 30°C, above which there was a steady decline. It was observed in the present study that there was gradual decline in the activity of the lipase from conophor nut with successive increases in temperature, from 30 to 80°C. However, the enzyme was fairly active at higher temperatures. The basis for sustained activity at relatively

higher temperatures is still unclear, but it might be linked to the fact that the lipases are highly hydrophobic. Enujiugha et al. (2004) reported that there was a decline in activity above 30°C, though there was a substantial lipolysis, even at 80°C, indicating a fairly high thermo-stability of the enzyme. The present study agrees with this observation using lipase obtained from *T. conophorum*.

A comparison of activity profile of the lipase at different pH is shown on Figure 2. The lipase was found to have highest lipolysis between pH 7.0 and 8.0. Past studies indicated different pH optima for lipases from different plant sources. The pH optimum for *Pentaclethra* lipase is near neutrality (Enujiugha et al., 2004). Ory et al. (1962) found out that some other lipases are optimally active at acid pH (e.g. pH 5.0 for ungerminated castor bean lipase); while oil palm mesocarp lipase has pH 4.2 (Abigor et al., 1985). The results of the present study revealed a peak at pH 8.0, followed by a slight drop at pH 9.0. The advantage of alkaline pH range lies in the absence of corrosion problems associated with acidic environments in industrial processes.

Conclusion

An alkaline lipase (optimum pH 8.0) with a fair thermo-activity was isolated from the cotyledons of raw conophor nut (*T. conophorum*) and the substrate specificity examined. The conophor nut lipase could prove useful in industrial biocatalytic hydrolysis. It could also be inferred from the present preliminary characterization that the conophor nut lipase could prove useful in processes that require lower cooling costs and minimal corrosion problems. Lipolysis was more pronounced in soybean and groundnut oils. Ca^{2+} and Hg^{2+} enhanced the enzyme activity, while Na^+ and EDTA caused various degrees of inhibition. The results show that the conophor nut lipase could be exploited in industrial processes.

REFERENCES

Abigor RD, Opute FI, Opoku AR, Osagie AU (1985). Partial purification and some properties of the lipase present in oil palm (*Elaeis guinensis*) mesocarp. J. Sci. Food Agric. 36: 599-606.

Abigor RD, Uadia PO, Foglia TA, Haas MJ, Scott K, Savary BJ (2002). Partial purification and properties of lipase from germinating seeds of *Jatropha curcas* L. J. Am. Oil Chem. Soc. 79:1123-1126.

Enujiugha VN (2003). Chemical and Functional characteristics of conophor nut; Pak. J. Nutr. 2(6): 335-338.

Enujiugha VN, Ayodele-Oni O (2003). Evaluation of nutrients and some antinutrients in lesser known under-underutilised oilseeds. Int. J. Food Sci. Technol. 38: 525-528.

Enujiugha VN, Thani FA, Sanni TM, Abigor RD (2004). Lipase activity in dormant seeds of the African oil bean (*Pentaclethra macrophylla* Benth). Food Chem. 88: 405-410.

Haas MJ, Cichowicz DJ, Bailey DG(1992). Purification and characterization of an extracellular lipase from the fungus *Rhizopus delemar*. Lipids 27: 571-576.

Hassanien FR, Mukherjee KD (1986). Isolation of lipase from germinating oilseeds for biotechnological processes. J. Am. Oil Chem. Soc. 63(7): 893-897.

Hills MJ, Kiewett I, Mukherjee KD (1990). Esterification reactions catalyzed by immobilized lipase from oil seed rape (*Brassica napus* L.). Biochem. Soc. 17: 478

Huang AHC, Moreau RA (1978). Lipases in the storage tissues of peanut and other oilseeds during germination. Planta 14: 111-116.

Innis SM (1991). Essential fatty acids in growth and development. Prog. Lipid Res. 30: 39-103.

Khor HT, Tan NH, Chua CL (1986). Lipase-catalyzed hydrolysis of palm oil. J. Am. Oil Chem. Soc. 63(4): 538-540.

Mukherjee KD (1990). Lipase-catalyzed reactions for modification of fats and other lipids. Biocatal. 3: 277-293.

Mukundan MK, Gopakumar K, Nair MR (1985). Purification of a lipase from the hepatopancreas of oil sardine (*Sardinella longiceps* Linnaeus) and its characteristics and properties. J. Sci. Food Agric. 36: 191-203.

Ogunsua AO, Adebona MB (1983). Chemical composition of *Tetracarpidium conophorum* (conophor nut). Food Chem. 10: 173-177

Ory RI, St. Angelo AJ, Altschul AM (1962). The acid Lipase of the castor bean: properties and substrates specificity. J. Lipid Res. 3:99-105.

Sanders TH, Pattee HE (1975). Peanut alkaline lipase. Lipids 10(1):50-54.

Wetter LR (1957). Some properties of the lipase present in germinating rapeseed. J. Am. Oil Chem. Soc. 34: 66-70.

Chemical composition and biological activities of essential oil of *Plectranthus tenuicaulis* (Hook f.) leaves from Gabon

Bikanga Raphaël[1]*, Obame-Engonga Louis-Clément[1], Agnaniet Huguette[1], Makani Thomas[1], Anguile Jean Jacques[1], Lebibi Jacques[1], and Menut Chantal[2]

[1]Laboratory of Natural Substances and Organometallic synthesis, University of Sciences and Technology of Masuku, P. O. Box 901 Franceville, Gabon.
[2]Team "Glyco and nanovecteurs for therapeutic targeting", Institute of biomolecules Mousseron, Faculty of Pharmacy, 15 avenue Charles Flahault, 34093 Montpellier, France.

Essential oils were obtained by hydro distillation of dried leaves from two kinds of *Plectranthus tenuicaulis* (Hook f.) J. k. Morton grown in Gabon and were analyzed using GC-FID, GC-MS, ^{1}H and ^{13}C NMR. The main constituent was *(R), (E)*-6,7-epoxyocimene [*(E)*-myroxide] (72.2 to 83.8%) with small amounts of *(Z)*-6,7-epoxyocimene (0.6 to 1.3%), *(E)* and *(Z)*-β-ocimene (1.1 to 2.4% and 0.6 to 1.7%). The oils were examined for their antioxidant, antimicrobial and antibacterial activities. The antioxidant activity was tested by using an *in vitro* radical scavenging activity test and β-carotene-linoleic acid assays. The essential oils showed antioxidant and DPPH radical scavenging activities, and they displayed the inhibition of lipid peroxidation. The oils were less effective than butylated hydroxytoluene (BHT). The antimicrobial activity of oils was studied by means of agar disc diffusion and broth microdilution methods. The essential oils of *P. tenuicaulis* showed antibacterial activity against microorganisms tested; however, they were unable to inhibit the growth of *Streptococcus pyogenes* and fungal strains. The reference bacteria were the most sensitive to the essential oils. We also observed a significant anticandidal effect.

Key words: *Plectranthus tenuicaulis*, essential oil, GC-MS, antimicrobial activity; antioxidant activity.

INTRODUCTION

Plectranthus tenuicaulis (Hook f.) J. K. Morton [syn. *Coleus tenuicaulis* Hook. f; *Plectranthus minimus* (Gürke)], is one of 14 species of the genus *Plectranthus* found in the flora of West Africa (Hutchinston et al., 1963). The plant is a small erect, branched, slender pubescent herb 26 to 36 inches high with dominant purple evergreen.

The aim of this study was to characterize the volatile extract of the species *P. tenuicaulis* which is widely spread in Gabon and traditionally used by local populations. During plant collection, it was noted that an infusion of the leaves is used to treat indigestion, sleep disorders and also by young mothers to promote lactation.

Despite the many papers that have been published on the chemistry of *Plectranthus* (Abdel-Mogib et al., 2002) before our works (Agnaniet et al., 2011; Agrebi et al., 2012), to our knowledge, there are no previous reports on either the biological or chemical activities of the volatile (or non-volatile) compounds of *P. tenuicaulis* (Hook).

In this paper, we report the chemical composition, along

*Corresponding author. E-mail: brbikanga@hotmail.fr.

Abbreviations: **AME,** Antifungal mixture elios; **BHT,** butylated hydroxytoluene; **DPPH,** 1,1-diphenyl-2-picrylhydrazil; **MBC,** minimum bactericidinal concentration; **MFC,** minimum fungicidal concentration; **MIC,** minimum inhibitory concentration.

with the antimicrobial and antioxidant activities of the essential oil obtained by hydro distillation of dried leaves of *P. tenuicaulis* collected in Gabon.

MATERIALS AND METHODS

P. tenuicaulis leaves were collected in Mopia (South-eastern Gabon) and analyzed using Gas Chomatography (GC-FID), Gas Chromatography-Mass Spectrometry (GC-MS), ^{1}H and ^{13}C NMR (Agnaniet et al., 2011; Agrebi et al., 2012).

Microbial strains

The reference strains used were: *Bacillus cereus* LMG 13569, *Enterococcus faecalis* CIP 103907, *Escherichia coli* CIP 11609, *Listeria innocua* LMG 1135668, *Salmonella enterica* CIP 105150, *Shigella dysenteria* CIP 5451, *Staphylococcus aureus* ATCC 9244, *Proteus mirabolis* 104588 CIP, *S. aureus* ATCC 25293 BHI and *Staphylococcus camorum* LMG 13567 BHI. The clinical strains were *Pseudomonas aeruginosa*, *S. aureus*, *E. faecalis*, *Streptococcus pyogenes* and *Candida albicans* vaginal were isolated in the Laboratory of Medical Biology (Burkina-Faso).

Microbiological methods

The agar disc diffusion method was employed for the screening of antimicrobial activities of the essential oils (NCCLS, 1997). The test was carried out in sterile Petri dishes (90 mm in diameter) containing solid and sterile Mueller-Hinton agar (Becton, Dickinson, USA). The oil absorbed on sterile paper discs (5 µl by Whatman disc of 6 mm diameter), were placed on the surface of the media previously inoculated with microbial suspension 0.1 ml (1 µg per Petri dish). One filter paper disc was placed per Petri dish in order to avoid a possible additive activity, exposed via phase steam, of the components from more than one disc. Each dish was sealed, to prevent evaporation, with a film laboratory and incubated aerobically at 30°C (Gram-negative) or 37°C (Gram-positive) depending on the strain for 24 h, followed by the measurement of the zone diameter of inhibition expressed in mm. Tetracycline (BIO-RAD, Marnes-la coquette, France) and ticarcilline (BIO-RAD, Marnes, la coquette, France) were used as antibiotic reference products. Flucanozole (100 µg) and cetoconazole (100 µg) were used as antifungal reference for the vaginal strain. Antifungal Mixture Elios (AME): Rosemary essential oil, geranium essential oil and lavender essential oil were used as the antifungal compound. All tests were conducted in triplicate.

A broth micro dilution method was used to determine the Minimum Inhibitory Concentration (MIC) and the Minimum Bactericidal Concentration (MBC) (Agnaniet al., 2009). All tests were performed in Mueller-Hinton broth (Becton Dickinson, USA). A serial doubling dilution of each essential oil was prepared in 96 wells plates over the range of 0.03 to 8% (v/v). The broth was supplemented by tween 80 (Merck, Germany), at a concentration of 0.1%, to enhance the solubility of the essential oils.

Overnight cultures of each broth strain were prepared in Nutrient Broth (Diagnosis Pasteur, France) and the final concentration in each well was adjusted to 5×10^5 CFU/ml after inoculation. Each inoculum concentration was confirmed by the viable count on Plate Count Agar (Merck, Germany). Positive and negative growth controls were included in each test. The tray was incubated aerobically at 30°C (reference Gram-negative strain) or 37°C (reference and isolated Gram-positive) and MICs were determined. The MICs defined as the lowest concentration of essential oil in which the microorganism tested does not show visible growth. To determine the MBCs, a suspension of 10 µl was taken from each well and inoculated in Mueller-Hinton Agar (Becton Dickinson, USA) during 24 h at 30 or 37°C. The MBCs are defined as the lowest concentration of essential oil (Michel-Briand et al., 1986). Minimum inhibitory concentration (MIC) and minimum fungicidal concentration (MFC) were determined for *C. albicans* according to the same protocol.

Antioxidant activity

1,1-Diphenyl-2-picrylhydrazyl (DPPH) assay

The hydrogen atoms or electron-donating ability of corresponding extracts and butylated hydroxytoluene (BHT) was determined by the laundering of purple-colored DPPH methanol solution. This spectrophotometric assay uses the stable radical DPPH as reagent (Cuendet, 1997; Burits, 2000). Briefly, 0.5 mM DPPH radical solution in methanol was prepared, and then 1 ml of this solution was mixed with 3 ml of the solution of the sample in a final concentration of ethanol. After incubation for 30 min in the dark, the absorbance was measured at 517 nm. Decreasing the absorbance of the DPPH solution indicates an increase in DPPH radical scavenging activity. This activity is given as percentage DPPH radical scavenging, which is calculated with the equation:

% DPPH radical cleaning = [(sample absorbance - control absorbance)] / control absorbance]

DPPH radical scavenging activity was an average of the three repeated measurements.

β-carotene-linoleic acid assay

In this trial, antioxidant capacity was determined by the extent of the inhibition of volatile organic compounds and the formation of the conjugated hydroperoxide diene from linoleic acid oxidation (Dapkevicius, 1998). A solution of β-carotene-linoleic acid mixture was prepared as follows: 0.5 mg of β-carotene dissolved in 1 ml of chloroform (HPLC grade); 25 µl of linoleic acid and 200 mg of tween 80 were added as emulsifier because β-carotene is not soluble in water. Chloroform was completely evaporated using a vacuum evaporator. Then, 100 ml of distilled water saturated with oxygen was added with vigorous agitation at a rate of 100 ml/min for 30 min; 2500 µl of the reaction mixture was then dispersed in test tubes and different concentrations of essential oil were added. The emulsion system was incubated up to 48 h at room temperature. The same procedure was repeated with a positive control BHT and a blank. After this period of incubation, the absorbance of the mixture was measured at 490 nm. Antioxidant capacity of the extracts was compared with those of the BHT and the blank. Tests were carried out in triplicate.

RESULTS

The hydro distillation of young green clear and light green mature stem purple leaves of *P. tenuicaulis* (*Coleus tenuicaulis*) (Hook) of Gabon produced essential oil with 1.3 and 1.7% (w/w), respectively. The major constituents identified by GC and GC-MS analyses are given in Table 1.

The essential oils were subjected to screening for possible antioxidant activity by two complementary test systems to DPPH free radical scavenging and β-carotene/linoleic acid systems.

Table 1. Percentage composition of *Plecthrantus tenuicaulis* essential oils obtained by hydrodistillation.

RI	Compound	Composition (%)	
		Green leaves	Purple leaves
970	octen-3-ol + sabinene	0.5	0.4
986	myrcene	0.2	0.3
1031	(Z)- β-ocimene	1.1	2.0
1036	(E)- β-ocimene	0.6	0.9
1085	terpinolene	0.4	0.1
1089	6,7-epoxiocimene	0.1	0.4
1090	linalol	0.2	0.4
1094	ipsenol	0.5	0.3
1121	(Z)-6,7-epoxiocimene	1.3	1.3
1141	(E)-6,7-epoxiocimene	83.8	76.0
1160	Rose furane epoxide	0.3	0.2
1172	borneol	0.2	0.2
1182	3-(Z)-hexenyl butanoate	1.3	1.8
1187	M=154	1.0	0.9
1199	(E)-2,6-dimethylocta-3,5,7-trien-2-ol	0.1	0.2
1301	(Z)-2,6-dimethylocta-5,7-dien-2,3-diol	0.5	0.4
1306	(E)-2,6-dimethylocta-5,7-dien-2,3-diol	0.1	0.2
1359	eugenol		0.1
1372	α-copaene	0.1	0.1
1388	β-elemene	0.1	0.2
1432	β-caryophyllene	2.6	4.9
1445	(Z)-β-farnesene	1.1	2.2
1464	α-humulene	0.2	0.4
1485	γ-muurolene		0.1
1490	germacrene D	1.3	2.2
1495	γ-amorphene	0.1	0.4
1497	valencene	0.1	0.1
1518	γ-cadinene		0.1
1523	δ-cadinene		0.2
1530	Selina-3,7-11-diene	0.2	0.1
1604	Caryophyllene oxide	0.1	0.1
1668	intermediol	0.4	0.6
Total		98.5	97.8

BHT was used as positive controls in two test systems. The value of $IC_{[50]}$ essential oil was 18.35 mg/ml. The capacity of essential oil for inhibiting the lipid peroxidation was evaluated by β carotene bleaching test (Figure 1).

Table 2 presents the average inhibition zones. Antimicrobial activity of essential oils from the leaves of *P. tenuicaulis* was first evaluated against 17 strains of bacteria and 4 fungi of different origins by the agar disc diffusion method. Diameter of the inhibition zone of essential oil varies from 8 to 35 mm and 6 to 12 mm respectively for the bacteria and fungi. Other strains were sensitive between 8 to 14 mm: *S. pyogenes*, (14 mm), *E. coli* CIP 11602 (12 mm), and *S. dysenteria* CIP 5451 (11 mm).

The MICs, MFCS and MBCs of *P. tenuicaulis* for tested microorganisms were determined by a broth microdilution method (Table 3). The MIC was equivalent to the MBC, indicating a bactericidal action of oils.

DISCUSSION

The DPPH radical scavenging activity of *P. tenuicaulis* essential oil was high but relatively lower than that of BHT (Figure 2). The capacity of essential oil for inhibiting the lipid peroxidation evaluated by β-carotene bleaching test showed that lipid peroxidation is effectively inhibited by essential oil, RAA value = 94% (30 μg/ml).

The best antioxidant activity of the essential oils was obtained on β carotene/linoleic acid systems. The

Table 2. Antimicrobial activities of *Coleus tenuicaulis* essential oils: Screening by inhibition zone diameters (mm).

Reference strain	Origin	Leaves essential oil	Te[b]	Ti[b]
Bacillus cereus LMG13569	LMG	35	18	50
Enterococcus faecalis CIP103907	CIP	25	19	30
Escherichia coli CIP NCTC11602	CIP	12	22	8
Listeria innocua LMG1135668	LMG	35	14	50
Salmonella enterica CIP105150	CIP	35	16	50
Shigella dysenteria CIP5451	CIP	11	21	31
Staphylococcus aureus ATCC9244	ATCC	30	17	48
Staphylococcus camorum LMG13567	LMG	20	20	16
Proteus mirabilis CIP 104588	CIP	8	15	15
Hospital strain				
Enterococcus faecalis	Foecal	25	20	28
Pseudomonas aeruginosa	Vaginal liquid	11	21	19
Staphylococcus aureus	Vaginal liquid	12	21	27
Streptococcus pyogenes	Vaginal liquid	13	20	24
Fungal strain			Fluc	Gris
Candida albicans ATCC10231	ATCC	10	10	15
Candida albicans ATCC90028	ATCC	12	13	10
Candida albicans (n = 2)	Uro-vaginal liquid	8	13	11

Clinical strains	HE	(E)-ep	AME	HER	HEG	L
Pseudomonas aeruginosa	10	8	8/10	10/15	10	<6
Staphylococcus aureus	10	10	20	15	40	35
Escherichia coli	8	12	11	17	20	16
Aspergillus niger	<6	<6	18/21	<6	60	14/15

a) Tested at a concentration of 5 µl/disc. HE: *Coleus tenuicaulis* essential oils, (E)-ep: (E)-6,7-epoxiocimene; b) Te: Tetracycline , Ti : Ticarcilline; Fluc.: Fluconazole; Gris: Griseofulvine, AME: Antifungal Mixture Elios, HER: Rosemary, HEG: Géranium Bourdon, L : Lavander.

Table 3. Antibacterial (MIC, MBC) and antifungal parameters (MIC, MFC) of *Coleus tenuicaulis* essential oils.

Reference strain	Origin	MIC (%)	MBC (%)
Bacillus cereus LMG13569	LMG	1	1
Enterococcus faecalis CIP103907	CIP	0.5	0.5
Escherichia coli CIP 11602	CIP	>8	ND
Listeria innocua LMG1135668	LMG	1	1
Proteus mirabolis 104588 CIP	CIP	>8	ND
Salmonella enterica CIP105150	CIP	0.5	0.5
Shigella dysenteria CIP5451	CIP	4	8
Staphylococcus aureus ATCC9244	ATCC	0.5	0.5
Staphylococcus camorum LMG13567	LMG	1	1
Hospital strain			
Enterococcus faecalis	Foecal	1	1
Pseudomonas aeruginosa	Vaginal liquid	>8	ND
Staphylococcus aureus	Vaginal liquid	>8	ND
Streptococcus pyogenes	Vaginal liquid	1	1
Fungal strain		MIC	MIF
Candida albicans ATCC10231	ATCC	>8	ND
Candida albicans ATCC90028	ATCC	8	>8
Candida albicans (n = 2)	Uro-vaginal liquid	>8	ND

ND: not determined, values given as percentage.

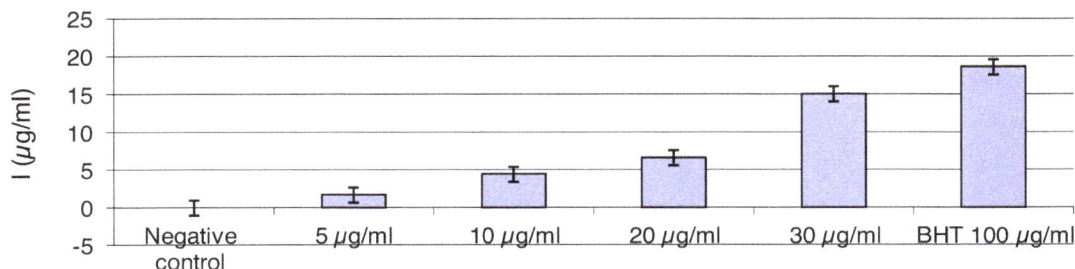

Figure 1. DPPH radical scavenging activity of *Coleus tenuicaulis* essential oils. $Y = 13.68 \log C - 8.3411$, $R^2 = 0.9018$; $IC_{[50]} = 18.35$ mg/ml. $I = \%$DPPH radical scavenging = [(control absorbance - sample absorbance)]/control absorbance] × 100.

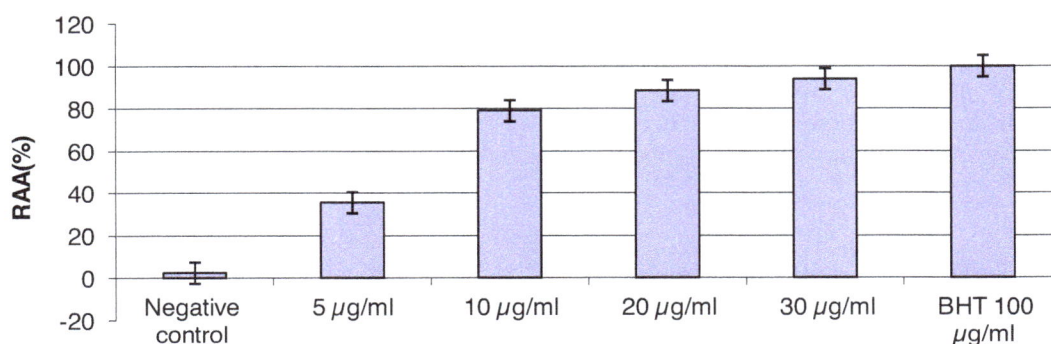

Figure 2. Antioxidant activity by ß-carotene bleaching test of *P. tenuicaulis* essential oils. RAA: Relative Antioxidant Activity = (sample absorbance/BHT absorbance) × 100. BHT: 2,6-di-tert-butyl-4-methylphenol or 2,6-Bis (1,1-dimethylethyl)-4-methylphenol; RAA = 94% (30 µg/ml).

sensitivity to the essential oil was obtained on *P. mirabolis* 104588 PAC (8 mm) and *E. coli* (8 mm) for clinical strains. The different reference strains were less sensitive to the essential oil than tetracycline, and less sensitive to the essential oils than tircacilline.

The essential oils exhibited more activity on *B. cereus* LMG 13569 BHI, *S. enterica* CIP 105150, *E. faecalis* CIP 103907, *E. coli*, *L. innocua* LMG 1135668 BHI, *S. enterica* CIP 105150, *S. aureus* ATCC 9244 and *S. camorum* LMG 13567 BHI for reference strains and *E. faecalis*, *S. aureus* than tetracycline. The essential oils presented less antifungal activity with an inhibition zone diameter varying from 8 mm for hospital strain candidal, 8 to 12 mm for *C. albicans* ATCC 10231 and *C. albicans* ATCC 90028 and 6 mm for *Aspergillus niger*. The growth of the fungal specie was weak inhibited by *P. tenuicaulis* oil, reference origin *C. albicans* was more sensible to the essential oil than clinical strains of *C. albicans*. It is also interesting to note that the inhibition effect of fluconazole and griseofluvin against *C. albicans* were higher than *P. tenuicaulis* oil. Furthermore, under the experimental conditions, the efficiency was higher for the essential oil than for *(E)*-6,7-epoxyocimene. Two commercial antifungal compounds (fluconazole and griseofluvin),

AME, Rosemary essential oil, geranium essential oil and lavender essential oil were more active than the *P. tenuicaulis* essential oil in the case of the clinical strain.

P. tenuicaulis essential oil failed to inhibit *S. enterica* CIP 105150 and *S. aureus* ATCC 9244 (MIC = 0.5%). *P. tenuicaulis* essential oil showed antimicrobial activity (MIC = 1%) on *B. cereus* LMG 13569, *L. innocua* LMG 1135668, *S. camorum* LMG 13567 BHI and *S. pyogenes* (MIC = 1%) for clinical strains.

The result of MBC and MFC demonstrated bactericidal and fungicidal effect. *P. tenuicaulis* essential oil were bactericidal for *B. cereus* LMG 13569 BHI, *S. enterica* CIP 105150, *E. faecalis* CIP 103907, *E. coli* CIP, *L. innocua* LMG 1135668 BHI, *S. enterica* CIP 105150, *S. aureus* ATCC 9244, *S. camorum* LMG 13567 BHI for the reference strains and *E. faecalis*. *P. tenuicaulis* essential oils showed bacteriostatic activity for *S. dysenteria* CIP 5451. The most resistant strains are *E. coli* and *P. aeruginosa*, the latter is known to be resistant to several essential oils (Pattnaik et al., 1995; Agnaniet et al., 2009). This confirms the first screening by the agar diffusion method and is also in agreement with previous reports on the resistance of *P. aeruginosa* (Sonboli, 2005) to several essential oils. Comparison between our data and anticandidal previously reported data (Leeja et al., 2007) show similar results. Fungal strains were less sensitive to the essential oils of *P. tenuicaulis* leaves.

Conclusion

The present work shows high *in vitro* antimicrobial activity and antioxidant activity of the essential oil of *P. tenuicaulis* leaves. It was bactericidal and low fungicidal for most of the reference and some clinical strains. It is more effective against reference bacteria than clinical bacteria. The essential oils also exhibit an antioxidant activity.

These results indicate that the essential oils from the leaves of *P. tenuicaulis* could be used as a natural antimicrobial agent for infectious diseases and food conservation. Furthermore, the development of natural antimicrobial agents will help to reduce the negative effects (pollution of the environment, resistance), of synthetic chemicals and drugs. The interesting antimicrobial effects support the use of this plant, especially by the local population which has use for good traditional medicine. Nevertheless, it is important to take into account the chemical and biological variability observed, depending on the part of the plant used to make the oil. Further investigation on different sites and periods of harvest are needed to confirm this observation in order to optimize its traditional use.

REFERENCES

Abdel-Mogib M, Albar HA, Batterjee SM (2002). Chemistry of the genus *Plectranthus*. Molecules 7(2):271-301.

Agnaniet H, Agrebi A, Bikanga R, Makani T, Lebibi J, Casabianca H, Morere A, Menut C (2011). Essential Oils of *Plectranthus tenuicaulis* Leaves from Gabon, Source of *(R), (E),* 6-7-epoxyocimene. An unusual Chemical Composition within the Genus *Plectranthus*. Natl. Prod. Commun. 6(3):409-416.

Agnaniet H, Makani T, Bikanga R, Obame LC, Lebibi J, Menut C (2009). Chemical composition and antimicrobial activity of the Essential Oils of leaves, roots and bark of *Glossocalyx staudtii*. NPC 4(8):1127-1132.

Agrebi A, Agnaniet H, Bikanga R, Makani T, Anguile JJ, Lebibi J, Casabianca H, Morere A, Menut C (2012). Essential Oils of *Plectranthus tenuicaulis* for flavour and fragrance: synthesis of derivatives from natural and synthetic 6,7-epoxyocimenes. Flavour Fragr. J. 27(2):188-195.

Burits M, Bucar F (2000). Antioxidant activity of *Nigella sativa* essential oil. Phytother. Res. 14:323-328.

Cuendet M, Hostettmann K, Potterat O, Dyatmyko W (1997). Iridoid glucosides with free radical scavenging properties from *Fagraea blumei*. Helv. Chim. Acta. 80:1144-1152.

Dapkevicius A, Venskutonis R, Van Beek TAk, Linssen JPH (1998). Antioxidant activity of extracts obtained by different isolation procedures from some aromatic herbs grown in Lithuania. J. Sci. Food Agric. 77:140-146.

Hutchinston J, Dalziel JM (1963). Flora of West Tropical Africa. 2nd Ed., Hepper FN, London. 1:458-460.

Leeja L, Umesh BT, Thoppil JE (2007). Antimicrobial screening of essential oil of *Coleus malabaricus* Benth. Var. mollis (Benth) Hook. F. Int. J. Essent. Oil Therapeut. 1:49-50.

Michel-Briand Y (1986). Mécanismes moléculaires de l'action des antibiotiques. Collection de Biologie moléculaire. p. 370.

NCCLS (National Committee for Clinical Laboratory Standards) (1997). Performance Standards for Antimicrobial Disk Susceptibility Test, sixth ed. Approved standard. M2-A6, Wayne, PA.

Pattnaik S, Subramanyam VR (1995). Antibacterial activity of essential oils from *Cymbopogon*: Inter- and intra-specific differences. Microbios 84(341):239-245.

Sonboli A, Saleli P, Kanani MR, Ebrahimi SN (2005). γ-Terpinene, p-cymene antibacterial and antioxidant activities. Zeitschrift für Naturforschung, 60c:534-538.

3

Immunopurification of a rape *(Brassica napus* L.) seedling lipase

H. Belguith[1]*, S. Fattouch[2], T. Jridi[1] and J. Ben Hamida[1]

[1]Department of Biology, Faculty of Science, Bizerte, Tunisia.
[2]Biological Engineering Laboratory, INSAT, Tunis, Tunisia.

Lipase or triacylglycerol acylhydrolase (E.C.3.1.1.3) was purified to homogeneity from rapeseed-germinated cotyledons (*Brassica napus* L.). The purification scheme involved homogenization, centrifugation, ultracentrifugation and affinity chromatography using polyclonal antibodies raised against porcine pancreatic lipase. The purified rapeseed lipase was homogenous and did not contain contaminating proteins detectable by SDS-PAGE and HPLC analysis. The specific activity of the purified preparation was increased about 1950 times, with an overall yield of 35%. The rapeseed lipase was found to be a cytosoluble, glycosylated and heat-labile serine-hydrolase. It was monomeric with a molecular mass of 38 kDa and a pI of 6.6. The purification method used in the present work is rapid, simple and yields highly purified lipase. It may therefore be applicable in the purification of other uncharacterized plant lipases.

Key words: *Brassica napus* L., immuno-affinity, lipase, purification, triacylglycerol acyl hydrolase.

INTRODUCTION

Interest in lipases (triacyl glycerol acylhydrolases: E.C.3.1.1.3) from micro-organisms, animals and plant sources has markedly increased in the last decade owing to their novel and multifold applications in industry, oleochemistry, organic synthesis, detergent formulation, nutrition and medicine. Lipases are unique in catalyzing the hydrolysis of fats into fatty acids and glycerols at the water-lipid interface and reversing the reaction in non-aqueous media. The comparison of the properties of these proteins provided valuable information on their evolutionary relationships. In addition, their structure and their function analysis may help us to understand the fundamental mechanisms in this family (Kanaya et al., 1998). Recently, the alignment of animal, bacterial and fungal lipase sequences suggested the presence of sequence homologies including a significant conserved region, Gly-X-Ser-X-Gly as the catalytic moiety (Kanaya

et al., 1998; Saxena et al., 2003; Osterlund et al., 1996). This consensus sequence, found in the substrate binding site, contains a serine residue suspected to be essential for binding to lipid substrates.

Lipases from a large number of bacterial (Gotz, 1991; Tyski et al., 1983; Vicente et al., 1990; Aires-Barros and Cabral, 1991), yeast (Veeraragavan and Gibbs, 1989), fungal (Suzuki et al., 1986; Kundu et al., 1987; Sugihara et al., 1988; Isobe and Nokihara, 1991) and animal sources (Kanaya et al., 1998; Osterlund et al., 1996; Cambillau and Bourne, 1991; Cheng et al., 1985; Carriere et al., 1994) have been purified to homogeneity, but not from plants at our knowledge. Compared to bacterial, fungal and animal lipases, little is known about plant lipases. This lack of information about plant lipases could be due to their complicated and laborious purifycation (Weselake and Jain, 1992). The difficulties involved in carrying out research on plant lipases, particularly from oil seeds, have been essentially attributed to their low abundance, their instability and the loss of activity using the traditional purification strategies. These procedures were generally based on clarification, precipitation, differential and density gradient centrifugations followed by sequential combinations of different chromatographic steps resulting in a low final yield (Ben-Hamida and Mazliak, 1985). Many different methods have been

---

*Corresponding author. E-mail: hatem.belguith@fsb.rnu.tn.

Abbreviations: BHT; Butylhydroxytoluene **DTT;** Dithiotriethol, **EC-lipase;** Euphorbia characias lipase, **IRS-lipase;** Immunopurified rapeseed lipase, **HPL;** Human pancreatic lipase, **PPL;** Porcine Pancreatic Lipase, **TAGs;** Triacylglyceols.

applied to purify plant lipases without reaching homogeneity (Huang, 1993).

To overcome these difficulties, we have tried to apply a new technique essentially based on an immunopurification step to prepare the rapeseed lipase. Antibodies immobilized onto support matrices have been already used to immunopurify a wide variety of biological compounds such as bacterial proteins (Sjoberg and Holmgren, 1973), enzymes (Melchers and Messer, 1970), hormones (Murphy et al., 1973), viral proteins (Diaco et al., 1986) and virus particles (Kenyon et al., 1973).

We report here the immunopurification and characterrization of a rapeseed lipase by affinity chromatography using an immunosort prepared by covalently coupling polyclonal antibodies raised against the commercial porcine pancreatic lipase (anti-PPL) to activated CNBr-Sepharose 4B. In a previous work, we found that these anti-PPL antibodies cross-react with the rapeseed lipase (Belguith et al., 2001).

MATERIALS AND METHODS

Plant material

Rapeseed (*Brassica napus* L.), cultivated locally, were soaked for 24 h, then, surface sterilized with 5% (v/v) $CaCl_2O_2$ in distilled water for 5 min. Sterilized seeds were allowed to germinate on filter paper moistened with distilled water in darkness at 26°C.

Homogenization and fractionation

The rape (*Brassica napus* L.) seedling lipase was prepared as described in a previous work [26]. Briefly, cotyledons (20 to 30 g) from 3-days old seedlings were excised from the hypocotyls and homogenized using a pestle and a mortar in grinding buffer containing 0.15 M Tris, pH 7.5, 0.6 M sucrose, 1 mM DTT, 1 mM benzamidine and 10 mM KCl at a ratio of 1:5 (w/v). The resulting homogenate was filtered through a Miracloth layer and fractionated by centrifugation at 10,000 g for 30 min into a yellow fatty layer (removed carefully with a spatula), a supernatant solution (S10) and a pellet. The S10 fraction was centrifuged at 100,000 g for 3 h. The resulting supernatant (S100) was stored at -20°C. All the homogenization and centrifugation steps were performed at 4°C.

Lipase activity assay

Colorimetric method

Lipase activity was determined by measuring free fatty acids produced by triacylglycerol (TAGs) hydrolysis using a colorimetric method (Duncomb, 1962). Experiments were carried out in a Teflon screw-top glass test tube, in a total volume of 1 ml. The reaction mixture contained 50 mM triolein emulsified in 5% (w/v) arabic gum, 1 M Tris-HCl pH 7.5 (Lin and Huang, 1983). Reactions were started by addition of 50 µl enzymatic solution and allowed to proceed for 20 min in a shaking bath at 30°C. Appropriate controls were included and reactions were stopped with 5 ml of cold chloroform. Fatty acids released were extracted and converted to copper soap using 0.1% (w/v) sodium diethyl dithiocarbamate. The copper complex was subsequently

estimated spectrophotometrically at 440 nm.

Radioactive method

In radiometric assays, lipase activity was measured using TAGs containing radiolabelled acyl chains (243 cpm/µl) as a substrate (Beisson et al., 1999). 20 µl of the rapeseed lipase preparation was incubated at pH 7.5 in the presence of 10 µl of radiolabeled triolein (22 Ci/mM) (Perkin-Elmer), 1% sodium taurodeoxycholate (NaTDC) and 7.6 µl of 4 M $CaCl_2$, in a final volume of 200 µl. After each 15 min, 50 µl of the reaction mixture were added to 1 ml of the stopping buffer. Then, a mixture of methanol/chloroform/heptane (21/18/15; v/v/v) was added to extract the radiolabeled free fatty acids. After a centrifugation at 13,000 g for 2 min, 200 µl of the aqueous upper phase were taken and mixed with 8 ml of scintillation liquid (Hionic FluorTM Packard BioScience B.V.). The radioactivity of the tritium resulting of the hydrolysis of radiolabeled TAGs was counted on a Beckman LS 1801 apparatus.

Fluorimetric method

Fluorescent TAGs were extracted from the seed kernels of *Parinari glaberrimum*. 5 mg of the crude lipidic extract was dissolved in 1 ml of diethylether containing 0.01% (w/v) butylhydroxytoluene (BHT) as antioxidant. The TAGs were isolated by preparative TLC under an argon atmosphere and the purity was checked by TLC. Purified TAGs were stored in an ethanol solution in the presence of 0.01% (w/v) BHT (stock solution) and stored in the dark at −20 °C under an argon atmosphere.

After evaporating ethanol under a nitrogen stream, 3 mg of purified TAGs was placed in a 0.5-ml polypropylene microtube, 100 µl of the following buffer was added: 50 mM Tris-HC1 (pH 8) containing 3% gum arabic, 4 mM NaTDC, 100 mM NaC1, 6 mM $CaCl_2$, 0.001% (w/v) BHT. The microtube was kept closed under argon and the mixture was sonicated for 30 s in a sonicating bath (35 kHz and 30 W) (Beisson et al., 1999).

The incubation buffer (990 µl) and the TAGs stock solution (10 µl) were added consecutively to a quartz cuvette of 1.5 ml (optic path-length 1 cm) containing a magnetic stirring bar 8 mm. The final TAGs concentration was 18 µg/ml. The above mixture is slightly turbid and requires continuous stirring to ensure homogeneity. The cuvette was kept under nitrogen using a teflon cap. After gently shaking the cuvette, it was left to equilibrate at 25°C.

Lipase solution was injected and the fluorescence due to the free fatty acids released was read at regular intervals under continuous stirring in a spectrofluorimeter (SFM 25 from Kontron). Excitation was at 324 nm and emission at 420 nm (Beisson et al., 1999).

The following standard incubation buffer (pH 8) was used: 4 mM NaTDC, 100 mM NaCl, 6 mM CaC1$_2$, 0.001% (w/v) BHT.

Electrophoresis

SDS-PAGE (Laemmli, 1970) was carried out on a discontinuous 6% stacking and 12% resolving polyacrylamide gel with Tris-Glycine running buffer and staining with Coomassie Brilliant Blue R-250 or silver nitrate. Standards (Sigma, St. Louis, MO, USA): lactalbumin (14.4 kDa), trypsin inhibitor (20.1 kDa), carbonic anhydrase (29 kDa), albumin (45 kDa), bovine serum albumin (66 kDa), phosphorylase (97 kDa). The native-PAGE was done in the same gel and buffer conditions without SDS. Total proteins were quantified assayed according to the Bradford method (Bradford, 1976) using bovine serum albumin (Sigma) as standard.

HPLC analysis

Reversed-phase HPLC (C8 column Ultra sphere octyl 5 μm; 25 x 0.46 cm), using a linear gradient of acetonitrile in 0.1% (v/v) aqueous trifluoroacetic acid, was used to analyze the active immunopurified fraction.

Polyclonal antibody production

500 μl (0.5 mg protein) of porcine pancreatic lipase type VI-S (Sigma, St. Louis, MO) mixed with an equal volume of complete Freund adjuvant was injected into rabbits. 2 further injections were given fortnight intervals with the same amount of the immunogen emulsified in incomplete Freund adjuvant. 2 weeks after the booster injection, sera samples were collected in order to evaluate the immune response of the rabbits, using ring test and ELISA as described later. Another injection was carried out after 5 weeks and the animals were bled 2 weeks after that last injection.

Direct-binding ELISA test

Enzyme-linked Immunosorbent assay was performed using the method of Kang et al. (1988). For all ELISA procedures the following buffers were used. Coating buffer: PBS. Wash medium: PBS-Tween-20 (0.5 g/L). Saturating buffer: PBS-Tween-20 containing bovine serum albumin (BSA; 5%). Substrate solution: O-phenyllene-diamine (0.4 g/L) (Sigma, St. Louis, MO) dissolved in 0.05 M sodium phosphate/citrate buffer, pH 5, containing hydrogen peroxide. Stop solution: 3 M HCl.

Immunoprecipitation

To immunoprecipitate lipases (IRS-lipase and HPL), 20 μl of lipase was added to 20 μl of anti- EC-A lipase immune serum at a 1:500 dilution. The total mixture were incubated at 37 °C for 1 h, immunoprecipitate was centrifuged at 10,000 x g for 10 min and residual lipase activity remaining in the supernatant was determined by the fluorimetric and radiolabeled method.

Immunoaffinity chromatography column

A CNBr-Sepharose 4B (Pharmacia Biotech AB) pre-activated gel was used to prepare the immunoaffinity column. 20 mg of the purified IgG anti-PPL was dissolved in the coupling buffer 0.1 M Na HCO3 pH 8.3 containing 0.5 M NaCl.

The anti-PPL antibodies were first coupled to the CNBr-Sepharose 4B pre-activated gel according to the manufacturer's instructions and all free antibodies were removed by washing and BSA (3%) was used to block unreacted gel groups. To control the efficiency of the adsorbance of antibodies to the gel, we measured the absorbance at 280 nm of the eluted fraction. Proteins of the S100 fraction (10 ml) concentrated by acetone precipitation were taken up in binding buffer (0.1 M NaHCO₃, pH 8.2), applied to the immunoaffinity column equilibrated with the same buffer and allowed to react overnight at 4 °C with the immunoadsorbent to optimize the fixation of the rapeseed lipase by the antibodies Anti-PPL. The column was washed with binding buffer (4 x 5 mL), to remove unbound proteins.

To elute bound proteins, the column was washed with glycine buffer 0.25 M pH 2.2 and in order to make the pH close to the neutrality, each eluted volume was supplemented with 50 μl of Tris-Hcl 1 M (pH 8). The eluted fractions were submitted to a lipase activity test and a protein quantification test (measurement of the absorbance at 280 nm).

RESULTS

Enzyme purification

Rapeseed lipase activity was found in the soluble fraction of 3-day old seedlings. After ultracentrifugation at 100,000 x g approximately 70% of rapeseed lipase activity was remained in the supernatant S100 (Belguith et al., 1999). In the present work we used an immunoaffinity chromatography to purify rapeseed lipase from the soluble fraction S100.

The S100 proteins concentrate was subjected to immunoaffinity chromatography on immobilized polyclonal antibodies against PPL, which had been demonstrated in a previous work to cross react with the rapeseed lipase (Belguith et al., 2001).

Figure 1 shows the elution profile of rapeseed lipase with 0.2 M glycin buffer, pH 2.2, from the antibodies Sepharose column. Protein quantification and lipase activity test were performed for each fraction, only in the fractions 4, 5, 6 and 7 a lipase activity was detected.

The overall purification protocol is summarized in the Table 1. The most striking aspect of the procedure is that the immunopurification step yielded relatively little protein, but with a specific activity of 91.83 nkat.mg-1. The enzyme preparation eluted from the immunoaffinity column which corresponded to a yield of about 35% and a 1953-fold enrichment of lipase activity of the original fraction (Table 1), was homogeneous as proved by SDS-PAGE analysis. The polypeptide eluted from the immunoaffinity column migrated as a single peptide protein species upon SDS-PAGE at approximately 39 kDa (Figure 2). This result suggests that the IRS-lipase is a monomeric protein with a molecular weigh about (39 kDa) similar to the Euphorbia characias lipase (Moulin et al., 1994).

In order to evaluate the purity of the immunopuri-fied rapeseed lipase (IRS-lipase), the active immuno-purifiedfraction was analyzed by reverse phase HPLC. A single and symmetric peak was eluted at about 50% acetonitrile and eluted at 19.72 min (Figure 3). The HPLC profile confirms that IRS-lipase was homogeneous. Our results demonstrate the efficiency of the immunoaffinity chromatography to purify plant lipases. The immunoaffinity column allowed us to purify the immunoaffinity column allowed immunoaffinity column allowed us to purify the rapeseed lipase with a high purification factor.

This type of strategy may be useful in the purification of low abundance plant proteins such as lipases.

Biochemical characteristics of cytosolic rapeseed lipase

In a first step, different tests were used to evaluate and to confirm that the IRS-lipase is a true triacylglycerol acylhydrolase (E.C.3.1.1.3) of germination. In a second step we have characterized the IRS-lipase. All experiments were carried out in triplicate.

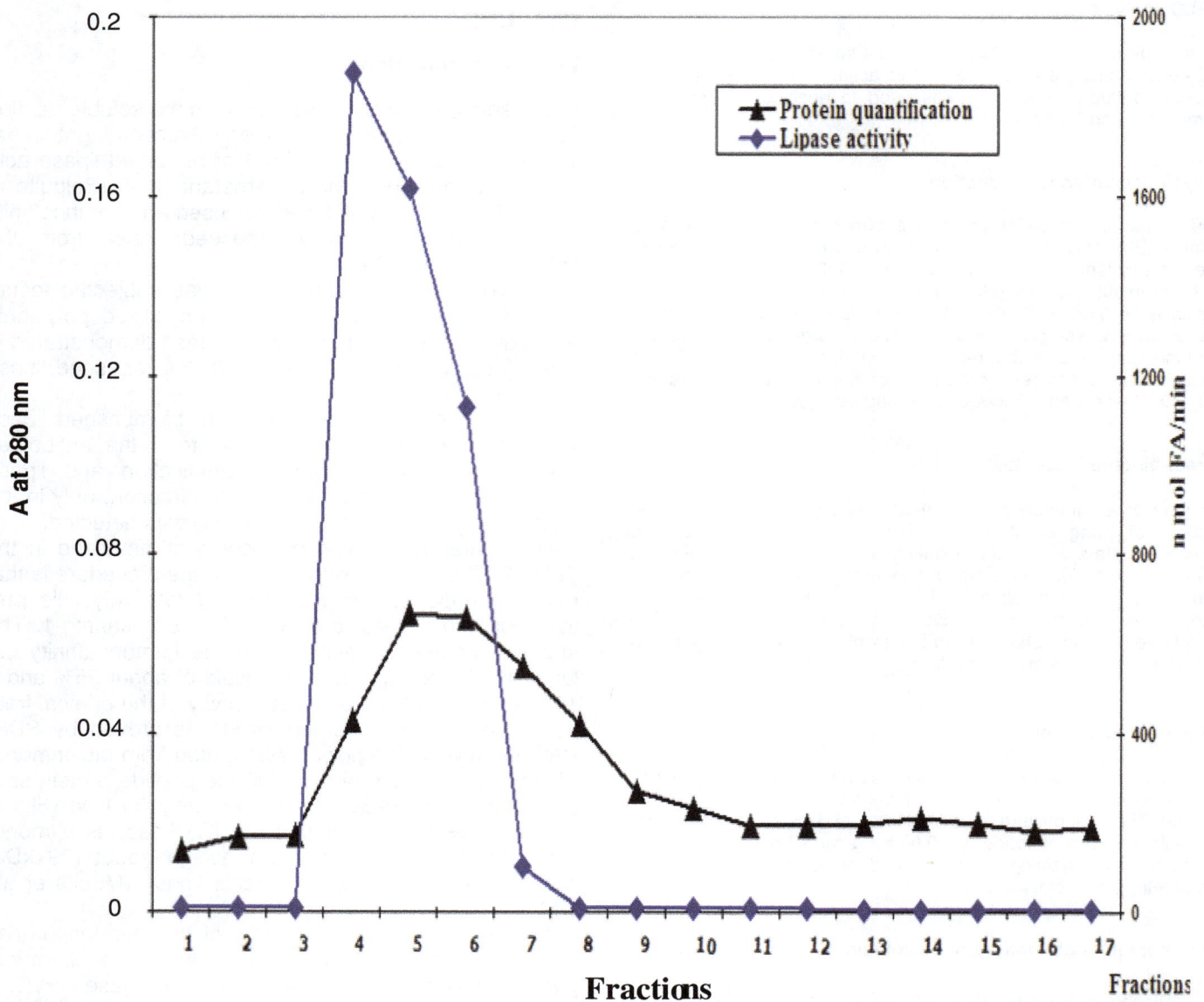

Figure 1. Elution profile from affinity chromatography on CNBr-activated sepharose-4B column. Enzyme preparation (S100) was applied onto the column with stepwise elution with 0.2 M glycin buffer, pH 2.2. Protein quantification and lipase activity test were performed for each fraction.

Table 1. Effect of treatments on separation of seeds from wild pomegranate and their chemical characteristics.

| Treatments | Time taken for separation (min) | Labour saving (%) | | Titratable reducing ascorbic acid | | | |
		Excluding treatment time	Including treatment time	pH	Acidity (%)	Sugar (%)	(mg/100 g)
CT	13.6 (±5.4)	–	–	6.63 (±0.39)	2.8 (±0.20)	7.2 (±0.2)	19.20 (±0.65)
SR1	8.3 (±3.7)	39.0	3.40	6.16 (±0.16)	3.0 (±0.20)	7.8 (±0.4)	17.70 (±0.71)
SR2	5.8 (±2.6)	57.0	6.02	6.13 (±0.11)	3.1 (±0.23)	7.9 (±0.4)	17.20 (±0.38)
HW1	3.4 (±2.0)	75.0	60.0	6.03 (±0.10)	3.1 (±0.21)	7.8 (±0.2)	19.60 (±0.60)
HW2	2.5 (±1.7)	80.0	74.0	6.10 (±0.13)	3.0 (±0.10)	7.8 (±0.1)	19.47 (±0.55)
WS1	10.0 (±3.0)	–	–	6.70 (±0.16)	2.9 (±0.15)	7.2 (±0.3)	19.30 (±0.75)
WS2	12.1 (±4.1)	–	–	6.80 (±0.20)	2.8 (±0.20)	7.1 (±0.1)	19.56 (±0.45)

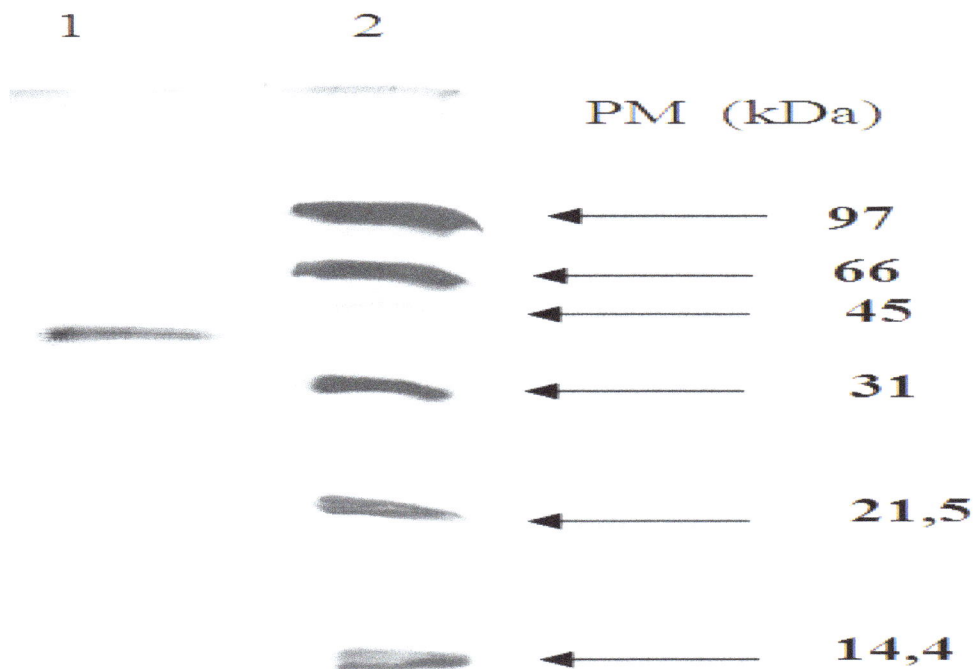

Figure 2. SDS-PAGE analysis of the immunopurified rapeseed lipase on a 12% gel, stained with silver nitrate. Lane 1: 12 µg of IRS-lipase, lane 2: molecular mass markers.

Lipase activity tests

In order to confirm that the IRS-lipase is a true triacylglycerol acylhydrolase (E.C.3.1.1.3) of germination, we tested the lipase activity using 2 different methods.

Results obtained using radiolabeled TAGs

Lipase activity was measured using TAGs containing radiolabelled acyl chains (243 cpm/µl) as a substrate. We quantify the cpm resulting of the hydrolysis of radiolabeled TAGs. Figure 4a clearly shows that the IRS-lipase presents a high activity level with a specific activity of about 94 nkat.mg^{-1}.

We observe a linear kinetic for 15 min, after that it reaches a plateau. No activity was observed in the heat-treated IRS-lipase for 5 min at 90°C used as a control, these results also suggest that the IRS-lipase is a heat-labile enzyme.

Result obtained using fluorimetric method

Lipase activity was measured using a *Parinari glaberrium* TAGs as a substrate, the fluorescence was recorded every 10 min for 3 h. Figure 4b shows that the increase of relative fluorescence was linear with time during 120 min of incubation, with a specific activity of about 92 nkat.mg^{-1}, after that it reaches a plateau. These results are identical of the obtained ones using the radiolabeled method.

Esterase activity test

This test was performed to investigate if the IRS-lipase presents or not an esterase activity such as some described partially purified plant lipases. Figure 5 shows the result of the separation by native-PAGE of the IRS-lipase and the crude 3-days germinating cotyledons extract, submitted after that to a specific revelation using the α-naphtol acetate as a substrate. The obtained electrophoregramme revealed three esterase isoforms in the crude extract (lane 2 and 3) as found in the sunflower germinating crude extract (Ben Hamida, 1986) and in the excised sunflower cotyledons crude extract. No esterase activity was detected in the IRS-lipase fraction (lane1). This result indicates that the IRS-lipase did not exhibit any esterase activity.

The different methods used to test lipase activity confirm that the immunopurified rapeseed lipase is a true lipase of germination and not an esterase.

Some structure and molecular properties of IRS-lipase

Effect of antibodies anti-*Euphorbia characias* lipase on rapeseed lipase activity

In order to study the effect of the polyclonal anti-*Euphorbia characias* (anti-EC-lipase) antibodies (made by the laboratory of LLE-CNRS Marseille-France) on the IRS- lipase activity, we measured the residual lipase acti

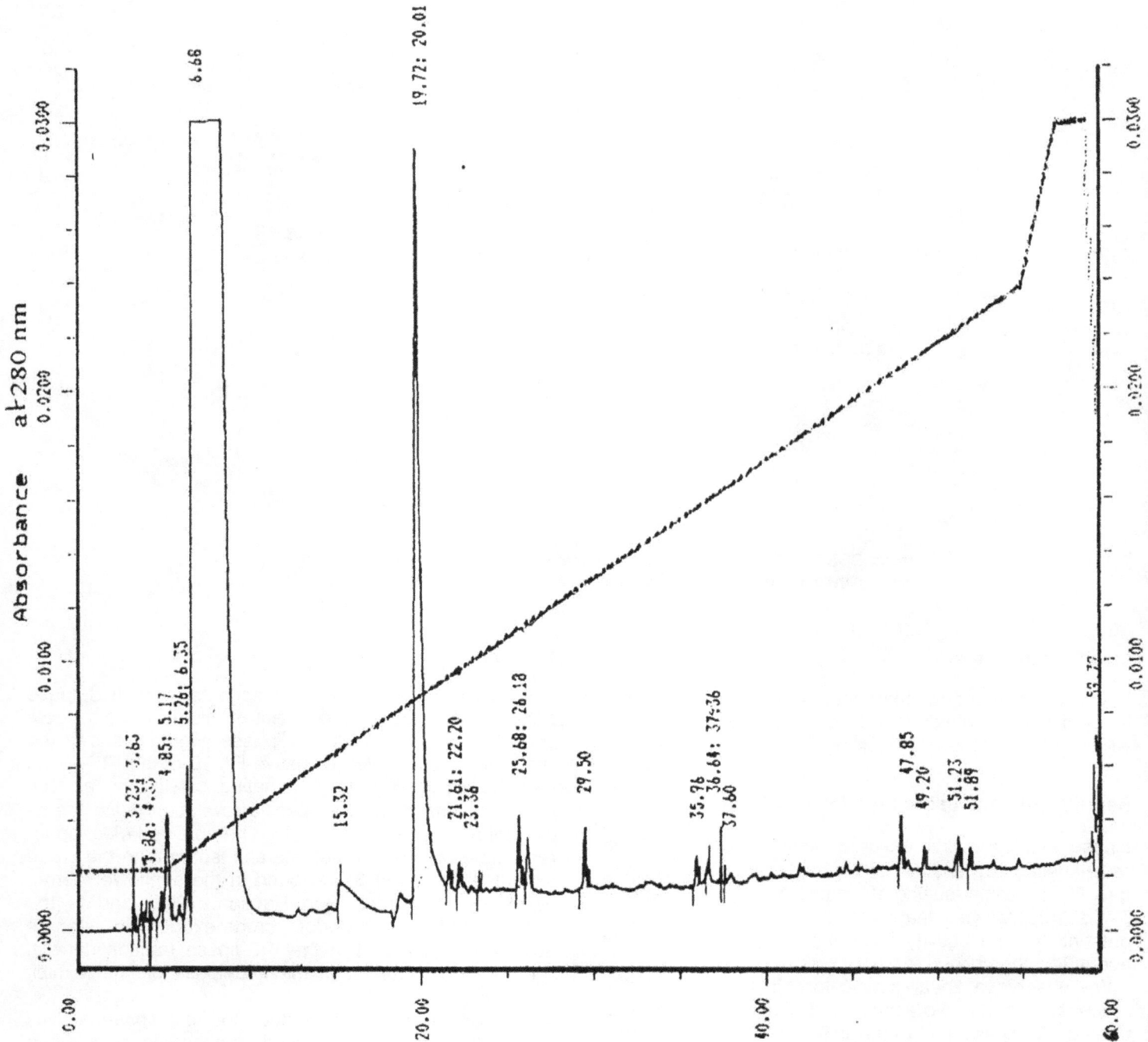

Figure 3. Reversed-phase HPLC analysis of the immunopurified rapeseed lipase. The separation was carried out on C8 column Ultra sphere octyl 5 mm (25 x 0.46 cm), using a linear gradient of acetonitrile in 0.1% (v/v) aqueous trifluoroacetic acid.

vity using 2 different methods (radiolabeled and fluorimetric) as described in the material and method section. 2 control tests were carried out using pre-incubated IRS-lipase with a pre-immune rabbit serum and a heat-treated IRS-lipase for 5 min at 90℃. In these cases, no change in the lipase activity amount was observed, but a very low residual lipase activity was found in the recuperated supernatant (S10) with a 97% inhibition rate (Figure 6). These results show evident immunochemical cross-reactivity between EC-lipase and rapeseed lipase, suggesting that these antibodies recognized the native IRS-lipase and bound to some residues located in the catalytic site

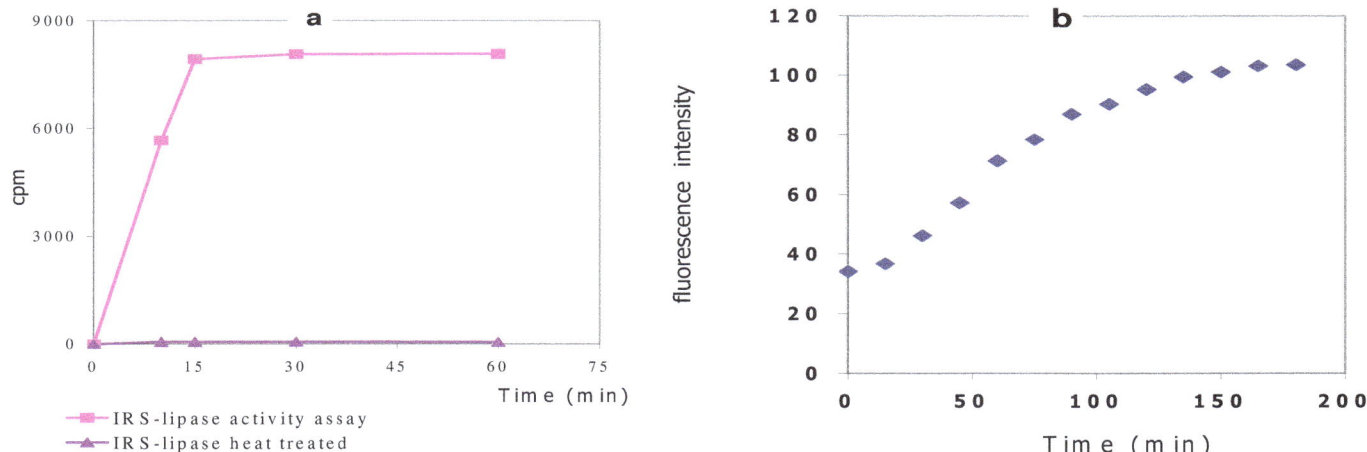

Figure 4. Test of the IRS-lipase activity using 2 methods. (a): Radiolabeled assay, using 20 μl of native and heat treated IRS-lipase were incubated at pH 7.5 and 37°C, in presence of 10 μl of radiolabeled triolein. (b): Kinetics of hydrolysis of naturally fluorescent TAGs from *P. glaberrimum* upon incubation with 20 μl of IRS-lipase incubated in I ml of the standard reaction medium (pH 8), containing 16 μg of fluorescent TAGs, 50 mM Tris, 100 mM NaTDC, 6 mM CaCl$_2$, and 0.001% BTH.

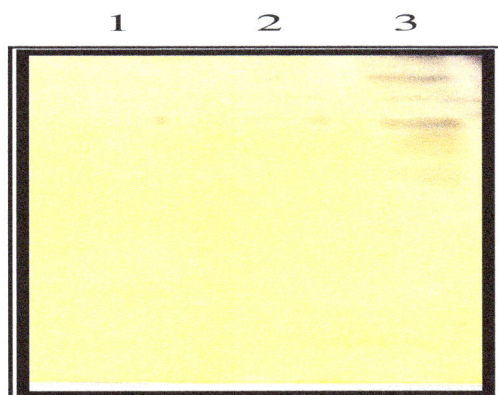

Figure 5. Electrophoregramme of esterases specific revelation. 6 to 12 μg of proteins were separated by native PAGE, the gel was submitted after that to an esterase specific revelation, using α-naphtol acetate as a substrate. Lane 1: 12 μg of ISR-lipase, lane 2 and 3: 6 and 12 μg of the rapeseed crude extract (S10).

or are related to it.

Evidence of a cross-reactivity with *Euphorbia characias* lipase

To confirm the last results, we assessed the immunoblotting technique, using the anti-EC-lipase polyclonal immune serum at a 1: 500 dilution (Belguith et al., 2001). The immunoblot presented in Figure 7 demon-strates clearly that the anti-ECL antibodies cross-react with IRS-lipase. This result suggests an antigenic relationship be-

tween the 2 lipases.

Effect of the tetrahydrolipstain on rapeseed lipase activity

The tetrahydrolipstain (THL) is considered as the first irreversible and selective inhibitor of lipases (Hadvary et al., 1991). These authors had demonstrated that the THL bind specifically to the serine- residue of the catalytic site. We had tested the THL effect on rapeseed lipase activity using the fluorimetric method. 2 cuvette assays were prepared, in the first we incubated the IRS-lipase alone as a positive control and in the second we added 5 μl of THL before 20 min of incubation. We observed (Figure 8) that the increase of the fluorescence intensity measured at 420 nm can be stopped readily by the addition of THL and it reaches a plateau at about 70 (arbitrary units). This inhibition by THL suggests the presence of a serine-residue in the catalytic site of IRS-lipase and that it is a serine triacylglycerol acyl hydrolase as the major of lipases.

Mass spectrometry analysis

Mass spectrometry is considered as an indispensable technique to analyze and characterize the primary structure of proteins. This technique allows us to evaluate the purity and to get the exact molecular weight of proteins. The MALDI-TOF mass specter of the rapeseed lipase shows a single peak with a MW of 38,0387 (Figure 9), this MW is very close to that of EC-lipase described by Moulin et al. (1994).

A short protein sequence of the IRS-lipase was obtained by Edman degradation and the sequence generated was

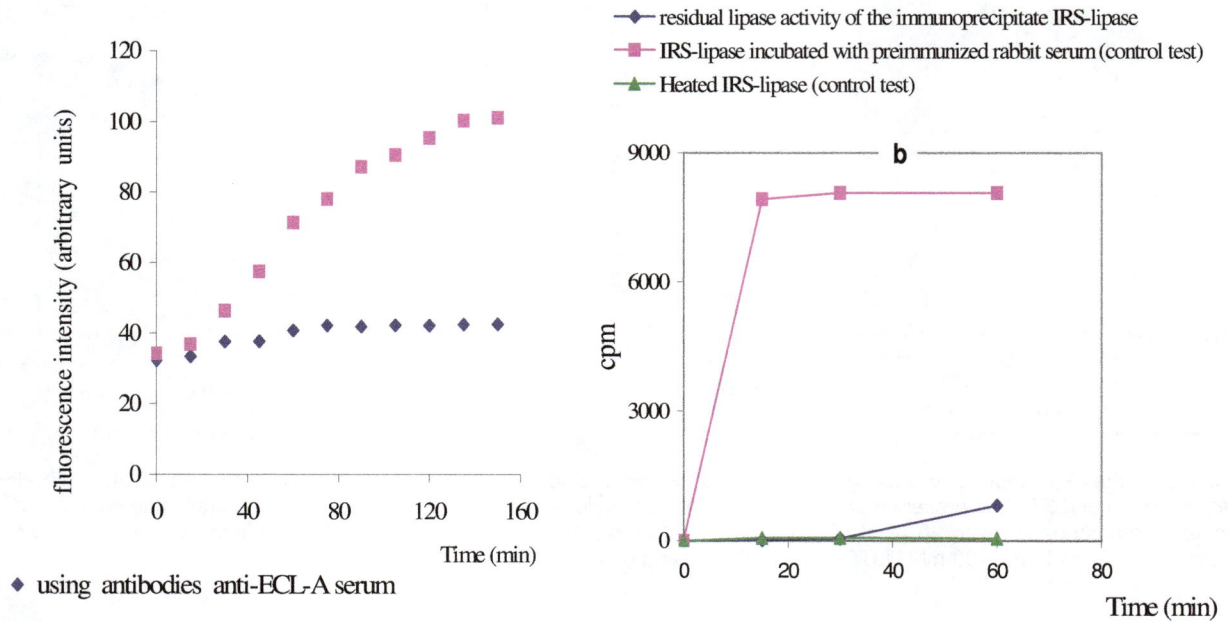

Figure 6. Immunoprecipitation of IRS-lipase and HPL by antibodies anti-EC-A lipase. Residual lipase activity was measured using Fluorimetric method (a) and the radiolabeled method (b).

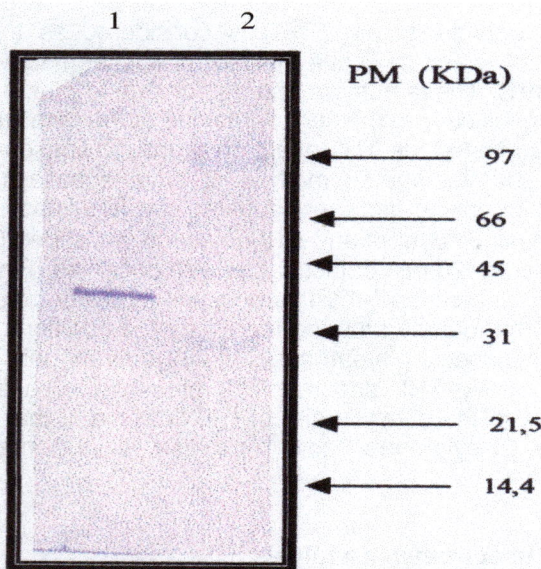

Figure 7. Immunoblot analysis of the IRS-lipase separated by SDS-PAGE, showing a cross reactivity with *E. characias* lipase. The western blot was probed with an anti-EC-lipase polyclonal immune serum at a 1:500 dilution. Lane 1, 15 µg of the IRS-lipase; 2, Protein markers.

(D, S)INGGXATLPQX.

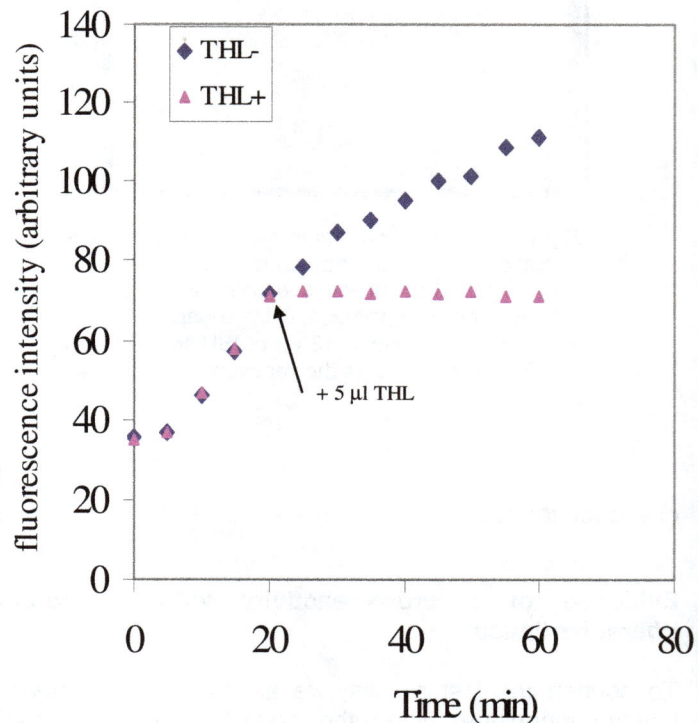

Figure 8. Effect of the addition of tertahydrolipstain (THL) on the IRS-lipase activity. 5 µl of THL was added in the cuvette assay. Lipase activity was measured using the fluorimetric method.

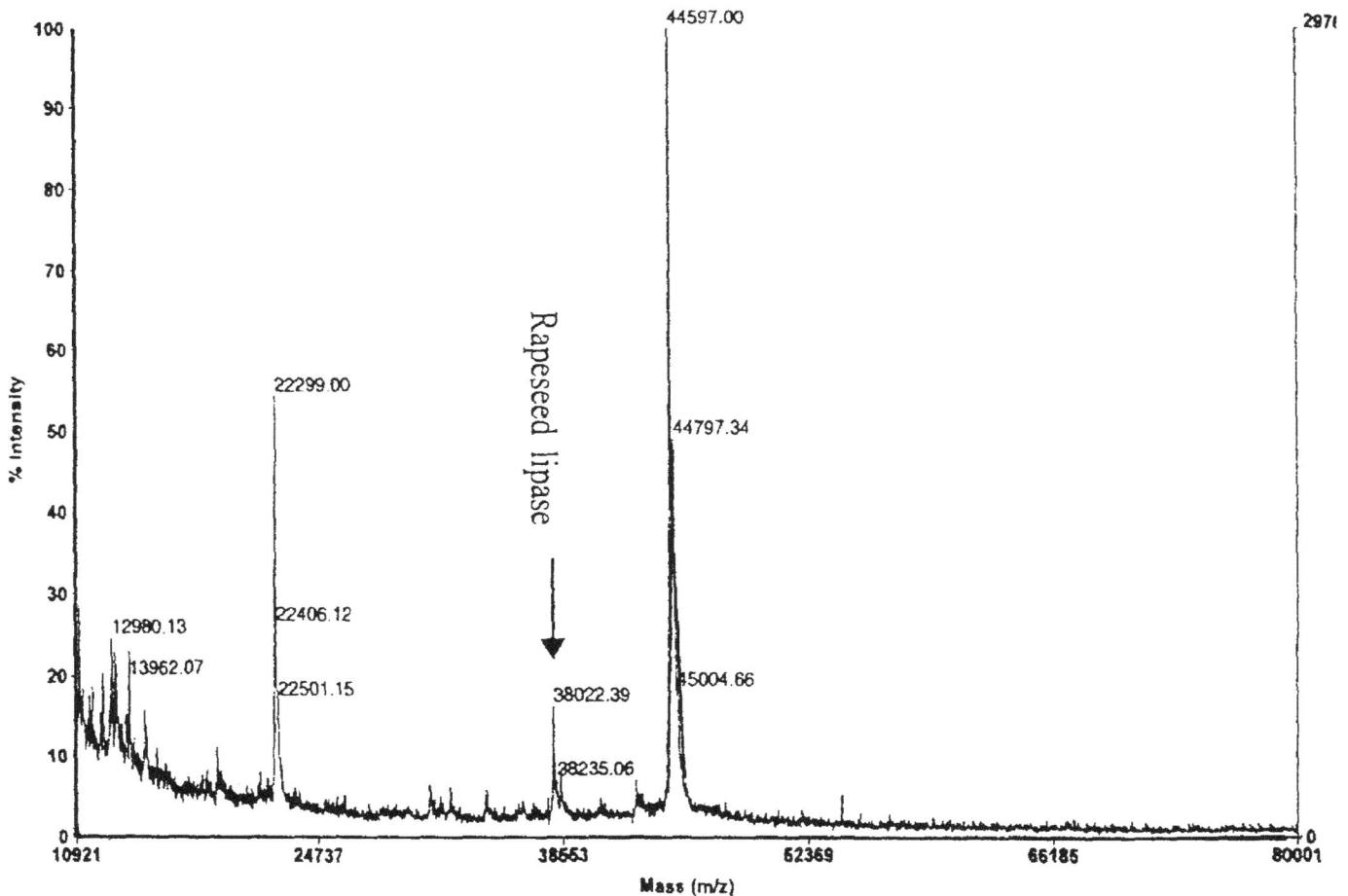

Figure 9. The MALDI-TOF spectrum of the immunopurified rapeseed lipase.

DISCUSSION

We have performed a rapid, high yielding, inexpensive and reproducible procedure for rapeseed lipase purification, it seems to be efficient to purify low abundant plant lipases since the overall obtained yield in our case is about 35.19% with a 1950-fold purification factor. The purity of the immunopurified lipase was firstly tested by SDS-PAGE, a single band was stained in the polyacrylamide gel with a MW of about 39 kDa, secondly by HPLC where a single and symmetric peak was obtained. Finally, using mass spectrometry analysis, we determined that the rapeseed lipase is a monomeric protein having an exact MW of 38.0387 kDa, molecular weight identical to that of EC-lipase. Our results confirm that the immunopurification is one of the most selective and powerful affinity techniques for protein purification leading to purification of 1000 to 10,000-fold in a single step (Harlow and Lane, 1988). Very few examples of immunopurification of lipase are found in the literature.

This method is recommended to be used as a first step on purification procedures leading to high recovery yields. It can be considered as a highly selective purification technique, that require less number of steps leading to an efficient separation and high recovery purification. Immunopurification techniques can be used in industries, because they are rapid, efficient, inexpensive, powerful affinity techniques for protein purification and amenable to large-scale operations.

The different methods used to test lipase activity confirm that the immunopurified rapeseed lipase is a true lipase of germination and not an esterase.

Generally plant proteins are glycosylated and characterized by the presence of a glycane structure related to the asparagines residues. A positive reaction was obtained after incubation of the transferred IRS-lipase nitrocellulose membrane with the biotine conjugated hydrazide and with the alkaline phosphatase conjugated avidine. This result suggests that the IRS-lipase is a glycoprotein (data not shown).

We also demonstrate that the rapeseed lipase is glycosylated and it was inhibited by THL. It is worth noting that the total inactivation of the rapeseed lipase by THL supports the existence of an activated serine in the catalytic site and that is a true serine triacyl glycerol acylhydrolase.

This result suggests a structure similarities and a chemical relationship between lipases from different organisms. This suggestion is consistent with the presence in the primary rapeseed lipase amino acid sequence of the consensus motif Gly-X-Ser-X-Gly. Nearly, all lipases that undergo interfacial activation are somehow akin to a molecular structure covering the catalytic site. Using anti-EC lipase antibodies, we demonstrate an evident immunochemical cross reactivity between EC-lipase and IRS-lipase, reflecting a considerable degree of structure homology between these enzymes. Furthermore, these antibodies were able to total inhibit rapeseed lipase activity suggesting that they bind to some residue located in the catalytic site.

Moulin et al. (1994) had demonstrated that the EC-lipase presents the consensus sequence Gly-X-Ser-X-Gly found in the B chain of the ricin lipase. It seems that the rapeseed lipase as all the purified lipases from different organisms present the consensus sequence as described by Moulin et al. (1994) and that these plant lipases are antigenically very similar.

ACKNOWLEDGMENTS

The authors gratefully acknowledge financial support from the Tunisian Ministry of Higher Education, Scientific Research and Technology and the University of 7 November at Carthage.

REFERENCES

Aires-Barros MR, Cabral JMS (1991). Purification and kinetic studies of lipases using reversed micellar systems. in: Alberghina L, Schmid RD, Verger R(Eds.), Lipases: Structure, Mech. Genet. Eng. GBF Monographs, Braunschweig, Germany 16: 407-416.

Beisson F, Ferte N, Nari J, Noat G, Arondel V, Verger R (1999). Use of naturally fluorescent triacylglycerols from *Parinari glaberrimum* to detect low lipase activities from *Arabidopsis thaliana* seedlings. J. Lip. Res. 40: 2313-2321.

Belguith H, Khodjet El, Kill H, Fattouch S, Jridi T, Ben-Hamida J (2001). Contribution of blotting techniques to the study of rapeseeds (*Brassica napus* L.) lipases. Electrophoresis 22: 18-22.

Belguith H, Jridi T, Ben Hamida J (1999). Subcellular repartition and characterization of lipase during rape seeds (*Brassica napus* L.) germination. Ann. de l'INRAT 72: 197-211.

Ben-Hamida J, Mazliak P (1985). Les lipases des graines oleagineuses. Ann. Biol. T XXIV, Fasc 3: 201-232.

Ben-Hamida J (1986). Etude des lipides, des lipases et d'un inhibiteur des lipases, chez le tournesol (*Helianthus annuus* L.) Ph-D dissertation, Sciences University, Tunis, Tunisia.

Bradford MM (1976). A rapid and sensitive method for the quantitation of microgram quantities of protein utilizing the principal of protein of dye binding 72: 248-254.

Cambillau C, Bourne Y (1991). Crystallographic studies of the pancreatic lipase/colipase system. in: Alberghina L, Schmid RD, Verger R (Eds.), Lipases: Structure, Mechanism and Genetic Engineering. GBF Monographs, Braunschweig, Germany 16: 31-34.

Carriere F, Thirstrup K, Hjorth S, Boel E, Verger R, Thim L (1994). Structure-function relationships in naturally occurring mutants of pancreatic lipase. Protein Eng. 7: 563-569.

Cheng CF, Bensadoun A, Bersot T, Hsu JST, Melford KH (1985). Purification and characterization of human lipoprotein lipase and hepatic triglyceride lipase. J. Biol. Chem. 260: 10720-10726.

Diaco R, Hill H, Durand DP (1986). Purification of soybean mosaic virus by affinity chromatography using monoclonal antibodies 67: 345-351.

Duncomb WC (1962). The colorimetric determination of long chain fatty acids in the 0.05-0.5 mole range. Biochem. J. 83: 6.

Gotz F (1991). Staphylococcal lipases and phospholipases. In: Alberghina L, Schmid RD, Verger R (Eds.), Lipases: Structure, Mechanism and Genetic Engineering. GBF Monographs, Braunschweig, Germany 16: 285-292.

Hadvary P, Sidler W, Meister W, Velter W, Wolf H (1991). Inhibition of pancreatic lipase in vitro by the covalent inhibitor tetrahydrolipstain. J. Bio. Chem. 266: 2021-2027.

Harlow E, Lane D (1988). Antibodies Cold Spring Harbour Publications, USA.

Huang AHC (1993) Lipases in: lipid Metabolism in plants. Moore TS, (Eds). CRC Press Inc., Boca Raton pp. 473-502.

Isobe K, Nokihara K (1991). Physicochemical properties of mono- and diacylglycerol lipase from *Penicillium camembertii. in*: Alberghina L, Schmid RD, Verger R (Eds.), Lipases: Structure, Mechanism and Genetic Engineering. GBF Monographs, Braunschweig, Germany 16: 345-348.

Kanaya S, Koyanagi T, Kanaya E (1998). An esterase from *Escherichia coli* with a sequence similarity to hormone-sensitive lipase. Biochem. J. 322: 75-80.

Kang AS, Abott SJ, Harris N (1988). Manipulating secondary metabolism in culture, Cambridge University Press, Cambridge.

Kenyon AJ, Gander JE, Lopez C, Good RA (1973). Isolation of Aleutian mink disease virus by affinity chromatography. Sci. 179: 187-189.

Kundu M, Basu J, Guchhait M, Chakrabarti P (1987). Isolation and characterization of an extracellular lipase from the conidia of *Neurospora crassa*. J. Gen. Microbiol. 133: 149-153.

Laemmli UK (1970). Clivage of structural proteins during the assembly of the head of bacteriophage T4. Nature 277: 680-685.

Lin YH, Haung ACH (1983). Lipase in lipid bodies of cotyledons of rape and mustard seedlings. Arch. Biochem. Biophys. 225: 360- 369.

Melchers F, Messer W (1970). The activation of mutant β-galactosidase by specific antibodies. Euro. J. Biochem. 17: 267-272.

Moulin A, Tiessere M, Bernard C, Pieroni G (1994). Lipases of the Euphorbaceae family: Purification of a lipase from *Euphrbia characias* latex and structure-function relationships with the B chain of ricin. Biochem. 91: 11328-11332.

Murphy RF, Buchanan KD, Elmore DT (1973). Isolation of glucagon-like immunoreactivity of gut by affinity chromatography on anti-glucagon antibodies coupled to sepharose 4B. Bioch. Biophy. Acta. 303(1): 118-127.

Osterlund T, Danielsson B, Degerman E, Contreras JA, Edgern G, Davis RC, Schotz MC, Holm C (1996). Domain-structure analysis of recombinant rat hormone-sensitive lipase. Biochem. J. 319411-420.

Saxena RM, Holmgren A (1973). Purification of thioredoxin from *Escherichia coli* and bacteriophage immunosorbent affinity chromatography. Bioch. Biophy. Acta. 315: 176-180.

Sugihara A, Shimada Y, Tominaga Y (1988). Purification and characterization of *Aspergillu niger* lipase. Agric. Biol. Chem. 1591-1592.

Suzuki M, Yamamoto H, Mizugaki M (1986). Purification and general properties of a metal-insensitive lipase from *Rhizopus japonicus* NR 400. Japanese Biochem. Soc. 100: 1207-1213.

Tyski S, Hryniewicz W, Jeljaszewicz J (1983). Purification and some properties of the coccal extracellular lipase. Biochim. Biophys. Acta. 749: 312-317.

Veeraragavan K, Gibbs BF (1989). Detection and partial purification of two lipases from *Candida rugosa*. Biotechnol. Lett. 11: 345-348.

Vicente MLC, Aires-Barros MR, Cabral JMS (1990). Purification of *Chromobacterium viscosum* lipases using reversed micelles. Biotechnol. Tech. 4: 137-142.

WeselaKe R, Jain JC (1992). Strategies in the purification of plant proteins. Physiol. Plant. 84: 301-309.

4

Changes in physiological and biochemical indicators associated with salt tolerance in cotton, sorghum and cowpea

Valdinéia Soares Freitas[1], Nara Lídia Mendes Alencar[1], Claudivan Feitosa de Lacerda[2], José Tarquinio Prisco[1] and Enéas Gomes-Filho[1]*

[1]Department of Biochemistry and Molecular Biology, National Institute of Science and Technology in Salinity (INCTSal), Federal University of Ceará, Fortaleza-CE, Brazil.
[2]Departament of Agronomy Engineering, Center of Agricultural Sciences, Federal University of Ceará, Fortaleza-CE, Brazil.

The aim of this study was to evaluate the interaction between salinity, growth, and enzymatic antioxidants in cotton, cowpea and sorghum. Salt stress significantly reduced plant growth, especially in cowpea. Na^+ and Cl^- concentrations increased in leaves and roots of these three species, especially for cotton at 4.0 and 8.0 dS/m, in relation to control. Salinity significantly increased lipid peroxidation levels in cowpea, whereas in cotton these levels were reduced. Superoxide dismutase (SOD), catalase (CAT) and ascorbate peroxidase (APX) enzyme activities in leaves were not changed by saline treatments at 4.0 and 8.0 dS/m. However, reductions were observed in the SOD activity in cotton at 8.0 dS/m NaCl and in CAT activity in sorghum at 4.0 and 8.0 dS/m. The growth results obtained confirmed the highest cotton tolerance and the highest cowpea susceptibility to salinity, while antioxidant enzyme activities changes suggest that cotton constitutive enzyme system seems to be more efficient than the others.

Key words: Salt stress, antioxidant enzymes, inorganic solutes, *Gossipium hirsutum*, *Vigna unguiculata*, *Sorghum bicolor*.

INTRODUCTION

Abiotic stress may be caused by numerous factors such as drought (Simova-Stoilova et al., 2009), cold (Van Heerden et al., 2003), high temperature (Reynolds-Henne et al., 2010), salinity (Meloni et al., 2003), heavy metals (Smeets et al., 2009) and ultraviolet radiation (Gao and Zhang, 2008). Among the various abiotic stresses to which plants are constantly exposed, salt stress is the one that most affects growth and productivity of plants around the world (Vaidyanathan et al., 2003; Veeranagamallaiah et al., 2007), reaching more than 800 million land hectares around the globe (FAO, 2005).

In general, salt stress consists of two components: osmotic and ionic. The first is a result of the high concentration of salts in the root environment that leads to decreased in soil water potential and reduces the availability of water for the plant. The ionic component arises from the accumulation of certain ions (usually Na^+ and Cl^-) and can cause nutritional imbalance, toxicity or both (Greenway and Munns, 1980; Munns, 2002; Munns and Tester, 2008). In addition to these effects, salt stress also causes an imbalance in redox status of cells, generating an oxidative stress through overproduction of reactive oxygen species (ROS) such as superoxide ($O_2\bullet^-$) and hydroxyl (HO•) radicals and hydrogen peroxide (H_2O_2) (Vaidyanathan et al., 2003). These ROS are highly reactive and can alter normal cellular metabolism through the oxidation of important biomolecules such as proteins and nucleic acids (McKersie and Leshem, 1994; Mittler, 2002).

*Corresponding author. E-mail: egomesf@ufc.br.

To minimize oxidative damage caused by ROS, plants have developed a complex antioxidant system that includes both non-enzymatic antioxidants of low molecular weight (mainly ascorbate and glutathione) and ROS scavenging enzymes such as superoxide dismutase (SOD), catalase (CAT) and peroxidases (McKersie and Leshem, 1994). The superoxide radical is eliminated mainly by SOD, resulting in the formation of H_2O_2 (Alscher et al., 2002). The H_2O_2 produced is then removed by CAT and peroxidases activities (McKersie and Leshem, 1994). The generation of ROS and increased activity of many antioxidant enzymes during salt stress have been reported in cotton (Meloni et al., 2003), sorghum (Costa et al., 2005), tomato (Mittova et al., 2002) and mangrove (Parida et al., 2004). However, until date, biochemical and physiological studies involving plants with different responses to salinity in the same experimental conditions are scarce.

In this study, we examined the interactions between salinity, growth, lipid peroxidation and antioxidant enzyme activities in three species with contrasting tolerance to salt (cotton, sorghum and cowpea) and aiming a better understanding of these salt tolerance indicators in these species.

MATERIALS AND METHODS

Plant materials and growth conditions

Seeds of cotton [*Gossypium hirsutum* (L.) Mast.], cultivar BRS 113 7MH, cowpea [*Vigna unguiculata* (L.) Walp.], cultivar Epace 10, and sorghum [*Sorghum bicolor* (L.) Moench.], genotype CSF 20, were sown in plastic cups containing vermiculite moistened with half-strength Hoagland's nutrient solution (½ HNS) and irrigated daily with distilled water. Fifteen day old seedlings were transferred to hydroponic media in plastic bowls (12 L) containing ½ HNS and after six days they were transferred to plastic buckets (5 L) and subjected to three saline treatments with electrical conductivity (EC) of 0.9, 4.0 and 8.0 dS/m. The lowest EC treatments consisted of plants in ½ HNS (low salinity - control), while 4.0 (moderate salinity) and 8.0 dS/m (high salinity) treatments consisted of plants in ½ HNS with the addition of NaCl, which resulted in these conductivities. They were kept aerated and their pH was checked daily and adjusted between 5.5 to 6.5 with 0.1 N NaOH or 0.1 N HCl when necessary. Every seven days, the nutrient solution was exchanged for a new one. The experiment was carried out under greenhouse conditions. Temperature and relative humidity mean values were 28.5 °C and 65.5%, respectively.

Growth measurements

Plants were harvested 25 days after the commencement of the addition of salt, when leaves, culms and roots were separated. Leaf area was determined using a Li-Cor area meter LI-3000 (Li-Cor., Inc., Lincoln, Nebraska, USA). The first and second fully expanded leaves from the base and apex roots (one third of the apical extremity) were frozen in liquid nitrogen and kept in a freezer (-80 °C) for further analyses. The rest of the plant material was weighed and after drying in a forced air circulation oven at 60 °C for 48 h for dry mass (DM) measurement.

Ion determination

Second leaf and root fresh matter were macerated in a mortar and pestle, thereafter pressed and filtered in a disposable syringe, using a muslin cloth. The juice extracted from the plant material was centrifuged at 12,000 g for 10 min. The supernatant (cell juice) was used for determining Na^+ and K^+ concentration using flame photometry and Cl^- concentration was determined according to the method of Gaines et al. (1984). The inorganic ion concentrations were expressed in mM.

Extract preparation

First leaf (1 g) and root (1 g) were homogenized in a mortar and pestle with 4 mL of ice-cold extraction buffer (100 mM potassium phosphate buffer, pH 7.0, 0.1 mM EDTA, 2 mM ascorbate). The homogenate was filtered through muslin cloth and centrifuged at 12,000 g for 15 min. The supernatant fraction was used as a crude extract for enzyme activity assays. All operations were carried out at 4 °C.

Enzyme assays

SOD (EC 1.15.1.1) activity was determined by measuring its ability to inhibit the photochemical reduction of nitroblue tetrazolium chloride (NBT), as described by Giannopolitis and Ries (1977). The SOD activity results were expressed as U/g fresh mass (FM), being one unity of SOD activity (U) defined as the enzyme amount required to cause 50% inhibition of the NBT photoreduction rate. CAT (EC 1.11.1.6) activity was measured according to Havir and McHale (1987), by H_2O_2 decrease that was monitored by reading the absorbance at 240 nm. The difference in absorbance was divided by the H_2O_2 molar extinction coefficient (36/M.cm) and enzyme activity expressed as µmol of H_2O_2/min.g FM. APX (EC 1.11.1.1) activity was performed according to Nakano and Asada (1981), being the ascorbate oxidation monitored by reading at 290 nm absorbance. The absorbance difference was divided by the ascorbate molar extinction coefficient (2.8/mM.cm) and enzyme activity expressed as µmol of H_2O_2/min.g FM, considering that 2.0 mol of ascorbate is required for reducing 1.0 mol of H_2O_2 (McKersie and Leshem, 1994).

Lipid peroxidation in leaves

Lipid peroxidation was performed using the thiobarbituric acid according to Cakmak and Horst (1991) method. The malondialdehyde (MDA) concentration was calculated using a 155/mM.cm extinction coefficient and results expressed as nmol/g FM.

Experimental design and statistical analysis

The experimental design was completely randomized, corresponding to factorial three (species) × three (salt levels), with five replicates of two plants each. Enzyme activities, MDA concentration and ion determination were assayed in duplicate. Data were analyzed using a two-way analysis of variance (ANOVA). The values were compared using Tukey´s test ($p < 0.05$).

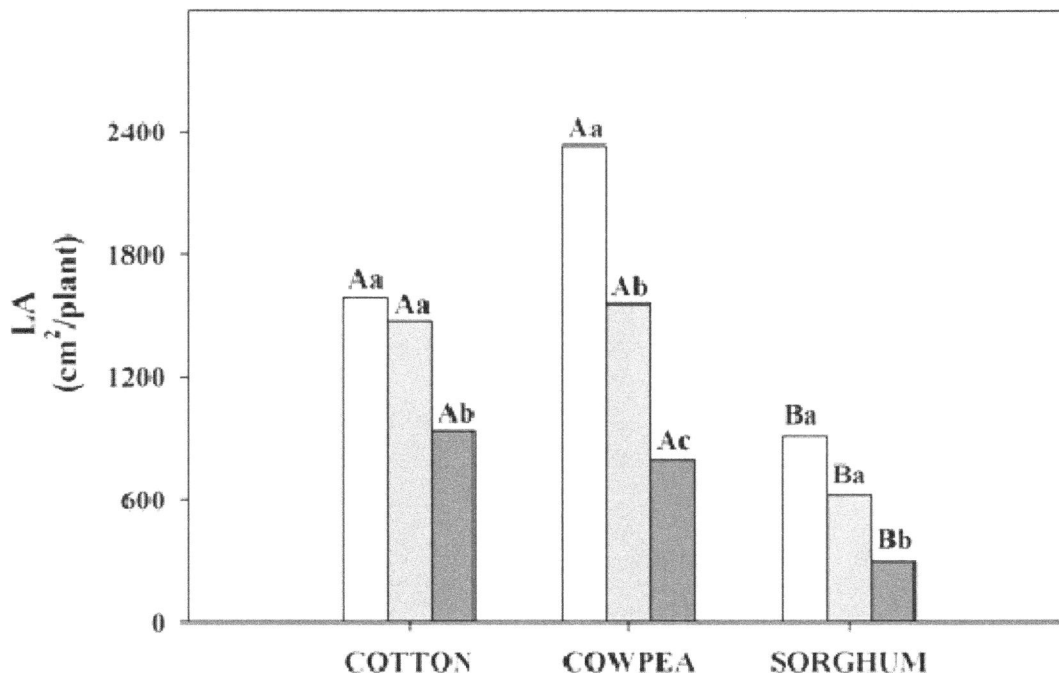

Figure 1. Leaf area (LA) of cotton, cowpea and sorghum plants subjected to salinity (low- ☐, medium - ▨ and high - ▧ levels). Columns followed by the same capital letters within each treatment and with the same lowercase letters in the same species are not significantly different ($p \leq 0.05$) by Tukey test.

RESULTS

Plant growth

The leaf area (LA) of these three species was affected by salt stress, but this reduction was more pronounced in cowpea (Figure 1). In cowpea, at the moderate (4.0 dS/m) and high salinity (8.0 dS/m) levels the reduction of LA reached values of 33 and 66%, respectively, in relation to control, whereas for sorghum and cotton plants LA reductions were observed at 8.0 dS/m, corresponding to 67 and 41%, respectively, compared to the lowest level.

In cotton and sorghum plants exposed to a moderate level (4.0 dS/m), there was no significant reduction in shoot dry mass (SDM) compared with the lowest salinity level (0.9 dS/m) (Figure 2a). However, reductions were observed at 41 and 58% for the two species, respectively, when exposed to the highest salinity level (8.0 dS/m). In cowpea, this parameter was significantly affected by the salinity moderate level, which remained unchanged even with higher salinity, showing average reductions of 50%, in these two salinity levels, in relation to plants subjected to 0.9 dS/m (Figure 2a). Sorghum showed a significant reduction in root dry mass (RDM) at 8.0 dS/m when compared to the low level of salinity, while cotton and cowpea did not suffer significant changes in this parameter (Figure 2b). In cotton and sorghum plants, the SDM/RDM ratio did not differ

statistically among the three treatments; however, this relationship was significantly reduced in cowpea, about 39% at moderate and high salinity treatments in relation to the low level of salinity (Figure 2c).

Inorganic solutes

In leaves, Na^+ concentration did not show differences between the species at the lowest salinity level (Figure 3a). In cotton, a significant increase ($p < 0.05$) was verified in Na^+ concentration, at 4.0 and 8.0 dS/m. Plants exposure to the highest level of salinity, when compared to the lowest, resulted in leaf, Na^+ concentration increases for cotton, cowpea and sorghum, corresponding respectively to 32.7, 4.4 and 20.6 times (Figure 3a). In roots, high salinity (8.0 dS/m) showed Na^+ concentration increase of 46.4, 24.1 and 20.6 times in cotton, cowpea and sorghum, respectively, when compared to the lowest salinity level (Figure 3b). In comparison with 0.9 dS/m treatment, K^+ concentration only increased in cotton and cowpea leaves at the highest salinity level (Figure 3c). In contrast, in sorghum plants, K^+ concentration in leaves was progressively reduced by salt stress (Figure 3c). In roots, K^+ concentration decreased significantly for cotton (16.9%) and cowpea (20.5%) with NaCl increase from 0.9 to 8.0 dS/m (Figure 3d). In sorghum roots subjected at 4.0 dS/m treatment, K^+ concentration increased 17.8%, when

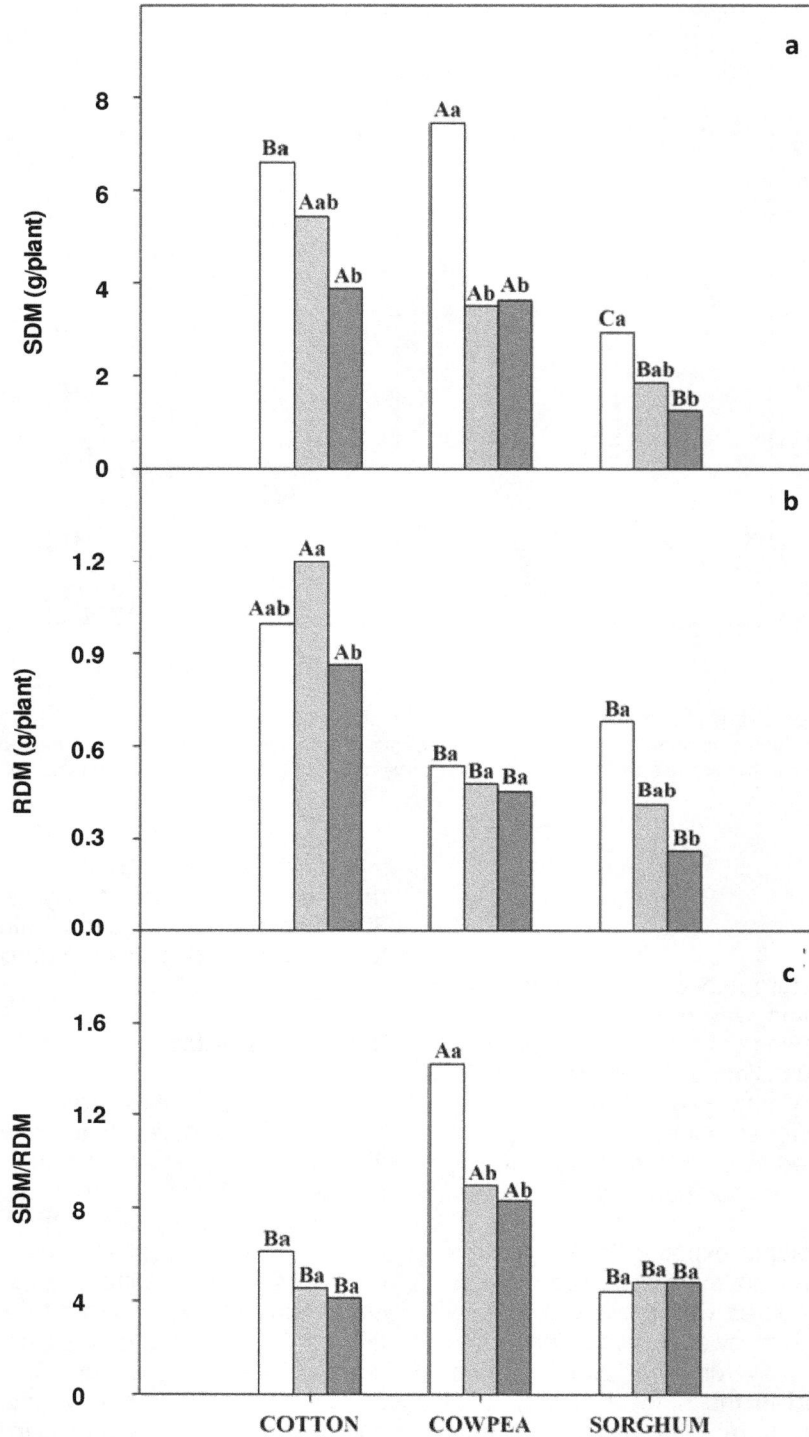

Figure 2. Shoot dry mass (SDM, a), root dry mass (RDM, b) and SDM/RDM ratio (c) of cotton, cowpea and sorghum plants subjected to salinity (low - ☐, medium - ▨ and high - ■ levels). Columns followed by the same capital letters within each treatment and with the same lowercase letters in the same species are not significantly different ($p \leq 0.05$) by Tukey test.

compared with 0.9 dS/m treatment (Figure 3d). Leaf Cl⁻ concentration increased with the salinity increment in the

three species, however in cotton and cowpea the Cl⁻ accumulation was more pronounced than sorghum

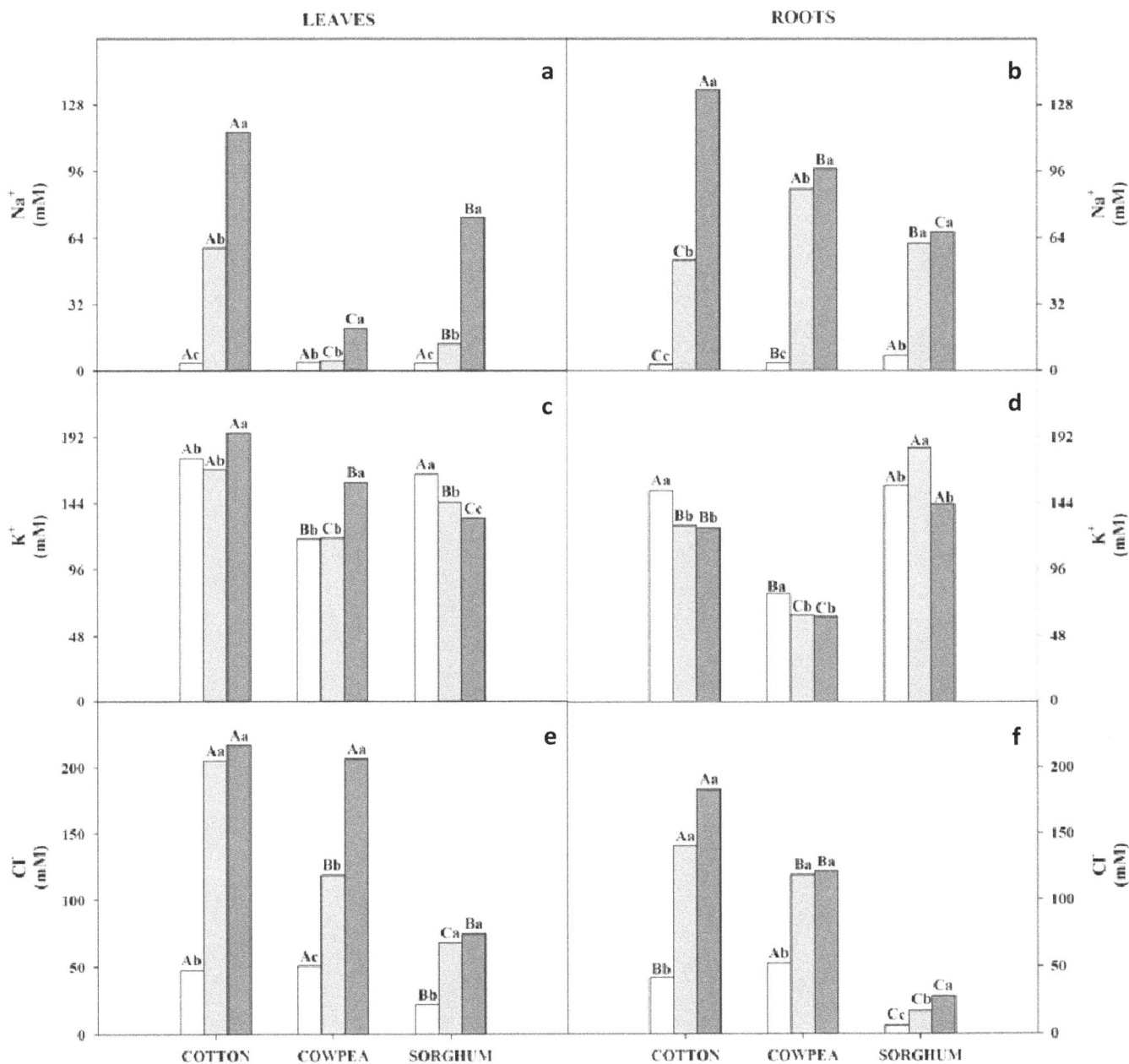

Figure 3. Na$^+$,K$^+$ and Cl$^-$ concentrations in leaves and roots of cotton, cowpea and sorghum plants subjected to salinity (low - ☐, medium - ▨ and high - ▩ levels). Columns followed by the same capital letters within each treatment and with the same lowercase letters in the same species are not significantly different ($p \leq 0.05$) by Tukey test.

(Figure 3e). Root Cl$^-$ concentration increased significantly in these species, being higher in cotton (Figure 3f).

SOD, CAT and APX activities

Constitutive SOD activity in leaves was higher in sorghum followed by cotton and cowpea (Figure 4a). In general, 4.0 and 8.0 dS/m treatments compared to 0.9 dS/m did not strongly affect SOD activity in leaves and

roots (Figures 4a and b), except by reductions of SOD activities in cotton leaves (Figure 4a). Constitutive CAT activity was markedly higher in cotton and sorghum leaves than in cowpea, whereas in roots the constitutive CAT activity was higher in cotton than in cowpea and sorghum (Figure 4d). In leaves and roots of cotton and cowpea, CAT activity remained practically constant under all NaCl levels (Figures 4c and d). However, CAT activity in sorghum leaves decreased about 37% with the salinity increase. APX activity was not significantly affected by

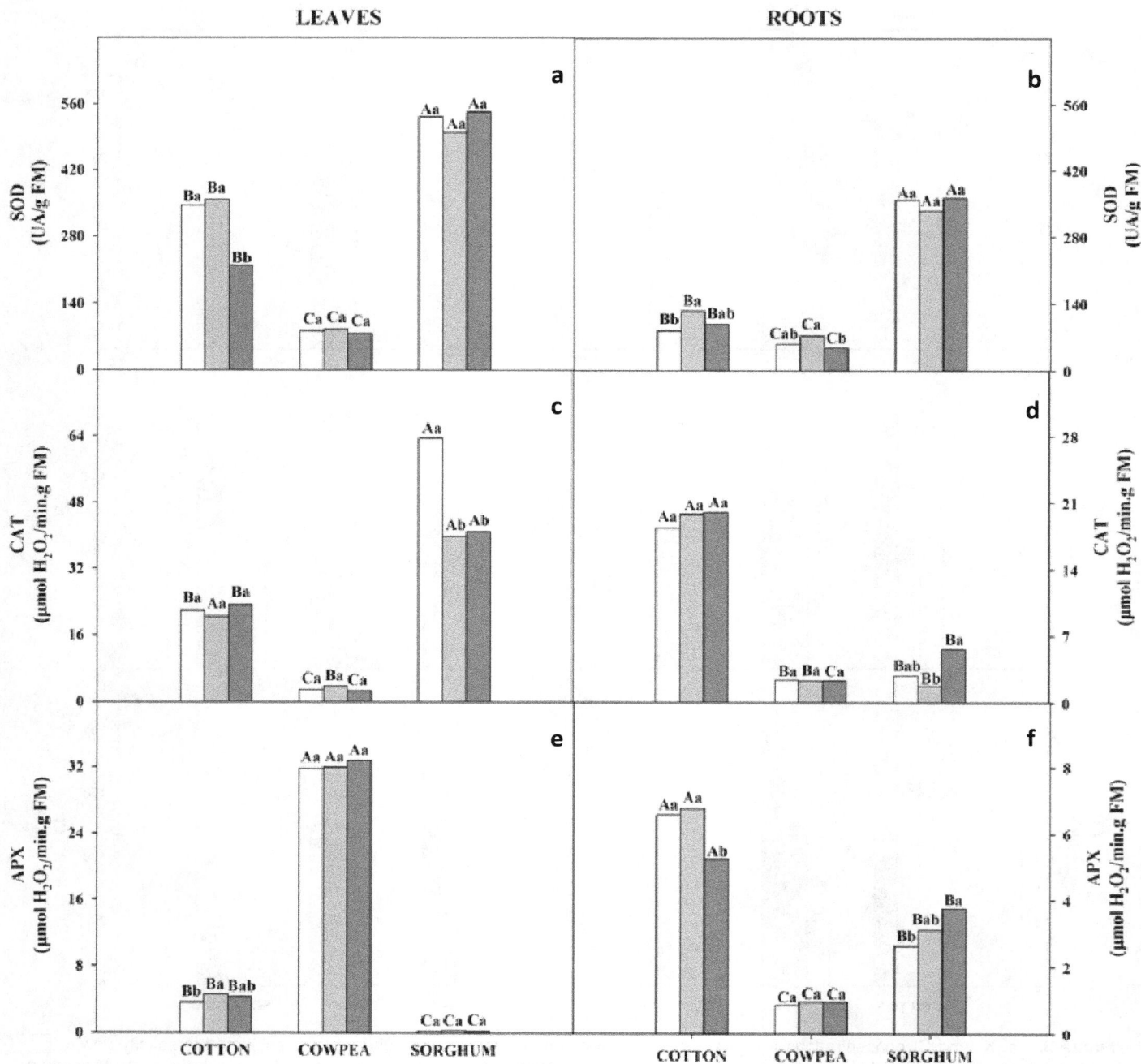

Figure 4. Superoxide dismutase (SOD), catalase (CAT) and ascorbate peroxidase (APX) activities in leaves and roots of cotton, cowpea and sorghum plants subjected to salinity (low - ☐, medium - ▨ and high - ▧ levels). Columns followed by the same capital letters within each treatment and with the same lowercase letters in the same species are not significantly different ($p \leq 0.05$) by Tukey test.

salt treatments in cowpea leaves and roots and in sorghum leaves (Figures 4e and f). However, APX activity in sorghum roots increased 42.6% at 8.0 dS/m treatment. In cotton leaves, APX activity increased in the treatment with moderate salinity, when compared to that with low salinity (Figure 4e), whereas in roots, the 8.0 dS/m treatment promoted a 20.4% decrease in total APX activity in comparison to 0.9 dS/m treatment (Figure 4f).

Lipid peroxidation in leaves

The salt stress (4.0 and 8.0 dS/m) caused a marked increase in MDA content in leaves only in cowpea (Figure 5), showing a medium increase of about 39.6% as compared to the lowest level of salinity. In sorghum, there were no significant differences in MDA content between the treatments, whereas in cotton, there were significant

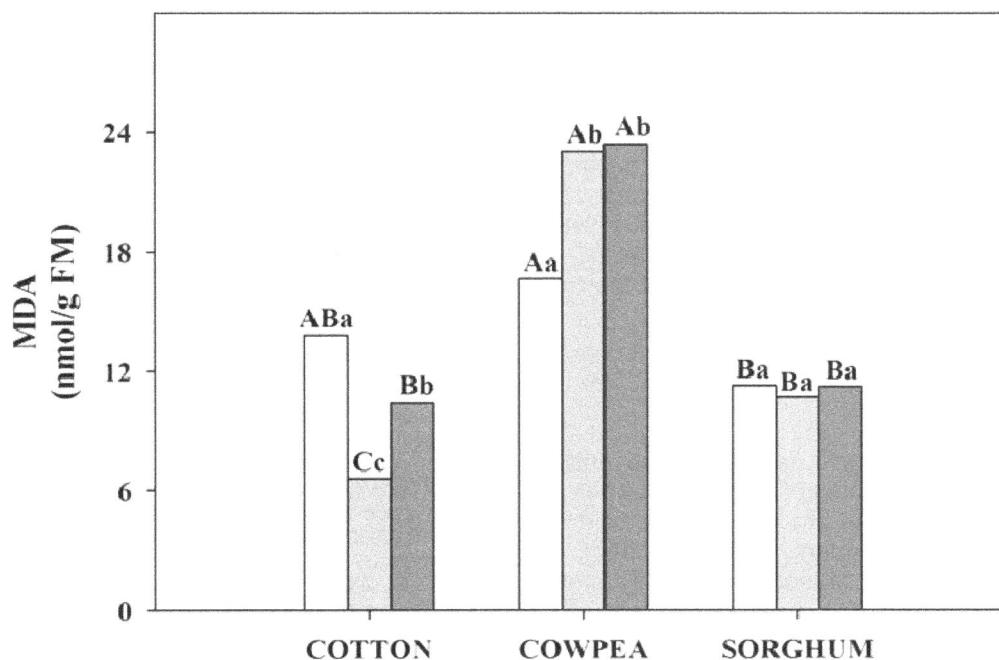

Figure 5. Malondialdehyde (MDA) content in leaves of cotton, cowpea and sorghum plants subjected to salinity (low - ☐, medium - ▨ and high - ▨ levels). Columns followed by the same capital letters within each treatment and with the same lowercase letters in the same species are not significantly different ($p \leq 0.05$) by Tukey test.

decreased at 4.0 and 8.0 dS/m treatments (Figure 5).

DISCUSSION

Salt stress can have many effects on plant growth being a complex syndrome that involves osmotic stress, ion toxicity, mineral deficiencies and formation of ROS such as $O_2\bullet^-$, H_2O_2 and $HO\bullet$ (Hasegawa et al., 2000; Munns, 2002; Vaidyanathan et al., 2003; Munns and Tester, 2008). In this study, cotton, cowpea and sorghum varied markedly in salt tolerance. The cowpea was the one that showed severe reductions in LA and SDM since at moderate salinity level (4.0 dS/m), whereas sorghum and cotton showed more severe reductions in these parameters only at the highest salinity level (8.0 dS/m), which clearly demonstrates cowpea's greater susceptibility to salinity compared with other species tested (Figures 1 and 2). This severe reduction in cowpea biomass may be due to the biggest sensitive to Na^+ and Cl^- accumulation in leaves and roots (Figure 3). Cotton seems to be a species that does not use mechanisms of exclusion of toxic ions (especially Na^+ and Cl^-) of photosynthetic tissues, as with other species of glycophytes (Lacerda et al., 2003). Therefore, depending on the specie, salt tolerance could be associated with the rate of ion transport to the shoot and/or plant capacity to compartmentalize toxic ions in different tissues or cells (Lacerda et al., 2001).

The biggest values for SOD and CAT activities in cotton and sorghum leaves, suggest that their constitutive antioxidant enzyme system is more efficient than in the cowpea (Figure 4). The SOD and CAT low activities in cowpea can at least be partially responsible for its bigger susceptibility to salt stress. This idea is supported by other studies that suggest greater tolerance to salinity in species that have a more efficient constitutive antioxidant enzyme system (Bandeoglu et al., 2004; Stepien and Klobus, 2005; Demiral and Türkan, 2005). Furthermore, the highest values of MDA (Figure 5) in any treatment were observed in cowpea leaves, indicating that, among the species, it has higher lipid peroxidation, thus corroborating the suggestion of being the species with the least effective constituent antioxidant system. APX is an important antioxidant enzyme, which acts by reducing H_2O_2 to water using ascorbate as a reducing agent (Asada, 1992). In this study, the constitutive activity of APX in sorghum roots was higher than that found in leaves, whereas with cowpea plants, the constitutive activity was higher in leaves than in roots (Figures 4e and f). Bandeoglu et al. (2004), while studying lentil under saline conditions found that APX activity was significantly increased in the leaves, while no changes were observed in roots and that their oxidative protection was due to the enzyme's high constitutive activity coupled with SOD increases.

Malondialdehyde (MDA) is a decomposition product of polyunsaturated fatty acids and their increases in plants have been widely used as an indicator of lipid peroxidation in salt stress, often serving to discriminate between species that are sensitive and tolerant to salinity (Meloni et al., 2003; Stepien and Klobus, 2005). In the present study, lipid peroxidation levels were significantly increased in medium and high salinity treatments in the most susceptible species (cowpea), while in the most tolerant (cotton) for the same treatments, reductions were observed (Figure 5). These results suggest that salinity does not induce lipid peroxidation in cotton plants, or even that these plants have high oxidative protection under saline conditions.

Conclusion

The growth results obtained confirmed the highest cotton tolerance and the highest susceptibility of the cowpea to salt stress, which can be related to the higher efficiency of cotton's constitutive antioxidant enzyme system, in reducing oxidative damage caused by salinity, than cowpea.

ACKNOWLEDGEMENTS

We are grateful to the Conselho Nacional de Desenvolvimento Científico e Tecnológico (CNPq) for their financial support.

REFERENCES

Alscher RG, Erturk N, Heath LS (2002). Role of superoxide dismutases (SODs) in controlling oxidative stress in plants. J. Exp. Bot., 53: 1331-1341.

Asada K (1992). Ascorbate peroxidase - a hydrogen peroxide-scavenging enzyme in plants. Physiol. Plant, 85: 235-241.

Bandeoglu E, Eyidogan F, Yücel M, Oktem HA (2004). Antioxidant responses of shoots and roots of lentil to NaCl-salinity stress. Plant Growth Regul., 42: 69-77.

Cakmak I, Horst WJ (1991). Effect of aluminium on lipid peroxidation, superoxide dismutase, catalase, and peroxidase activities in root tips soybean (Glycine max). Physiol. Plant, 83: 463-468.

Costa PHA, Azevedo-Neto AD, Bezerra MA, Prisco JT, Gomes-Filho E (2005). Antioxidant-enzymatic system of two sorghum genotypes differing in salt tolerance. Braz. J. Plant Physiol., 17: 353-361.

Demiral T, Türkan I (2005). Comparative lipid peroxidation, antioxidant defense systems and proline content in roots of two rice cultivars differing in salt tolerance. Environ. Exp. Bot., 53: 247-257.

FAO - FOOD AND AGRICULTURE ORGANIZATION OF THE UNITED NATIONS (2005). Land and Plant Nutrition Management Service. http://www.fao.org/ag/agl/agll/spush/topic2.htm#top. Accessed 10 December 2010.

Gaines TP, Parker MB, Gascho, GJ (1984). Automated determination of chlorides in soil and plant tissue by sodium nitrate. Agron. J., 76:371-374.

Gao Q, Zhang L (2008). Ultraviolet-B-induced oxidative stress and antioxidant defense system responses in ascorbato-deficient vtc 1 mutants of Arabidopsis thaliana. J. Plant Physiol., 165: 138-148.

Giannopolitis CN, Ries SK (1977). Superoxide dismutases. I. occurrence in higher plants. Plant Physiol., 59: 309-314.

Greenway H, Munns R (1980). Mechanisms of salt tolerance in nonhalophytes. Ann. Rev. Plant Physiol., 31: 149-190.

Hasegawa PM, Bressan RA, Zhu JK, Bohnert HJ (2000). Plant cellular and molecular responses to high salinity. Ann. Rev. Plant Physiol., 51: 463-499.

Havir E, McHale NA (1987). Biochemical and developmental characterization of multiple forms of catalases in tobacco leaves. Plant Physiol., 84: 450-455.

Lacerda CF, Cambraia J, Cano MAO, Ruiz HA (2001). Plant growth and solute accumulation and distribution in two sorghum genotypes, under NaCl stress. Braz. J. Plant Physiol., 13: 270-284.

Lacerda CF, Oliveira HPM, Oliveira TS, Gomes-Filho E (2003). Crescimento e acúmulo de íons em folhas de sorgo forrageiro submetido a soluções iso-osmóticas de sais (NaCl + KCl). Rev. Ciência Agron., 34: 1-6.

McKersie BD, Leshem YY (1994). Stress and stress coping and cultivated plants. Kluwer Academic Publishes, Dordrecht-London, 256 p.

Meloni DA, Oliva MA, Martinez CA, Cambraia J (2003). Photosynthesis and activity of superoxide dismutase, peroxidase and glutathione reductase in cotton under salt stress. Environ. Exp. Bot., 49: 69-76.

Mittler R (2002). Oxidative stress, antioxidants and stress tolerance. Trends Plant Sci., 7: 405-410.

Mittova V, Tal M, Volokita M, Guy M (2002). Salt stress induces up-regulation of an efficient chloroplast antioxidant system in the salt-tolerant wild tomato species Lycopersicon pennellii but not in the cultivated species. Physiol. Plant, 115: 393-400.

Munns R (2002). Comparative physiology of salt and water stress. Plant Cell Environ., 25: 239-250.

Munns R, Tester M (2008). Mechanisms of Salinity Tolerance. Ann. Rev. Plant Biol., 59: 651-681.

Nakano Y, Asada K (1981). Hydrogen peroxide is scavenged by ascorbate-specific peroxidase in spinash chloroplasts. Plant Cell Physiol., 22: 867-880.

Parida AK, Das AB, Mohanty P (2004). Defense potentials to NaCl in mangrove, Bruguiera parviflora: Differential changes of isoforms of some antioxidative enzymes. J. Plant Physiol., 161: 531-542.

Reynolds-Henne CE, Langenegger A, Mani J, Schenk N, Zumsteg A, Feller U (2010). Interactions between temperature, drought and stomatal opening in legumes. Environ. Exp. Bot., 68: 37-43.

Simova-Stoilova L, Demirevska K, Petrova T, Tsenov N, Feller U (2009). Antioxidative protection and proteolytic activity in tolerant and sensitive wheat (Triticum aestivum L.) varieties subjected to long-term field drought. Plant Growth Regul., 58: 107-117.

Smeets K, Opdenakker K, Remans T, Sander SV, Belleghem FV, Semane B, Horemans N, Guisez Y, Vangronsveld J, Cuypers A (2009). Oxidative stress-related responses at transcriptional and enzymatic levels after exposure to Cd or Cu in a multipollution context. J. Plant Physiol., 166: 1982-1992.

Stepien P, Klobus G (2005). Antioxidant defense in the leaves of C_3 and C_4 plants under salinity stress. Physiol. Plant, 125: 31-40.

Vaidyanathan H, Sivakumar P, Chakrabarty R, Thomas G (2003). Scavenging of reactive oxygen species in NaCl-stressed rice (Oryza sativa L.) - differential response in salt-tolerant and sensitive varieties. Plant Sci., 165: 1411-1418.

Van Heerden PDR, Krüger GHJ, Loveland JE, Parry MAJ, Foyer CH (2003). Dark chilling imposes metabolic restrictions on photosynthesis in soybean. Plant Cell Environ., 26: 323-337.

Veeranagamallaiah G, Chandraobulreddy P, Jyothsnakumari G, Sudhakar C (2007). Glutamine synthetase expression and pyrroline-5-carboxylate reductase activity influence proline accumulation in two cultivars of foxtail millet (Setaria italica L.) with differential salt sensitivity. Environ. Exp. Bot., 60: 239-244.

5

Comparative phenolic profile of Persian walnut (*Juglans regia L.*) leaves cultivars grown in Iran

Amir Jalili and Asra Sadeghzade*

Department of Biology, Islamic Azad University, Urmia Branch, Urmia, Iran.

In this study Walnut leaves from 11 different cultivars (Mayette, Fernor, Mellanaise, Elit, Orientis, Lara, Hartley, Franquette, Parisienne, Arco and Marbot) were studied for their phenolic compounds. The evolution of major phenolic compounds amounts was monitored from May to September. Two extractive procedures were assayed and the best results were obtained using acidified water (pH 2) and a solid phase extraction column purification step. Qualitative analysis was performed by HPLC-DAD/MS and, in all samples; nine phenolic compounds (3-caffeoylquinic, 3-p-coumaroylquinic and 4-p-coumaroylquinic acids, quercetin 3-galactoside, quercetin 3-arabinoside, quercetin 3-xyloside, quercetin 3-rhamnoside, quercetin 3-pentoside and kaempferol 3-pentoside) were identified. Quantification of phenolic compounds was performed by HPLC-DAD, which revealed that quercetin 3-galactoside was always the major compound while 4-p-coumaroylquinic acid was the minor one. All cultivars presented slightly higher values of total phenolic compounds in May and July.

Key words: *Juglans regia* L., phenolic profile, walnut leaf, high performance liquid chromatography -diode array detector (HPLC-DAD), high performance liquid chromatography/ diode array detector-electrospray ionization-mass spectrometry (HPLC/DAD/ESI-MS).

INTRODUCTION

Thirteen phenolic compounds are identified in walnut hulls: chlorogenic acid, caffeic acid, ferulic acid, sinapic acid, gallic acid, ellagic acid, protocatechuic acid, syringic acid, vanillic acid, catechin, epicatechin, myricetin, and juglone (Stampar et al., 2006). Phenolic compounds are constituents of both edible and non-edible parts of plants (Amarowicz et al., 2004). Several natural antioxidant compounds such as flavonoids, tannins, coumarins, curcuminoids, xanthons, and terpenoids are found in fruits, leaves, seeds, and oils of various plant products. Phenolic compounds are the most active natural antioxidants in plants (Bors et al., 2001). They are known to act as antioxidants (Cuvelier et al., 1992). Walnuts are rich in components that have anti-oxidant and anti-inflammatory properties (Muthaiyah et al., 2011). Walnut (*Juglans regia* L.) leaf has been widely used in folk medicine for treatment of venous insufficiency and haemorrhoidal symptomatology, and for its antidiarrheic, antihelmintic, depurative and astringent properties (Bruneton, 1993; Van Hellemont, 1986; Wichtl and Anton, 1999). Keratolytic, antifungal, hypoglycaemic, hypotensive, anti-scrofulous and sedative activities have also been described (Gırzu et al., 1998; Valnet, 1992). Walnuts are rich in components that have anti-oxidant and anti-inflammatory properties (Muthaiyah et al., 2011).

Juglone (5-hydroxy-1, 4-naphthoquinone) is the characteristic compound of *Juglans* spp., which was reported to occur in fresh walnut leaves (Bruneton, 1993; Gırzu et al., 1998; Wichtl and Anton, 1999). Nevertheless, because of polymerization phenomena, juglone was reported to occur in the drug (dry leaves) only in vestigial amounts (Wichtl and Anton, 1999), which means that the compound is not suitable for use in the quality control of the dry plant. Besides these, other phenolics, namely phenolic acids and flavonoids, have been reported in walnut leaves (Wichtl and Anton, 1999). The chemical characterization of the drug is described

---

*Corresponding author. E-mail: Asra.sadeghzade@yahoo.com.

in some pharmacopoeias, by thin layer chromatography (TLC) detection of quercetin 3-galactoside and quercetin 3-rhamnoside. However, due to their ubiquity in nature, these flavonoids do not guarantee the plant authenticity and, more identified compounds would be more useful for characterisation. In some European countries, dry walnut leaves are still largely used as an infusion. Because flavonoids and phenolic acids have already been successfully applied in the quality control of several foodstuffs (Andrade et al., 1997a, b; Areias et al., 2001; Ramos et al., 1999; Silva et al., 2000). In the present work, phenolics of walnut leaves have been studied by high performance liquid chromatography/ diode array detector-electrospray ionization- mass spectrometry (HPLC/DAD/ESI/MSMS). The evolution of phenolic compounds from May to September was monitored and a useful methodology for routine quality control, based on HPLC-DAD quantification of major phenolics was developed and applied to eleven different cultivars growing under the same agricultural, geographical and climatic conditions.

MATERIALS AND METHODS

Samples

Studies were carried out on walnut leaves from eleven cultivars (Mayette, Fernor, Mellanaise, Elit, Orientis, Lara, Hartley, Franquette, Parisienne, Arco, Marbot) grown in Iran. Fresh leaves were collected at "Saatloo", an orchard in Urmia, in the Northwest of Iran (37°44 N, 45°10 E, altitude 1338 m). The orchard has a planting density of 8×8 m, with all trees being more than eighteen years old. They are pruned when necessary and receive organic fertilization, but no phytosanitary treatments are applied. Fresh samples of all cultivars were collected on the same day, from May to September of 2010, at the end of each month. For each sample, about 100 g of leaves were manually collected from the middle third of branches exposed to sunlight, dried in a stove at 30°C for five days and stored in paper bags in order to protect them from light. Just before phenolic extraction, each sample was powdered to a maximum particle size of 0/5 mm.

Chemicals

The standards were purchased from Sigma (St. Louis, MO, USA) and extrasynthese (Genay, France). Methanol and hydrochloric and formic acids were obtained from Merck (Darmstadt, Germany). The water was treated in a Milli-Q water purification system (Millipore, Bedford, MA, USA).

Solid-phase extraction columns

The ISOLUTE C18 non-end-capped (NEC) solid phase extraction (SPE) columns (50 μm particle size, 60 A porosity; 10 g sorbent mass/70 ml reservoir volume) were purchased from International Sorbent Technology Ltd (UK).

Extraction of phenolic compounds

For analytical purposes, the sample (ca. 0.2 g) was thoroughly mixed with methanol until complete extraction of phenols (negative reaction with NaOH 20%). The extract was then filtered, evaporated to dryness under reduced pressure (40°C), and redissolved in 3 ml of methanol. A chloroformic extract was also prepared with the same sample: ca. 0.5 g of plant material was extracted three times with 100 ml of chloroform, with agitation, for 10 min. The extracts were pooled, taken to dryness under reduced pressure (40°C) and the residue dissolved in 3 ml of methanol. For quantification purposes, each sample (ca. 0.2 g) was thoroughly mixed with acidified water (pH 2 with HCl) until complete extraction of phenolic compounds (negative reaction to NaOH 20%) and filtered. The filtrate was passed through an ISOLUTE C18 (NEC) column, previously preconditioned with 60 ml of methanol, followed by 140 ml of water (pH 2 with HCl). The retained phenolic fraction was eluted with methanol (ca. 75 ml) and the methanolic extract obtained was filtered, evaporated to dryness under reduced pressure (40°C) and redissolved in methanol (3 ml).

High performance liquid chromatography/diode array detector/mass spectrometry (HPLC/DAD/MS/MS) for qualitative analysis

Chromatographic separation was carried out on a reversed-phase LiChroCART column (250×4 mm, RP- 18, 5 lm particle size; Merck, Darmstadt, Germany) using two solvents: trifluoroacetic acid (0.1%) (A) and methanol (B), starting with 30% methanol and installing a gradient to obtain 50% B at 30 min, 70% B at 32 min, 80% B at 33 min and 80% B at 35 min. The flow rate was 1 ml min^{-1}, and the injection volume was 5 μl. The HPLC system was equipped with a diode array detector (DAD) and mass detector in series (Agilent 1100 Series LC/MSD Trap). It consisted of an Agilent G1312A HPLC binary pump, an Agilent G1313A autosampler, an Agilent G1322A degasser and an Agilent G1315B photo-diode array detector controlled by Agilent software v. A.08.03 (Agilent Technologies, Waldbronn, Germany). Chromatograms were recorded at 280, 320 and 350 nm. The mass detector was an Agilent G2445A Ion-Trap Mass Spectrometer (Agilent Technologies, Waldbronn, Germany) equipped with an electrospray ionisation (ESI) system and controlled by Agilent Software v. 4.0.25. Nitrogen was used as nebulizing gas at a pressure of 65 psi and the flow was adjusted to 11 L min^{-1}. The heated capillary and voltage were maintained at 350°C and 4 kV, respectively. The full scan mass spectra of the phenolic compounds were measured from m/z 60 up to m/z 800. Collision-induced fragmentation experiments were performed in the ion trap using helium as the collision gas, with a voltage ramping to 0.3 up to 2 V. Mass spectrometry data were acquired in the negative ionisation mode. MS^2 data were acquired in the automatic mode.

High performance liquid chromatography/diode array detector (HPLC/DAD) for quantitative analysis

Chromatographic separation was achieved with an analytical HPLC unit (Gilson), using a reversed-phase Spherisorb ODS2 (250×4.6 mm, 5 lm particle size, Merck, Darmstadt, Germany) column. The solvent system used was a gradient of water/formic acid (19:1) (A) and methanol (B), starting with 5% methanol and installing a gradient to obtain 15% B at 3 min, 20% B at 5 min, 25% B at 12 min, 30% B at 15 min, 40% B at 20 min, 45% B at 30 min, 50% B at 40 min, 70% B 45 min and 0% B at 46 min. The flow rate was 1 ml min^{-1}, and the injection volume was 20 μl. Detection was accomplished with a DAD (Gilson), and chromatograms were recorded at 320 and 350 nm. Spectral data from all peaks were accumulated in the 200 to 400 nm range. Data were processed on Unipoint system software (Gilson Medical Electronics, Villiers le Bel, France). Phenolic compounds quantification was achieved by the

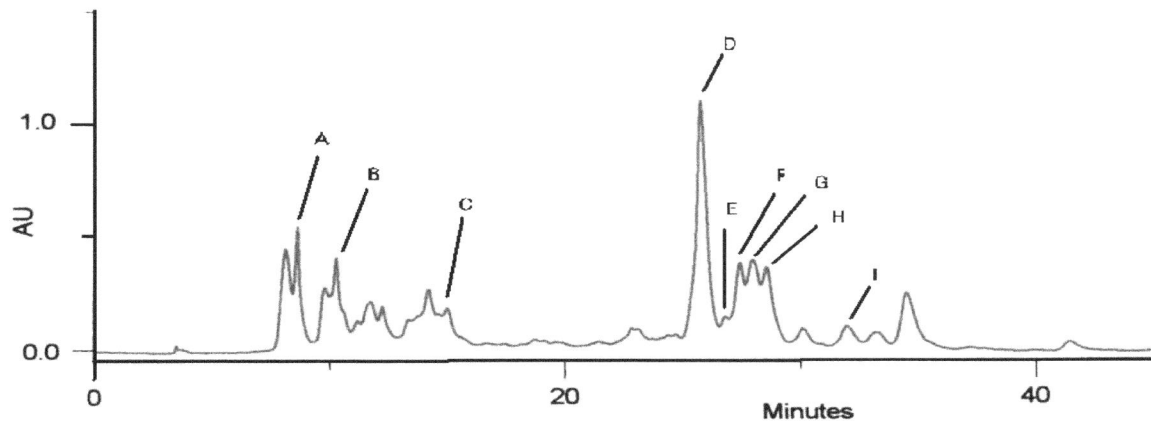

Figure 1. HPLC/DAD walnut leaf phenolic profile. Detection at 320 nm. (A) 3-caffeoylquinic; (B) 3-p-coumaroylquinic acid; (C) 4-p-coumaroylquinic acid; (D) quercetin 3-galactoside; (E) quercetin 3-pentoside derivative; (F) quercetin 3-arabinoside; (G) quercetin 3-xyloside; (H) quercetin 3-rhamnoside; (I) kaempferol 3-pentoside.

absorbance recorded in the chromatograms relative to external standards, with detection at 320 nm for phenolic acids and at 350 nm for flavonoids. 3-O-caffeoylquinic acid was quantified as 5-O-caffeoylquinic acid, 3-p-coumaroylquinic and 4-p-coumaroylquinic acids were quantified as p-coumaric acid; the quercetin3-pentoside derivative and quercetin 3-xyloside were quantified as quercetin 3-arabinoside. The other compounds were quantified as themselves.

RESULTS AND DISCUSSION

As dried leaves were used, juglone was not detected in any extract, which is in accordance with Wichtl and Anton (1999). Gırzu et al. (1998) have reported the isolation of juglone from a chloroformic extract of fresh walnut leaves. In this study, it was possible to detect this compound only in the chloroformic extract from a fresh sample. Bearing in mind that infusion is traditionally prepared with dry plant and that juglone is not detected in the water extract, a methodology based on phenolic compounds determination seemed to be useful for the quality control of walnut leaves.

With the development of electrospray ionisation mass spectrometry (ESI/MS), it has become technically and economically feasible to analyse polar compounds by liquid chromatography coupled with ESI/MS. As several authors have successfully used HPLC/DAD/MSESI in the identification of phenolic compounds in foodstuffs (Llorach et al., 2003; Zafrilla et al., 2001), this technique was applied to walnut leaf in order to identify the highest possible number of compounds. The UV spectra of the compounds obtained by high performance liquid chromatography/diode array detector (HPLC/DAD) analysis revealed that phenolic acids and flavonoids were the two main groups of compounds in walnut leaf extract. The first group, corresponding to peaks A, B, C (Figure 1), presented spectral characteristics of cinnamic acids, with two absorption maxima at 250 and 320 nm. HPLC-

MS data provided some interesting information about those compounds. Fragmentation of pseudomolecular ion [M–H]$^-$ at m/z 353.70, found for compound A, yielded the ion at m/z 191.47 ([M–H]$^-$-162), base peak corresponding to quinic acid by the loss of a caffeoyl radical from the pseudomolecular ion. Besides, in the MS^2 study, the ion at m/z 179.63 was also obtained with an abundance of 35% which, according to Clifford et al. (2003), characterizes 3- caffeoylquinic acid. A pseudomolecular ion [M–H]$^-$ at m/z 333.92 was found for compound B. Fragmentation of this ion yielded a base peak at m/z 163.2, corresponding to the loss of quinic acid radical, which is in accordance with literature data found for 3-p-coumaroylquinic acid (Clifford et al., 2003). Compound C also had a pseudomolecular ion at an identical m/z found for compound B; in the MS^2 study, the base peak was at m/z 173.2. According to Clifford et al. (2003) the compound was identified as 4-p-coumaroylquinic acid. The second group of compounds, corresponding to peaks D–I, showed UV spectra characteristic of flavonoids. Pseudomolecular ions [M–H]$^-$ at m=z 464.61, 434.39, 434.37 and 448.58 were found for peaks D, F, G and H, respectively (Figure 1). Fragmentation of these ions provided a characteristic m=z at 300.87, a typical mass in the negative mode of the quercetin aglycone. Injection of authentic standards of quercetin 3-galactoside, quercetin 3-arabinoside, quercetin 3-xyloside and quercetin 3- rhamnoside confirmed the occurrence of these compounds in walnut leaf extract. Compound E had a pseudomolecular ion [M–H]$^-$ at m/z 432.94 and fragmentation of this also provided a characteristic m/z at 300.9, suggesting the presence of a pentosyl quercetin derivative. In peak I, three compounds with pseudo-molecular ions at m/z 417.4, 475.4 and 489.4 were co-eluting in the same order. An extracted ion chromatogram (EIC), and MS^2 study were done for these ions. The MS of 417.4 yielded a main ion at m/z 284.9,

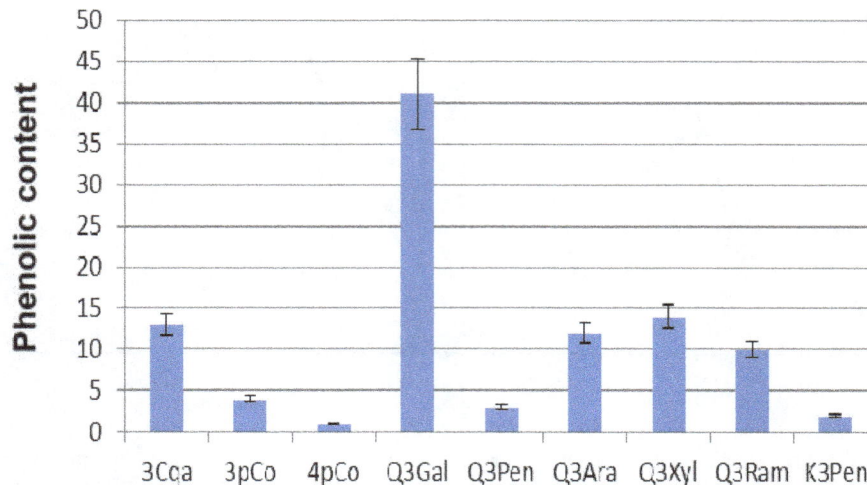

Figure 2. Phenolic fingerprint of Mayette, Fernor, Mellanaise, Elit, Orientis, Lara, Hartley, Franquette, Parisienne, Arco and Marbot cultivars. Results are the means of five samples collected for each cultivar. Identities of compounds are in Figure 1.

characteristic of kaempferol, suggesting the presence of a pentosyl kaempferol derivative. MS data from the all cultivars showed a common qualitative pattern, presenting nine identified phenolic compounds (Figure 1): 3-caffeoylquinic, 3-p-coumaroylquinic and 4-p-coumaroylquinic acids, quercetin 3- galactoside, quercetin 3-arabinoside, quercetin 3-xyloside, quercetin 3-rhamnoside, quercetin 3-pentoside and kaempferol 3-pentoside.

As far as we know, 3-p-coumaroylquinic and 4-p-coumaroylquinic acids are reported in this species for the first time. Wichtl and Anton (1999) described the existence of other phenolic acids in walnut leaves, namely; caffeic, ferulic, p-coumaric, p-hydroxyphenylacetic, gallic, salicylic, chlorogenic and neochlorogenic acids but, with the exception of 3-caffeoylquinic acid, those compounds were not detected in the present study. As no data concerning the cultivar and geographical origin of those samples were found, no correlation between the absence of those compounds and a given cultivar or origin can be inferred. Chloroform was the solvent that extracted a small number of phenolic compounds. Besides, when comparing methanolic extraction and extraction with acidified water and purification via the SPE columns, it was possible to observe that both led to the same qualitative phenolic profile but, as a general rule, the extraction with acidified water, leads to an extract with a higher amount of phenolic compounds. This last technique has the advantage of eliminating chlorophylls and allows the concentration of the extract in a shorter period of time. For quantification purposes all samples were subjected to this procedure. As with the qualitative profile, all the analysed samples showed a common quantitative pattern if the results are analysed as percentages (Figure 2).

In this profile, quercetin 3-galactoside was the major flavonoidic compound (42±2.18%) and 3-caffeoylquinic acid was the major phenolic acid (14±0.73%). A systematic analysis was carried out and samples were collected from May to September, in order to evaluate changes in the phenolic composition of the studied cultivars. All cultivars presented slightly higher values of total phenolic compounds in May and July. Likewise the maximum phenolic content in leave extract was 35.9 (g/kg, dry basis) for Lara cultivar in May and minimum phenolic content was 6.2 (g/kg, dry basis) for Hartley cultivar in June (Figure 3).

Mean values of phenolic acids, flavonoids and total phenolic contents seem to point to a decrease of compounds from May to June, an increase in July and a new decrease in September. The first decrease might be related to the rapid development of the fruit in June, when most of the nutrients and photoassimilates are employed for fruit growth (Charlot and Germain, 1988). The hypothesis that flavonoid content is related to sun exposure, because of their function as sun filters, may possibly explain their rise in July, since this was the month with a higher value of solar radiation (Figure 4).

In conclusion this study suggests that the technique herein described seems to be quite useful for analysis of walnut leaf phenolic compounds. This set of compounds, when examined qualitatively and quantitatively defines a fingerprint that may be suitable for assessing identity and quality. The nature of the cultivar and the month of collection do not seem to influence the mentioned phenolic fingerprint of walnut leaves. Bearing in mind that flavonoids and phenolic acids have been the subject of several studies because of their antioxidant potential (Aquino et al., 2001; Sakakibara et al., 2003; Valentao et al., 2002), the results obtained suggest that, for this

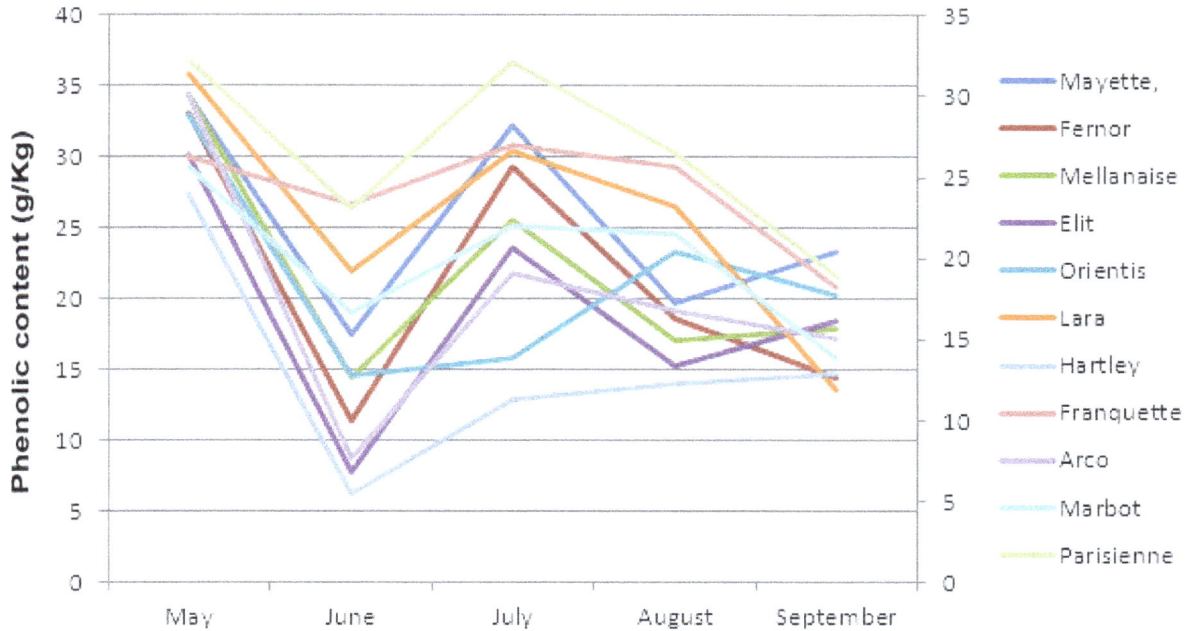

Figure 3. Total phenolics of walnut leaf samples from May to September.

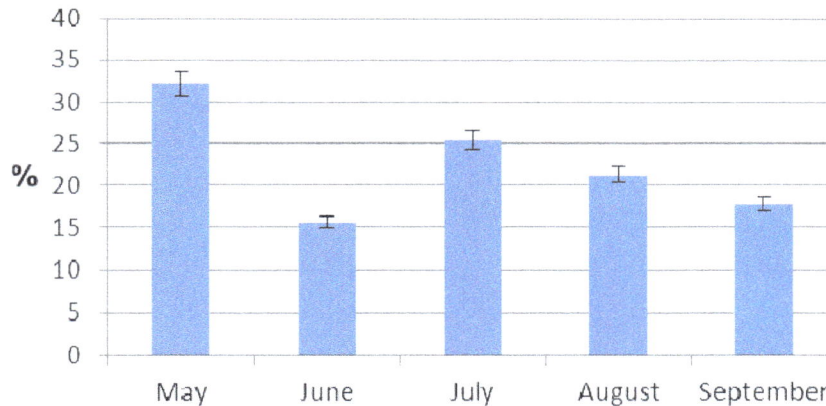

Figure 4. Total phenolics percentage, between May and September 2010, of the walnut leaf samples. Results are the means of the six analysed cultivars and standard error bars are on the top of each column.

purpose, walnut leaves should preferentially be collected in May or July, when phenolic content is higher. Besides, the phenolics present, mainly caffeic acid and quercetin derivatives, are excellent antioxidants since they bear the required structural characteristics for that, namely an ortho-dihydroxy group on an aromatic ring. In addition, Parisienne cultivar has the highest total phenolic compounds in comparism with other cultivars.

REFERENCES

Amarowicz R, Pegg RB, Rahimi-Moghaddam P (2004). Free radical scavenging capacity and antioxidant activity of selected plant species from the Canadian prairies. Food Chem., 84: 551-562.

Andrade P, Ferreres F, Amaral MT (1997a). Analysis of honey phenolic acids by HPLC, its application to honey botanical characterization. J. Liquid Chromatogr. Related Technol., 20: 2281–2288.

Andrade PB, Leitao R, Seabra RM, Oliveira MB, Freira MA (1997b). 3,4-Dimethoxycinnamic acid levels as a tool for differentiation of Coffea canephora var. robusta and Coffea arabica. Food Chem., 61: 511–514.

Aquino R, Morelli S, Lauro MR, Abdo S, Saija A, Tomaino A (2001). Phenolic constituents and antioxidant activity of an extract of Anthurium versicolor leaves. J. Nat. Prod., 64: 1019–1023.

Areias FM, Valentao P, Andrade PB, Ferreres F, Seabra RM (2001). Phenolic fingerprint of peppermint leaves. Food Chem., 73: 307–311.

Bors W, Michel C, Stettmaier K (2001). Structure–activity relationships governing antioxidant capacities of plant polyphenols. Methods Enzymol., 335: 166-180.

Bruneton J (1993). Pharmacognosie, phytochimie, medicinal plants. Paris: Tec & Doc-Lavoisier. p. 348.

Charlot G, Germain E (1988). The Walnut New Techniques. Paris: Ctifl.

Clifford MN, Johnston KL, Knight S, Kuhnert N (2003). Hierarchical scheme for LC-MSn identification of chorogenic acids. J. Agric. Food Chem., 51: 2900–2911.

Cuvelier ME, Richard H, Berset C (1992). Comparison of the antioxidant activity of some acid phenols: structure–activity relationship. Biosci. Biotechnol. Biochem., 56: 324-325.

Gırzu M, Carnat A, Privat AM, Fiaplip J, Carnat AP, Lamaison JL (1998). Sedative effect of walnut leaf extract and juglone, an isolated constituent. Pharmaceut. Biol., 36(4): 280–286.

Llorach R, Gil-Izquierdo A, Ferreres F, Tomas-Barberan FA (2003). HPLC-DAD-MS/MS ESI characterization of unusual highly glycosylated acylated flavonoids from cauliflower (Brassica oleracea L. var. botrytis) agroindustrial byproducts. J. Agric. Food Chem., 51: 3895–3899.

Muthaiyah B, Essa MM, Chauhan V, Chauhan A (2011). Protective Effects of Walnut Extract Against Amyloid Beta Peptide-Induced Cell Death and Oxidative Stress in PC12 Cells. Neurochem. Res., 36(11): 2096–2103.

Ramos R, Andrade PB, Seabra RM, Pereira C, Ferreira MA, Faia MA (1999). A preliminary study of non-coloured phenolics in wines of varietal white grapes (codega, gouveia and malvasia fina): Effects of grape variety, grape maturation and technology of winemaking. Food Chem., 67: 39–44.

Sakakibara H, Honda Y, Nakagawa S, Ashida H, Kanazawa K (2003). Simultaneous determination of all polyphenols in vegetables, fruits and teas. J. Agric. Food Chem., 51: 571–581.

Silva BM, Andrade PB, Valentao P, Mendes GC, Seabra RM, Ferreira MA (2000). Phenolic profile in the evaluation of commercial quince jellies authenticity. Food Chem., 71: 281–285.

Stampar F, Solar A, Hudina M, Veberic R, Colaric M (2006). Traditional walnut liqueur – cocktail of phenolics. Food Chem., 95: 627-631.

Valentao P, Fernandes E, Carvalho F, Andrade PB, Seabra RM, Bastos ML (2002). Antioxidative properties of cardoon (Cynara carduncunlus L.) infusion against superoxide radical, hydroxyl radical, and hypochlorous acid. J. Agric. Food Chem., 50: 4989–4993.

Valnet J (1992). Herbal treatment of diseases in plants. Paris: Maloine, pp: 476–478.

Van Hellemont J (1986). Compendium de phytotherapie. Bruxelles: Association Pharmaceutique Belge, pp. 214-216.

Wichtl M, Anton R (1999). Plantes therapeutiques. Paris: Tec. Doc., pp. 291-293.

Zafrilla P, Ferreres F, Tomas-Barberan FA (2001). Effect of processing and storage on the antioxidant ellagic acid derivatives and flavonoids of red raspberry (Rubus idaeus) jams. J. Agric. Food Chem., 49: 3651-3655.

6

Fatty acid composition and properties of skin and digestive fat content oils from *Rhynchophorus palmarum* L. larva

Edmond Ahipo Dué[1]*, Hervé César B. L. Zabri[2], Jean Parfait E.N. Kouadio[1] and Lucien Patrice Kouamé[1]

[1]Laboratoire de Biochimie et Technologie des Aliments de l'Unité de Formation et de Recherche en Sciences et Technologie des Aliments de l'Université d'Abobo-Adjamé, 02 BP 801 Abidjan 02, Côte d'Ivoire.
[2]Laboratoire de Chimie Bio-organique de l'Unité de Formation et de Recherche des Sciences Fondamentales Appliquées de l'Université d'Abobo-Adjamé, 02 BP 801 Abidjan 02, Côte d'Ivoire.

Skin and digestive fat content (DFC) oils from *Rhynchophorus palmarum* L. larva (Curculionidae) were extracted and their physicochemical properties were characterized. Water content (0.41 %) of skin oil was higher than the amount of DFC (0.04 %). While, the lipid fraction of the skin (35.16%) was slightly lower than the DFC (49.05%). The fatty acid compositions of the both oils were determined. Results showed that the most abundant fatty acids in skin and DFC oils were palmitic and oleic acids. In both oils, oleic fatty acid showed the highest percentage of composition of 45.62 and 46.71% for skin and DFC, respectively with palmitic acid followed close by 39.87 and 40.44%, respectively. In this study, saturated fatty acids accounted for 45.06 and 44.97% of total fatty acids, for skin and DFC oils, respectively. Myristic, myristoleic, stearic and linoleic acids were also detected in the both oils. Physicochemical properties of skin and DFC oils respectively include: iodine index, 51.22 and 48.35; acid value, 4.72 and 2.21; saponification value, 189.22 and 198.26; unsaponifiable matter, 0.97 and 0.98; peroxide index, 6.90 and 0; oleic acidity, 7.76 and 0.568; vitamin A, 0 and 12.04 and refractive index, 1.45440 and 1.45424. Results suggested that Skin and DFC oils from *R. palmarum* L. larva could deserve further consideration and investigation as a potential new multi-purpose product for nutritional, industrial, cosmetic and pharmaceutical uses.

Key words: Fatty acids, digestive fat content, skin of larvae, oil, *Rhynchophorus palmarum*.

INTRODUCTION

Insects have played an important role in history of human nutrition in Africa, Asia and Latin America (Bodenheimer, 1951). The larva of *Rhynchophorus palmarum* L., a Coleoptera of Curculionidae family is used as traditional food in several countries (Sanchez et al., 1996; 1997; Cerda et al., 1999). It is a delicacy in many part of Côte d'Ivoire and other countries in Africa where it is found. Generally, the *Rhynchophorus* larvae are strongly looked for the people who believe them to have high nutritive as well as certain pharmaceutical values.

In several countries, these larvae are consumed in their entireties prepared either by stewing, frying in oil with salt and pepper, adding to squash seed paste, or putting on brochettes grilled over coals. In Cameroon for example, these larvae called "FOS" are washed by lot of water and pierced in the abdomen with a sharp pierce of bamboo between each washing to let a white fatty liquid escape before any cooking (brochette, frying etc...) (Grimaldi et Bikia, 1985).

Traditionally, there are many claims that, the *Rhynchophorus* larvae have medicinal properties. In Delta state of Nigeria, the Itskiris believe that *R. phoenisis* larvae could cure a certain ailment in infants which presents such symptoms as the twitching of the hands and feet, restlessness between others (Ekpo, 2003). For certain Amazonian population who live in the forest, *R. palmarum* L. larva's oil is used for treating respiratory sickness (Bourdy et al., 2000). The biochemical basis for this treat-

*Corresponding author. E-mail: ahipoedmond@yahoo.fr.

ment is not known. The nutritional compositions of this larva used as a food by the Amazonian Indians and its palm substrata have been fairly well investigated by Cerda et al. (1999).

Several authors showed that some of the insects which are pests also have high nutritional qualities. Proximate composition of these insects have been studied from Central Africa (Richards, 1939), South Africa (Quinn, 1959; Dreyer and Weameyer, 1982) and South America (Dufour, 1987). Insects provide high quality of proteins and supplements (minerals and vitamins) (Banjo et al., 2006).

Nutritionally, a high level of saturated fatty acids in foods might be undesirable because of the linkage between saturated fatty acids and atherosclerotic disorders (Rahman et al., 1995). The presence of the essential fatty acids such as linoleic, linolenic and arachidonic acids in substantial amounts further points to the nutritional value of the larval oil. One implication of the high fat content in the insect larva is that it may increase susceptibility of the undefatted larva to storage deterioration via lipid oxidation (Greene and Cumuze, 1982). This may then be accompanied by increased browning reactions concurrent with reduced lysine availability (Pokorny, 1981).

However, fatty acid composition and properties of the oil from R. palmarum larvae are not yet known.
To evaluate the nutritional value of some parts of this larva eliminated by populations before consumption, we suggested studying in this work the characteristics of skin oil (obtained by extraction in Sohxlet) and digestive fat content (DFC) oil (obtained by heating) extracted from larvae of R. palmarum L.

MATERIALS AND METHODS

Materials

Hundred live larvae of R. palmarum L. were collected in Epimbé, Adzopé subprefecture (Côte d'Ivoire) palm grove. The species were specifically identified in the entomology department of Abobo-Adjamé University (Côte d'Ivoire).

Methods

Skin and digestive fat content (DFC) obtaining

Larval heads were cut, using a pair of chisel. Press then the abdomen upwards to bring out all the digestive fat content. Skin and digestive fat content are so separated.

Lipid extraction

Skin of larvae was dried at 70°C for seven days, and grinded for obtaining skin powder. Hexan was used as a solvent, and 140 g of skin powder of larva was dissolved in it, in a soxhlet extractor during four hours, to obtain skin oil of larva of the weevil.

Digestive fat content (DFC) oil was extracted from 158 g of DFC of R. palmarum L. for 30 min by heating. The extracted lipid was obtained by filtration of the extract.

Physicochemicals analysis of oils extracted from larva

The weights of oils extracted from 110 g of skin powder and 158 g of R. palmarum L DFC were determined to calculate the lipid contents. Result was expressed as the percentage of lipids in the dry matter of skin powder or in wet matter of DFC.

Acid index and acidity of R. palmarum L oils were determined according to AOCS Ca 5-40 official method with the morn ISO-9001. A volume of 100 ml ethanol was neutralised with a solution of NaOH (N/10). The titration was performed using a solution of KOH (1N) in the presence of phenolphthalein.

Iodine values of R. palmarum L oils were determined according to NF ISO 3961 (February 1990) method, and ISO 9001, norm. A volume of 30 ml carbon tetrachloride was used to dilute 0.4 g oil in the presence of 25 ml of Wijs reagent and 10 ml of acetate mercury. The titration was carried out using a solution of sodium thiosulfate (0.1 N).

The peroxide value was determined according to AOCS Cd-8-53/1960 and ISO 9001 norm. A solution made out of potassium iodine added to a mixture of acetic acid – chloroform: 3/2 (v/v) has been used. The titration was carried out using a solution of sodium thiosulfate (N/100).

Saponification values, unsaponifiables and oils moistures were determined according to official method AOCS-Ca-2b-38-T60-2. An amount of 2 g oil has been treated using alcoholic potash (0.5 N) and titrated hot with hydrochloric acid (0.5 N) in the presence of phenolphthalein.

Vitamin A was extracted according to the method described by Panfili et al. (1994).

Fatty acid composition was analysed by gas-liquid chromatography after derivatives changed to fatty methyl esters (FAMEs) with 2 M KOH in methanol at room temperature following IUPAC standard method (IUPAC, 1992). Analyses of FAMEs were carried out with a Hewlett-Packard 5890 series gas chromatograph GC system equipped with a hydrogen flame ionisation detector and a capillary column: HP-5 Cross-linked 5% PH ME Siloxane (30 m x 0.32 mm x 0.25 µm film). The column temperature was programmed from 160 to 325°C at 5°C/min and the injector and detector temperatures were set at 275 and 325°C, respectively.
Identification and quantification of FAMEs was accomplished by comparing the retention times of peaks with those of pure standards purchased from Sigma and analysed under the same conditions. The results were expressed as a percentage of individual fatty acid in the lipid fraction.

Refractive index of R. palmarum L oils was determined using a refractometer RFM 81, Multisecale Automatic from Bioblock Scientific.

Densities were determined using the NF T-60-214 method and ISO 9001norm.
All analytical experiments were repeated three times. Values of each parameter were shown as mean ± standard deviation (mean ± S.D.).

Statistical analysis was carried out by Student's t-test using SPSS Version 11.0 software and ANOVA (Duncan multiple range test) using SAS system Version 8e. $p < 0.05.$, was considered, as level of error.

RESULTS

Table 1 shows Skin and Digestive fat content (DFC) oils moistures and lipid value from R. palmarum L. larva. Analysis showed that the water content of skin and digestive fat content (DFC) oils from the larva were respectively 0.41 ± 0.020 and 0.04 ± 0.004%. It also showed that the lipid values obtained from skin (35.16) was lower than the DFC (49.05).

Table 1. Skin and Digestive fat content (DFC) oils water and lipid content (g/100 g of matter of weevil's larva).

	Skin	Digestive fat content (DFC)
Moisture (%H$_2$O)	0.41 ± 0,020[a]	0.04 ± 0,004[b]
Lipid value (g/100g of matter)	35.16 ± 0,0002[a]	49.05 ± 0,001[b]

All given values are means as three repeats for each experiment. Means for the determined values in the same line followed by the same superscript letter are not significantly different (p<0.05).

Table 2. Fatty acid composition (g/100 g of total fatty acid) of skin and digestive fat content oils from *Rhynchophorus palmarum* L larva. Values are all expressed as percentages.

Fatty acid	% composition	
	DFC	Skin
Myristic acid (C14:0)	2.54 ± 0.135[a]	3.02 ± 0.041[a]
Myristoleic acid (C14:1)	2.06 ± 0,002[a]	1.91 ± 0,005[a]
Palmitic acid (C16 :0)	40.44 ± 0.088[a]	39.87 ± 0.11[a]
Stearic acid (C18 :0)	1.99 ± 0,01[a]	2.17 ±0,009[a]
Oleic acid (C18 :1)	46.71 ± 0,008[a]	45.62 ±0,02[a]
Linoleic acid (C18:2)	6.24 ± 0,005[a]	7.37 ± 0,004[a]

All given values are means of three repeats for each experiment. Means for the determined values in the same line followed by the same superscript letter are not significantly different (p<0.05).

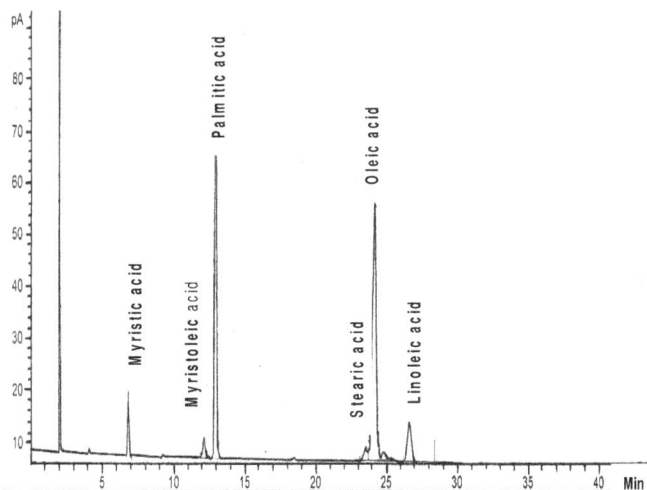

Figure 1. Fatty acid profile of skin oil from *Rhynchophorus palmarum* L larva. Fatty acid composition was analysed by liquid gas chromatography after derivatives changed to fatty methyl esters (FAMEs) with 2 M KOH in methanol at room temperature according to the IUPAC standard method (IUPAC, 1992).

Fatty acid composition determination was another important characteristic carried out on the studied oils from *R. palmarum* L. larva (Table 2). The must abundant fatty acids in the skin and DFC oils were the oleic, palmitic, linoleic followed by the myristic, myristoleic and stearic fatty acids. The major fatty acid were oleic fatty acid with 45.67% (skin oil) and 46.71% (DFC oil), followed by palmitic acid with 39.87 (skin oil) and 40.44% (DFC oil). Figures 1 and 2 show the fatty acid profiles of skin and DFC oils, respectively. There was no significant difference (p>0.05) in amounts of the fatty acids in the skin and DFC oil samples. In this study, saturated acids accounted for 45.06 and 44.97% of total fatty acids, for skin and DFC oils respectively (Table 3). However, these oils are enough unsaturated with a total monounsaturated fatty acids as most important in skin and DFC oils.

Table 4 shows the physical characteristics of skin and DFC oils extracted from weevil's larvae.

These oils were clear liquids with light brown and light yellow colours respectively for skin and DFC origins. Refractive index was studied and showed that there was no significant difference (p>0.05) in the values of the both oils (1.45440 for skin and 1.45424 for DFC oils).

The chemicals characteristics of skin and DFC oils are given in the Table 5. Peroxide value of DFC oil was nil contrary to that of skin that was relatively high (6.90 ± 0.57). The low values of iodine indexes (51.22 ± 0.25 for skin and 48.35 ± 0.55 for DFC oils) indicated that these oils could not be used mainly as unsaturated oils. The acid value of DFC oil (2.21 ± 0.02) was lower than the skin one (4.72 ± 0.06). This result was similar to oleic acidity of these oils. As a result of determining the saponification indexes, skin oil showed a lower number (189.22 ± 0.92) compared to the DFC one (198.26 ± 0.99) but these indexes were much higher in the both oils. There was significant difference (p<0.05) in the saponification numbers and also the density (0.79 ± 0.0 for skin and 0.77 ± 0.00046 for DFC oils) of the both oils.

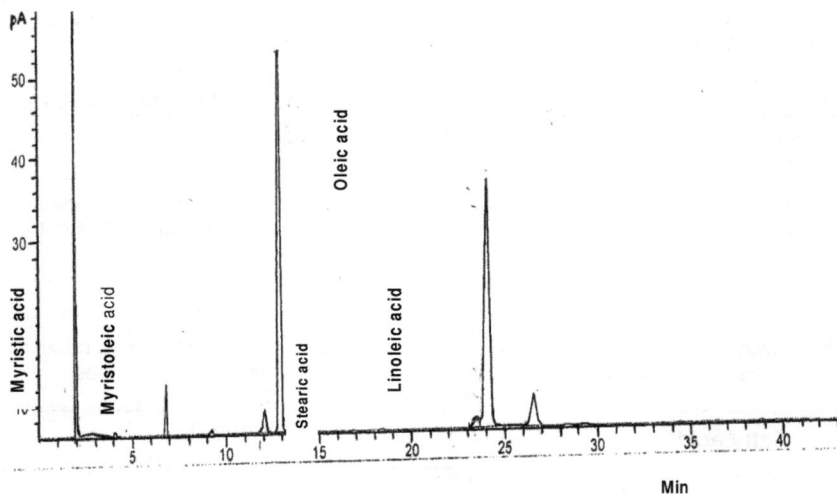

Figure 2. Fatty acid profile of digestive fat content (DFC) oil from *Rhynchophorus palmarum* L larva. Fatty acid composition was analysed by liquid gas chromatography after derivatives changed to fatty methyl esters (FAMEs) with 2 M KOH in methanol at room temperature according to the IUPAC standard method (IUPAC, 1992).

Table 3. Degree of saturation of skin and digestive fat content oils from *Rhynchophorus palmarum* L larva expressed, as percentages.

	DFC	Skin
TUFA	55.00	54.90
TSFA	44.97	45.06
MUFA	48.77	47.53
PUFA	6.24	7.37

TUFA, Total unsaturated fatty acid; TSFA, Total saturated fatty acid; MUFA, Monounsaturated fatty acid; PUFA, Poly unsaturated fatty acid.

Table 4. Physical characteristics of skin and DFC oils from *Rhynchophorus palmarum* L larva.

Physical characteristics	Values (DFC)	Values (Skin)
Density	0.77 ± 0.00046^a	0.79 ± 0.0^b
Refractive index	1.45424 ± 0.55^a	1.45440 ± 0.25^a
Colour	Light yellow	Light brown

All given values are means of three repeats. Means for the determined values in the same line followed by the same superscript letter are not significantly different ($p < 0.05$).

Table 5. Chemical characteristics of skin and DFC oils from *Rhynchophorus palmarum* L larva.

Chemical characteristics	Values (DFC)	Values (Skin)
Iodine index (g d'l$_2$ / 100 g of oil)	48.35 ± 0.55^a	51.22 ± 0.25^b
Saponification value (mg of KOH / g of oil)	198.26 ± 0.99^a	189.22 ± 0.92^b
Peroxide index (meq O$_2$ / kg of oil)	0^a	$6.90 \pm 0.57b$
Acid value (mg of KOH / g of oil)	2.21 ± 0.02^a	4.72 ± 0.06^b
Oleic acidity (%)	0.568 ± 0.02^a	7.76 ± 0.14^b
Unsaponifiable matter (%)	0.98 ± 0.12^a	0.97 ± 0.44^a
Vitamin A	12.04 ± 0.021^a	0.00^b

All values given are means of three determinations. Means for the determined values in the same line followed by the same superscript letter are not significantly different ($p < 0.05$).

The unsaponifiable matter of the *R. palmarum* L. larva oils were similar (0.97 ± 0.44 for skin and 0.98 0.12 for DFC oils). However, the vitamin A was present only in DFC oil. As the results of determining the iodine values, *R. palmarum* L. larva oils showed a much lowers numbers. There was also significant difference ($p<0.05$) in these low iodine indexes.

DISCUSSION

This study revealed that the oils from skin and DFC of *R. palmarum* L. larva possess high nutritional qualities.

Water content (0.41%) of skin was higher than the amount of DFC (0.04%) while, the lipid fraction of the skin (35.16 %) was slightly lower than the DFC one (49.05%). The high percentages of *R. palmarum* L larva oils make this insect a distinct potential for the oil industry. Variation in oils yield may be due to the different larval parts; the extraction method used and the larva nutritional metabolism. Skin lipid value was higher than those (dry matter basis) reported from other insects include 3.1 and 4.0%, respectively for the larva and adult beetle of *Lachnosterna* species (Davis, 1918), 7.21% for dried *Melanoplus* (Mc Hargue, 1970), 2.1% for the Japanese beetle, *Popillia japonica* nuwman (Fleming, 1968), 15.5% in the pupae of housefly *Musca domestica* (Calvert et al., 1969). Teotia and Miller (1974) reported the work done by Calvert et al. (1969) and reported the same results except that the lipids content was a little lower.

However, skin lipid value (35.16%) was lower than those reported by Leung (1972) for termites (55.24%); by Ukhun and Osasona (1985) for *Macrotermes bellicosus* (46.10%) and by Ekpo and Onigbinde (2007) for *M. bellicosus* (36.12%). Furthermore, the DFC lipid value (49.05% wet weight basis) was higher than those reported by Ekpo and Onigbinde (2004, 2005) for *R. pheonicis* (25.30 ± 0.20) and Oryctes rhinoceros larval (14.87 ± 0.33) oils, respectively. The fat content of these larvae could have contributed to its highly acceptable flavour when fried or roasted.

Fatty acid composition study revealed that the most abundant fatty acids in skin and DFC oils were palmitic and oleic acids. In both oils, oleic fatty acid showed the highest percentage of composition of 45.62 ± 0.02 and 46.71 ± 0.008 for skin and DFC, respectively with palmitic acid followed close by with 39.87 and 40.44%, respectively. In this study, saturated acids accounted for 45.06 and 44.97% of total fatty acids, for skin and DFC oils, respectively. Variances analysis showed that there was no significant difference ($p<0.05$) in these different type of fatty acids. Among them, the main saturated normal chain fatty acids were palmitic, myristic and stearic acids.

As shown in Table 2, Skin and DFC oils from *R. palmarum* L larva have unsaturated (TUFA) values similar to that observed for palm oil which is common household oil and the termite *M. bellicosus* oil (Ekpo and Onigbinde,

2007). Insect fatty acids are similar to those poultry and fish in their degree of insaturation, with some groups being higher in oleic and linoleic which are essential fatty acids (De Foliart, 1991). These oils can be used as source of palmitic acid which is an excellent energy-giving food. Industrially, it is used for margarine manufacture, and also hard soaps. These saturated fatty acids are also important for production of oils in paint industry (http://fr.wikipedia.org/wiki/Acide palmitique).

Physical characteristics as refractive index, density and colour of *R. palmarum* L oils were studied. There was no significant difference ($p>0.05$) in the refractive indexes of the both oils (1.45440 ± 0.25 for skin and 1.45424 ± 0.55 for DFC). These refractive indexes were similar to that of olive oils studied by Lalas and Tsakins (2002); that of *Nigella sativa* L. oil (Rouhou-Cheikh et al., 2007) and that those for Arachis and olive oils (Pearson, 1976). This implies that the oils from this insect are lighter as these oils that have been considered to be of high quality and as such find much use in the pharmaceutical industries. The chemical properties of oil are amongst the most important properties that determines its present condition. Free fatty acid of skin oil (4.72 ± 0.72) was higher than DFC oil one (2.21 ± 0.035) as well as oleic acidity (7.76 ± 0.14 and 0.568 ± 0.02, respectively). These characteristics in skin oil were significantly higher ($p<0.05$) than those of DFC oil. Furthermore, peroxide index of DFC oil was 0 meqO2/Kg and 6.90 meqO2/Kg ± 0.77 for skin oil. These results showed that DFC oil has high stability to oxidation and its lower acid value is also an indication of its lower susceptibility to rancidity compared to skin oil. The iodine indexes of *R. palmarum* L larva oils studied were lower than those reported for most insect lipids as reported for winged reproductives of the termite *M. Bellicosus* oil (108 ± 0.15) ; lepidopterous larvae (between 112-159); *Phytophagous chrysomefids* (106.6-118); *Rhynchophorus phoenicis* larvae oil (123.6); *Oryctes rhinoceros* larval oil (140) (EKPO and Onigbindé (2007); Wigglesworth (1976); Ekpo, (2003)).

As a result of determining the saponification values, *R. palmarum* L. larva oils showed 189.22 ± 0.92 and 198.26 ± 0.99 for skin and DFC, respectively. These parameters were higher than those reported for Mounga oleifera seed oil (164.26 ± 1.49 and 163 ± 0.98) by Abdulkarim et al. (2005). However, *R. palmarum* L larva oils saponification values were similar to those showed by Thiégang et al. (2004) for *Ricinodendron heudelotty* (Bail.) oils (between 193 and 195); and by Oomah et al. (2000) for Raspberry (*Rubus idaeus* L.) seed oils (between 191 and 192).

R. palmarum L. larva was shown to be rich in vitamin A (equivalent to 85.0 g of retinol) (Cerda et al., 1999). In the present work, the vitamin A was found in DFC oil (12.04 ± 0.02) but not in skin oil, which suggests that the vitamin A was produced in this part of *R. palmarum* L. larva.

This study showed that the *R. palmarum* L larvae oils have the potential to be developed either for food, pharmaceutical and chemical industries.

REFERENCES

Abdulkarim SM, Long K, Lai OM, Muhammad SKS, Ghazali HM (2005). Some physical-chemical properties of *Moringa oleifera* seed oil extracted using solvent and aqueous enzymatic methods. Food Chem. 93: 253-263.

Banjo AD, Lawal OA, Songonuga AE (2006). The nutritional value of fourteen species of edible insects in southwestern Nigeria. African. J. Biotechnol. 5: 298-301.

Bodenheimer FS (1951). Insect as Human Food, W. Suuk, The Hague, p.352.

Bourdy G, Dewalt S, Roca A, Chavez de Michel LR, Deharo E, Munoz V, Balderrama L, Quenevoc C, Gimenez A (2000). Medicinal plants uses of the Tacana, an Amazonian bolivian ethnic group. J. Ethnopharmacol. 70: 87-109.

Calvert CC, Martin RD, Morgan NO (1969). House fly pupae as food for poultry.J. Econ. Entomol. 62: 938-939.

Cerda H, Martinez R, Briceño N, Pizzoferrato L, Hermoso D, y Paoletti M (1999). Cria, Analysis nutricional y sensorial del picudo del cocotero R.P (Coleoptera: Curculionidea), insecto de la dieta traditional indigena Amazonica. *Ecotropicos* 12: 25-32.

Davis JJ (1918). Common white grubs.USDA Farmer's Bull. No, 940; 28.

DeFoliart GR (1991). Insect fatty acids: Similar to those of poultry and fish in degree of unsaturation, but higher in the polyunsaturates .*Food* Insects Newslet. 4.

Dreyer JJ, Weameyer AS (1982). On the nutritive value of mopanie worms. South Africa J. Sci. 78: 33-35.

Dufour DL (1987). Insects as food a case study from the Northwest Amazon. Am. Anthropol. 89: 383-397.

Ekpo KE (2003). Biochemical investigation of the nutritional value and toxicological safety of entomophagy in Southern Nigeria .ph.D thesis, Ambrose Alli university, Ekpoma. Edo state, Nigeria.

Ekpo KE, Onigbinde AO (2004). Pharmaceutical potentials of *Rhynchophorus phoenicis* larval oil. Nig. Annals of Natural Sci. 5: 28-36.

Ekpo KE, Onigbinde AO (2005). Nutritionnal potentials *oryctes rhinoceros* larva. Nig. J. Nut. Sci. 26: 54-59.

Ekpo KE, Onigbinde AO (2007). Characterization of lipids in winged Reproductives of the termite *macrotermes bellicosus*. Pakistan. J. Nutr. 6: 247-251.

Fleming WE (1968). Biological control of the Japanase beetle. USDA tech. Bull. N., 1383: 78.

Greene BE, Cumuze TH (1982). Relationship Between TBA Numbers and inexperienced panellists assessments of oxidised flavour in cooked beef. J. Food .Sci. 47-52.

Grimaldi J, Bikia A (1985). Le grand livre de la cuisine Camerounaise, p.136

http://fr.wikipedia.org/wiki/acide palmitique.

International Union of Pure and Applied Chemistry (IUPAC) (1992). In Paquot C, hautfenne A (Eds.). Standard method for analysis of oils, fats and derivatives (7th ed.) London; Blackwell Scientific Publications.

Lalas S, Tsakins J (2002). Characterization of *Moringa oleifera* seed oil variety "periyakulam 1". J. Food. Compos. Anal. 15: 65-77.

Leung WW (1972). Food consumption table for use in East Asia. FAO, U.S. Govt Printing Office. Washington D.C.

Mc Hargue JS (1970).A study of the proteins of certain insects with reference to their value as food for poultry.J. Agric. Res. 10: 633-637.

Oechlschlager AC, Chinchilla CM, Gonzalez LM, Jiron LF, Mexzon R, Morgan B (1993). Development of a pheromone-based trapping system for *rhynchophorus palmarum* (Coleoptera, Curculionidae). J. Econ. Entomol. 86: 1381-1392.

Oomah DB, Ladet S, Godfrey VD, Liang J, Girard B (2000). Characteristics of raspberry (*Rubus ideaus* L.) seed oil. Food Chem. 69: 187-193.

Panfili G, Manzi P, Pizzoferrato L (1994). HPLC simultaneous determination of tocopherol , carotenes, retinol and its geometric isomers in italian cheeses.analyst. 119: 1161-1165.

Pearson D (1976). The Chemical Analysis of Foods. 7th Edition. Churchill livingstone (publisher); pp. 491-516.

Pokomy J (1981). In: Progress in Food and Nutrition Science. Pergamon Press, Oxford, New york, 5: 1-6, 421.

Quinn PJ (1959). Foods and Feeding Habits of the Pedi, Witwatersrand, University, Johannesbury Republic of South Africa. pp. 278.

Rahman SA, Ituah TS, Hassa n O, David MN (1995). Fatty acid composition of some Malaysian freshwater fish. J. Food Chem. 54: 45-49.

Richards A (1939). Land Labour and diet in Northern Rhodesis, XV1-1-214, Oxford Univ, Press For, Int. Afr. Inst. London.

Rouhou-cheickh S, Besbes S, Hentati B, Blecker C, Deroanne C, Attia H (2007). *Nigella sativa* L. Chemical composition and physico-chemical characteristics of lipid fraction. Food Chem. 101: 673-681.

Sanchez P, Cerda H, Cabrera A, Caetano FH, Materán M, Sánchez F (1996). Secretory mechanisms for the male produced aggregation pheromone of the palm weevil *Rhynchophorus palmarum* L. (Coleoptera: Curculionidae). J. Insect. Physiol. 42 : 1113-1119.

Sanchez P, Jaffe K, Hevia P (1997). Comsumo de insectos : alternativa proteica del Neotropico. Boletin de Entomologia Venezolana 12 : 125-127

Tchiégang C, Oum NM, Dandjouma AA, kapsen C (2004). Qualité et stabilité de l'huile extraite par pressage des amandes de *Ricinodendron heudelotii* (Bail.) Pierre ex Pax pendant la conservation à température ambiante. J. Food. Engr. 62 : 69-77.

Teotia JS, Miller BF (1974). Nutritive content of housefly pupae and manure residue. Br. Poult. Sci., 15: 177-182.

Ukhun ME, Osasona MA (1985). Aspects of the nutrition chemistry of *Macrotermis Bellicosus*.Nutr. Rep. Int., 32: 1121-1129.

Wigglesworth VB (1976). The principles of insect physiology, 7th Edition. Methuen and Co Ltd. (publishers). London, p. 594.

7

Effect of boiling on the phytochemical constituents and antioxidant properties of African pear *Dacryodes edulis* seeds *in vitro*

T. Ogunmoyole[1]*, I. J. Kade[1], O. D. Johnson[1] and O. J. Makun[2]

[1]Department of Biochemistry, Federal University of Technology, P. M. B 704, Akure, Ondo State, Nigeria.
[2]Department of Science Laboratory Technology, Auchi Polytechnic, Auchi, Edo State, Nigeria.

African pear, *Dacryodes edulis,* seed extract has been used for the treatment of ailments in traditional medicine. However, the effect of boiling on its antioxidant properties is still poorly understood. Therefore, the present study investigates the effect of boiling on the antioxidant properties of *D. edulis* seed extract using *in vitro* parameters such as free radical scavenging ability against 2, 2- diphenyl -1-picrylhydrazyl (DPPH) radical, iron (III) reducing and iron (II) chelating ability. Furthermore, the ability of both extracts (boiled and unboiled) to offer protective benefit against lipid peroxidation in cerebral and hepatic tissues of rat was assessed. Moreover, the effect of boiling on the phytochemical constituents (total phenolics, flavonoids and vitamin C) of the seed extracts was measured. Results indicate that boiling significantly ($P < 0.05$) potentiates the total phenolic [(Boiled 60.1 ± 0.88 mg/g (GAE); Unboiled 30.2 ± 0.68 mg/g (GAE)] and flavonoid [(Boiled: 50.02 ± 0.12 mg/g (QE); Unboiled: 35.8 ± 0.15 mg/g (QE)] content but mildly depleted the vitamin C content [Boiled: (36.9 ± 0.44 mg/g; Unboiled: 40.1 ± 0.21 mg/g]. Similarly, boiling markedly increased the antioxidant properties (free radical scavenging, iron (II) chelating, iron (III) reducing and inhibitory effect against pro-oxidant-induced lipid peroxidation) of the seed extract. From the foregoing, the wide usage of African pear as remedy for ailment in folk medicine may be due to its phytochemical constituents which are potentiated by boiling. Hence, information from this study would create public awareness especially to traditional medical practitioners who are involved in the act of boiling the fruit to get the extract used for medicinal purposes.

Key words: African pear, phytochemical, degradation, proxidant, antioxidant, mechanism, lipid peroxidation.

INTRODUCTION

Dacryodes edulis is a dioecious shade loving species of non-flooded forests in the humid tropical zone (Leakey, 1999; Leakey et al., 2002; Waruhiu et al., 2004; Anegbeh et al., 2005) where its seed is widely cultivated for the production of its fruits which has vast economic and health-related benefits (Verheij, 2002). It consists of a seed surrounded by a pulpy butyraceous pericarp, which is the edible portion consumed either raw or cooked. Its fruit and seed is rich in oils, proteins, minerals and vitamins which makes it an excellent source of nutrition to consumers, stimulating its increased production and commercialization for decades (Sofowora, 1982). Its oil has been found suitable for cosmetics and food, while the flower nectar provides a good honey (Ayuk et al., 1999; Verheij, 2002).

Specifically, the seed oil is rich in arachidonic acid and other nutritionally beneficial fatty acids (Ajayi and Adesanwo, 2009). *D. edulis* is a versatile plant in African ethnomedicine, as its various parts are employed to treat several diseases. Its bark has long been used to cicatrize wound (Okunomo and Egho, 2010), and for the treatment of leprosy, dysentery, anaemia, spitting blood, debility, stiffness, tonsillitis and skin diseases (Dalziel, 1937; Hutchinson et al., 1963). The leaves are often crushed

---

*Corresponding author. E-mail: ogunmoyoledayo@yahoo.com.

and the juice released to treat generalized skin diseases such as scabies, ringworm, rash and wound, while the stem or stem twigs are employed as chewing sticks for oral hygiene (Igoli et al., 2005; Ajibesin et al., 2008). When chewed with kolanut, its leaves serves as an antiemetic, while its leaf sap could be used for treating ear infections, fever, headache, malaria and cephalgy (Bouet, 1980).

Recently, Jiofack et al. (2010) reported that the leaves are made into plaster to treat snakebite in Southwest Cameroon. Besides, Ajibesin (2011) had identified phenolics such as ethylgallate and quercitrin in the plant leaves. Flavonols such as quercitrin, isoquercitrin, isorhamnetin and rhamnoside, as well as anthocyanins such as petunidin and cyanidin were also reported to be present in the fruit skin zone and pulp of D. edulis during ripening (Missang et al., 2003). The stem exudates of the plant were reported to contain tannin, saponins, and alkaloids (Okwu and Nnamdi, 2008). The presence of bioactive compounds such as saponins, tannins, alkaloids and flavonoids identified in the plant has been suggested to be responsible for the various uses of D. edulis in traditional medicine to cure ringworm, wound, scabies, skin diseases and inflammation (Okwu and Nnamdi, 2008).

In addition, the potential health-related functions of dietary plants were found to include antibiosis, immunostimulation, nervous system action, detoxification, anti-inflammatory, antigout, antioxidant, glycemic and hypolipidemic properties (Johns, 2001). However, despite the widely reported pharmacological relevance of D. edulis, there is dearth of information on the effect of boiling on its pharmacopotency. Hence, there is dire need to unravel the effect of boiling on its bioactive constituents and antioxidant properties. This would furnish our traditional medical practitioners and the public with useful information that would guide them in the usage of the fruit for medicinal and nutritional purposes.

MATERIALS AND METHODS

Chemical reagents

Thiobarbituric acid (TBA) was obtained from Sigma (St. Louis, USA). DPPH (2, 2-diphenyl -1-picrylhydrazyl) and 1, 10 phenanthroline were obtained from Fluka Chemie (Buchs, Switzerland) and Merck (Germany). All other chemicals were obtained from standard chemical suppliers and were of analytical grade.

Plant material

Fruits of D. edulis were collected around the University campus of The Federal University of Technology, Akure, Nigeria, and were identified at the Crop Soil and Pest Management Department of the Federal University of Technology, Akure, Nigeria. The fruits were washed and opened to get its seeds which were air dried. The dried seeds were pulverized using a blender and the powdered seeds were stored in polythene bags and stored at room temperature until they were used.

Preparation of plant extracts

10 g each of powdered seeds were weighed in two separate extraction bottle. One of them was poured into a big Pyrex test tube containing 200 ml of distilled water and heated for 5 min at boiling temperature (100°C). This was allowed to cool, decanted and extract filtered using a Whatman's filter paper. The filtrate was stored in the refrigerator and used as stock of boiled sample for all determinations. On the other hand, 10 g of powdered seed was weighed into an extraction bottle containing 200 ml of distilled water and left for 24 h to allow for extraction. Thereafter, the sample was decanted and filtered. The filtrate was then kept in the refrigerator and used as stock of unboiled sample for all determinations.

Animals

Male adult Wistar rats (200 to 250 g) were used. The animals were used according to the standard guidelines of the committee on care and use of experimental animal resources, Federal University of Technology, Akure, Nigeria.

Determination of total phenol contents

The total phenol contents of the seeds of D. edulis were determined by mixing (0 to 1.0 ml) of the extracts with equal volume of water; 2.5 ml Folin - Ciocalteau's reagent and 2 ml of 7.5% sodium carbonate were subsequently added. The absorbance was measured at 765 nm after incubating at 45°C for 40 min. The amount of phenols in the both extracts was expressed as gallic acid equivalent (GAE).

Determination of total flavonoid content

The total flavonoid content of D. edulis was determined using quercetin as a reference compound. Briefly, (0 to 500 µl) of stock solution of both boiled and unboiled extract was mixed separately with 50 µl of aluminium trichloride and potassium acetate. The absorbance at 415 nm was read on (Spectrum Lab digital spectrophotometer) after 30 min at room temperature. Standard quercetin solutions were prepared from 0.01 g quercetin dissolved in 20 ml of ethanol. All determinations were carried out in triplicate. The amount of flavonoids in both extracts was expressed as quercetin equivalent (QE).

Vitamin C content

The level of vitamin C in D. edulis was determined colorimetrically as described by Jacques-Silva et al. (2001). An aliquot of both extracts (200 µl) was incubated for 3 h at 38°C then 1 ml H_2SO_4 65% (v/v) was added. The reaction product was determined using a color reagent containing 4.5 mg/ml dinitrophenyl hydrazine and $CuSO_4$ (0.075 mg/ml), and the absorbance of the colored product was measured at 520 nm. The ascorbic acid content was expressed as ascorbic acid equivalent (AscE).

Free radical scavenging ability

The free radical scavenging ability of D. edulis against DPPH (2, 2-diphenyl -1- picrylhydrazyl) free radicals were evaluated according to Gyamfi et al. (1999). 600 µl of extract was mixed with 600 µl, 0.3 mM methanolic solution containing DPPH radicals, the mixture was left in the dark for 30 min and the absorbance was measured at 516 nm.

Reducing property

The reducing property was determined by assessing the ability of both boiled and unboiled extract of *D. edulis* to reduce $FeCl_3$ solution as described by Pulido et al. (2000). Briefly, extract (0 to 250 µl of stock) was mixed with 250 µl, 200 mM sodium phosphate buffer (pH 6.6) and 250 µl of 1% potassium ferrocyanide, the mixture was incubated at 50°C for 20 min. Thereafter 250 µl, 10% trichloroacetic acid was added, and subsequently centrifuged at 650 rpm for 10 min, 1000 µl of the supernatant was mixed with equal volume of water and 100 µl of 0.1 g/100 ml ferric chloride. The absorbance was later measured at 700 nm, a higher absorbance indicates a higher reducing power.

Fe^{2+} Chelating assay

The Fe^{2+} chelating ability of both boiled and unboiled extract of *D. edulis* was determined using a modified method described by Puntel et al. (2005). Freshly prepared 500 µmol/L $FeSO_4$ (150 µl) was added to a reaction mixture containing 168 µl of 0.1 mol/L Tris-HCl (pH 7.4), 218 µl saline and the extract (0 to 100 µM). The reaction mixture was incubated for 5 min, before the addition of 13 µl of 0.25% (w/v) 1, 10 - phenanthroline. The absorbance was subsequently measured at 510 nm in a spectrophotometer. The Fe (II) chelating ability was subsequently calculated with respect to the reference (which contains all the reagents without seed extracts).

Lipid peroxidation

Rats were decapitated under mild ether anesthesia and the cerebral (whole brain) and hepatic (liver) tissues were rapidly dissected, placed on ice and weighed. Tissues were immediately homogenized in cold 50 mM Tris-HCl, pH 7.4 (1/10, w/v). The homogenates were centrifuged for 10 min at 4000 rpm to yield a pellet that was discarded and a low-speed supernatant (S1). An aliquot of 100 µl of S1 was incubated for 1 h at 37°C in the presence of both boiled and unboiled seed of African pear, with and without the prooxidants, iron (final concentration, 10 µM) and sodium nitroprusside (SNP) (final concentration, 30 µM). This was then used for lipid peroxidation determination. Production of thiobarbituric acid reactive species (TBARS) was determined as described by Ohkawa et al. (1979), excepting that the buffer of the color reaction has a pH of 3.4. The color reaction was developed by adding 300 µl 8.1% sodium dodecyl sulfate (SDS) to S1, followed by sequential addition of 500 µl acetic acid/HCl (pH 3.4) and 500 µl 0.8% thiobarbituric acid (TBA). This mixture was incubated at 95°C for 1 h. TBARS produced were measured at 532 nm and the absorbance was compared to that of the controls.

Statistical analysis

The results were expressed as mean ± SD of three-four independent experiments performed in triplicate and were analyzed by appropriate analysis of variance (ANOVA) followed by Duncan's multiple range test. Differences between groups were considered significant when $p < 0.05$.

RESULTS AND DISCUSSION

In traditional medicine, plants parts are subjected to various treatments to get their extracts which are used for the treatment of ailments. However, the effects of these treatments on the phytochemical constituents of such plants are greatly neglected. Meanwhile, such treatments may improve or deplete the phytochemical content and antioxidant activity of such plants. In fact, reports have shown that some phytochemicals which are insoluble at room temperature get solubilised and extracted at increased temperature (Kolodziej and Hemingway, 1991). Hence, the use of decoction as a method of extracting plants phytochemicals in herbal drug preparation has been used since antiquity in folkloric medicine because these phytochemicals form a major component responsible for the antioxidant properties of medicinal plants. Hence, some medicinal plants are better exploited when extracted with appropriate solvents at increased temperature (Hemingway et al., 1992).

Meanwhile, the onset of many degenerative diseases has been linked with oxidative stress (Valko et al., 2004) and efforts at arresting the menace of oxidative stress has been on the increase in recent times. One major potential solution to arresting oxidative stress is the application of phytochemicals (Agbor et al., 2007). In view of this, *D. edulis* has been used for curing diverse ailments in folkloric medicine. Hence, it is paramount to investigate the effect of boiling on its antioxidant constituents and activity.

Antioxidant constituents of *D. edulis*

The antioxidant constituents of *D. edulis* determined in the present study as shown in Table 1 include total phenols, flavonoids and vitamin C. The phenolic content of *D. edulis* was estimated to be 60.1 ± 0.88 mg/g (GAE) for boiled and 30.2 ± 0.68 mg/g (GAE) for unboiled aqueous extract, whereas the flavonoid content was estimated to be 50.0 ± 0.12 mg/g (QE) for boiled and 35.8 ± 0.15 mg/g (QE) for unboiled extract. In addition, the vitamin C contents were 36.9 ± 0.44 mg/g and 40.18 ± 0.21 mg/g, respectively for boiled and unboiled extract of dried *D. edulis* seeds.

Some authors have reported that *D. edulis* is rich in phenol and flavonoid including alkaloids and tannins (Sofowora, 2008). In agreement with these reports, phytochemical profile of *D. edulis* revealed that the plant's seed is rich in phenols, flavonoids and vitamin C (Table 1). However, it was discovered that boiling markedly increased the phenolics and flavonoid content of *D. edulis* but mildly depletes the vitamin C level (Table 1). Although, the reason for this observation still remains largely obscure, it could be suggested that boiling solubilises and releases some of the phenols and flavonoids that are insoluble at room temperature leading to an increase in its level. Meanwhile, vitamin C level was slightly altered, indicating that vitamin C is not stable at boiling temperature, hence, it probably gets oxidized at boiling temperature with consequent depletion in its content. Having identified the antioxidant constituents of

Table 1. Antioxidant constituent of *D. edulis*.

Parameter	Boiled	Unboiled
Total phenolics, mg/g (GAE)	60.1 ± 0.88*	30.2 ± 0.68
Total flavonoid, mg/g (QE)	50.0 ± 0.12*	35.8 ± 0.15
Vitamin C, mg/g(AsE)	36.9 ± 0.44	40.1 ± 0.21

GAE, Gallic acid equivalent; QE, Quercetin equivalent; AsE, Ascorbic acid equivalent. Each observation is a mean ± SD of 3 to 4 independent experiments.* Indicates a statistically significant difference at $p < 0.05$.

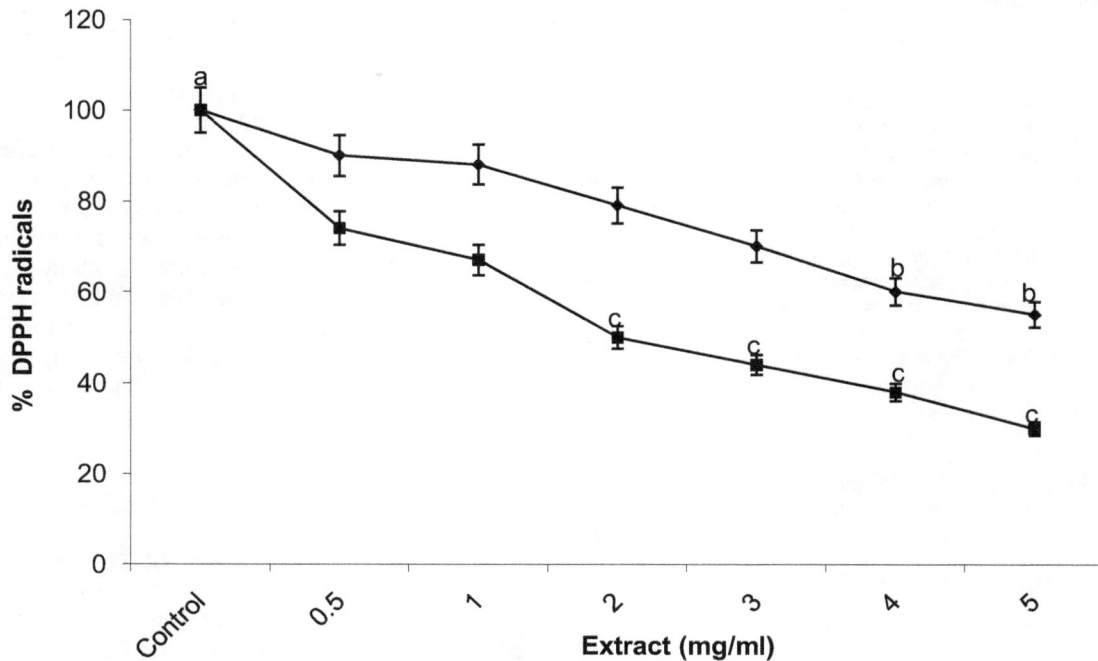

Figure 1. Free radical scavenging ability of boiled and unboiled seed extract of *D. edulis*. Data show means ± SEM values averages from 3 to 4 independent experiments performed in triplicate. 'b' and 'c' indicate a significant difference from the control 'a' at $p < 0.05$.

the seeds of *D. edulis*, it is rational to unravel the mechanisms of its antioxidant activity. Hence, some antioxidant parameters were determined in the present study.

Antioxidant mechanisms of *D. edulis*

In order to better ascertain the antioxidant potentials of *D. edulis*, several antioxidant mechanisms such as reducing property, metal chelating ability, free radical scavenging properties and inhibition of lipid peroxidation were employed. Generally, *D. edulis* seed extract exhibited potent antioxidant action in a concentration dependent. That is, the antioxidant activity of both boiled and unboiled extract increases with increasing concentration. However, boiled extract demonstrated better antioxidant Properties than unboiled extract in all parameters

determined.

Free radical scavenging ability

Figure 1 shows the free radical scavenging property of the seeds of *D. edulis*. Apparently, the boiled extract exhibited potent free radical scavenging activities which was significant (P < 0.05) than the unboiled extract at all dilutions of the stock solution used. One major routine *in-vitro* antioxidant parameters used for testing the potency of agents is their ability to scavenge 2, 2-diphenyl -1-picryl hydrazyl (DPPH) free radicals. The reaction involves protonation of the unstable 2, 2-diphenyl -1-picryl hydrazyl (DPPH) radicals turning it to stable diamagnetic molecule which is visually noticeable as a discoloration from purple to golden yellow.

Recently, Nguefack (2009) has reported the free radical

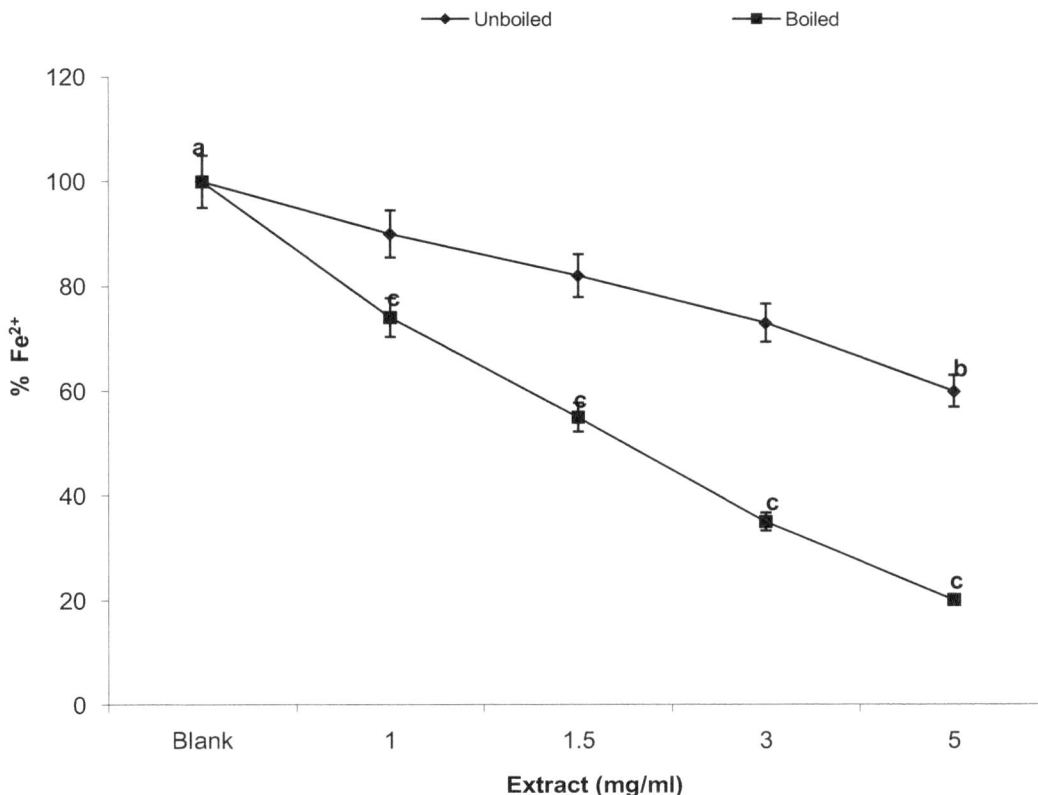

Figure 2. Fe^{2+}- chelating properties of boiled and unboiled seed extract of *D. edulis*. Data show means ± SEM values averages from 3 to 4 independent experiments performed in triplicate. 'b' and 'c' indicate a significant difference from the control 'a' at $p < 0.05$.

scavenging activity of ethanolic and aqueous extract of *D. edulis* seed extract. In line with this report, Figure 1 showed that African pear seed extracts demonstrated marked free radical scavenging activity (Figure 1). However, boiled extract showed a significantly higher radical scavenging effect than the unboiled aqueous extract. While the reason behind this observation is still not completely understood, it could be due to the higher phenolic and flavonoid content in the boiled extract. Interestingly, the antioxidant properties have been shown to have a direct relationship with their phytochemical constituents. For instance, phenolics and flavonoids are commonly known for their antioxidant activity. They modify the body's reactions to allergens, viruses, and carcinogens. They show anti-allergic, anti-inflammatory, antimicrobial and anticancer activity (Balch and Balchi, 2000), and may be useful in therapeutic remedy for ailments (Jisika et al., 1992). From the foregoing, it would be rational to expect that the extract with a higher content of these phytochemicals would exhibit a marked antioxidant activity. Since boiling increases the phytochemical (phenolics and flavonoid) content of the seeds of African pear, it is rational to attribute its higher free radical scavenging activity to its increased phytochemical content.

Fe^{2+} chelating ability

Figure 2 shows the Fe^{2+} chelating properties of *D. edulis*. One-way ANOVA followed by Duncan's test shows that the boiled extract of seeds of *D. edulis* was a better Fe^{2+} chelator than the unboiled extract. Antioxidants could elicit their effect by chelating and deactivating transition metals especially iron. Figure 2 revealed that boiling increases the iron chelating properties of *D. edulis* seed extract. This observation may be due to the variation in the phytochemical constituent of the extracts as observed for free radical scavenging activity (Figure 1).

Reducing property

The reducing property of *D. edulis* is as presented in Figure 3. One-way ANOVA revealed that *D. edulis* is rich in free electrons and readily supplies such electrons to Fe^{3+}, thereby reducing Fe^{3+} to Fe^{2+}. However, the boiled extract exhibited a better reductive ability than the unboiled. Antioxidants can also act by reducing transition metals specifically iron (III). Interestingly, boiling also increased the ability of *D. edulis* to reduce Fe^{3+} to Fe^{2+} (Figure 3). Meanwhile, reduction involves the addition of

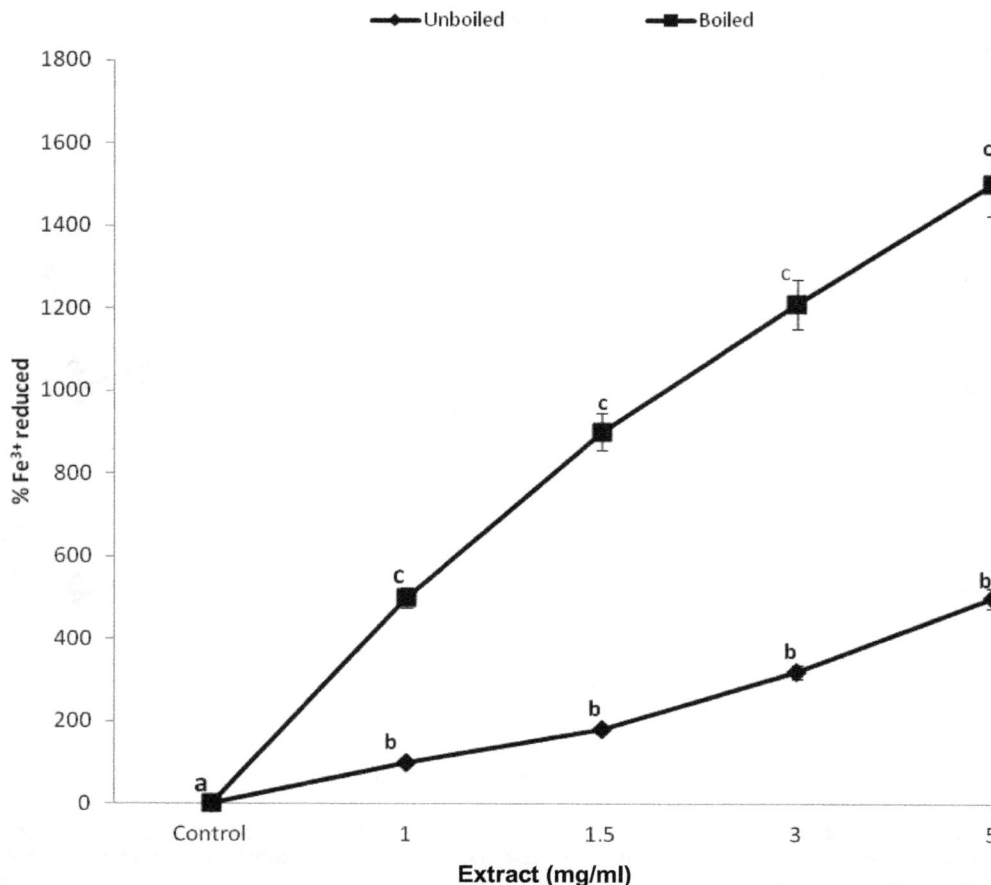

Figure 3. Ferric reducing properties of boiled and unboiled seed extract of *D. edulis.* Data show means ± SEM values averages from 3 to 4 independent experiments performed in triplicate. 'b' and 'c' indicate a significant difference from the control 'a' at p < 0.05.

hydrogen to substances. Hence, in this respect, hydrogen is being added to the unstable radicals leading to their stability and aversion of their deleterious effects. The fact that the boiled extract exhibited a higher reducing power may suggest that the extract is rich in constituents that are good nucleophiles which readily reduces Fe^{3+} to Fe^{2+}.

Lipid peroxidation

Figures 4 and 5 show the effect of *D. edulis* on lipid peroxidation subjected to oxidative assaults induced by iron and SNP, respectively. Figures 4a and 5a show that when brain lipids were subjected to stress-induced peroxidation either caused by Fe^{2+} or sodium nitroprusside in the presence of *D. edulis*, the extract exerted a significant inhibitory effect on the peroxidation processes. Similarly, Figures 4b and 5b show that when hepatic lipids were subjected to oxidative stress, *D. edulis* was able to significantly inhibit the peroxidation of hepatic lipids in a fashion similar to that observed when cerebral

lipids were used. One-way ANOVA revealed that irrespective of the prooxidant or lipid types, the inhibitory effect of *D. edulis* was significant at the lowest volume of extract tested (P < 0.05).

However, Figures 4 and 5 generally revealed that boiled extract was more potent than unboiled aqueous in the inhibition of prooxidant induced lipid peroxidation regardless of the tissue or prooxidant employed for oxidative assault. Furthermore, membrane lipids present in subcellular organelles are highly susceptible to free radical damage. Polyunsaturated lipids when reacted with free radicals can undergo oxidative degeneration which is a highly damaging chain reaction of lipid peroxidation leading to both direct and indirect detrimental effects. During lipid peroxidation, a large number of toxic by-products are also formed that can have effects at a site away from the area of generation, behaving as 'second messengers'. The damage caused by lipid peroxidation is highly detrimental to the functioning of the cell (Devasagayam et al., 2003). Hence, antioxidants are assessed based on their ability to offer protective shield

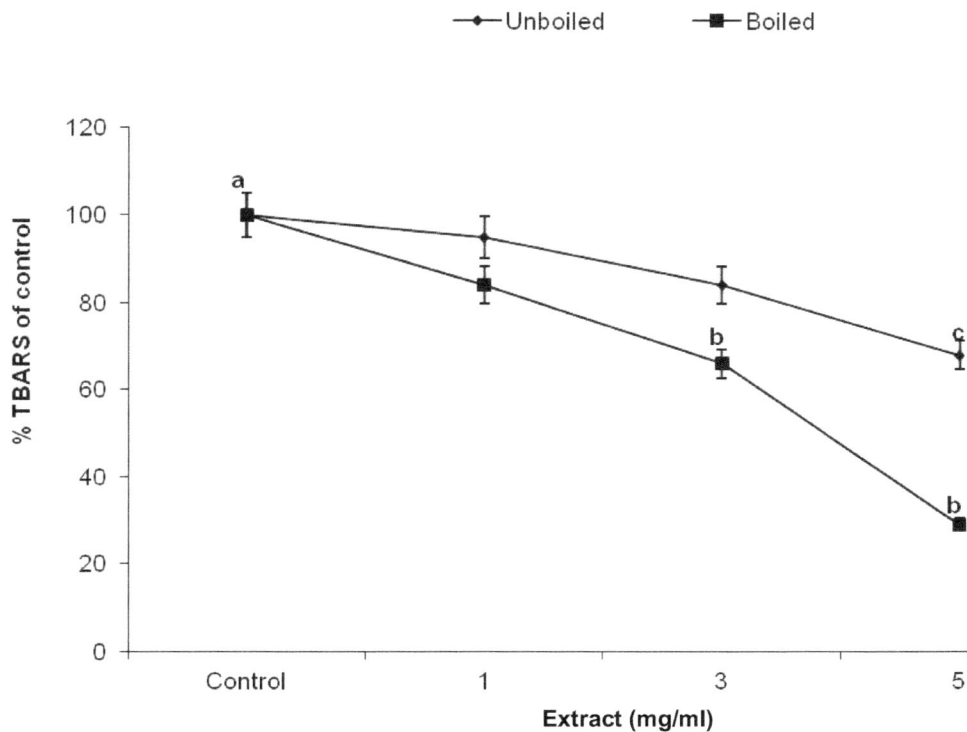

Figure 4a. Inhibitory effect of boiled and inboiled extracts of D. edulis on Fe^{2+}- induced lipid peroxidation in rat brain. Data show means ± SEM values averages from 3 to 4 independent experiments performed in triplicate. 'b' and 'c' indicate a significant difference from the control 'a' at $p < 0.05$.

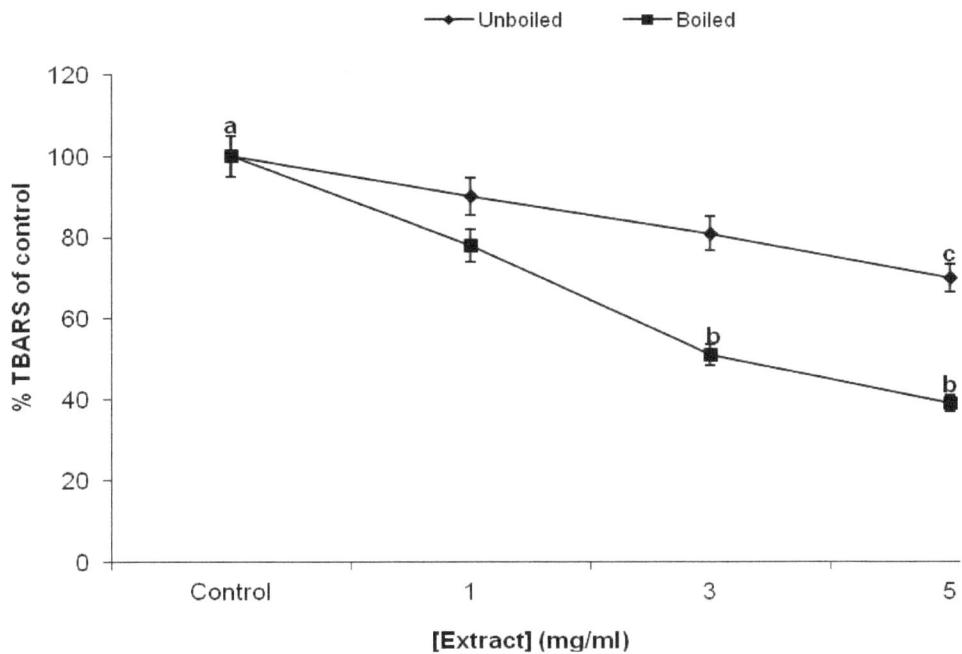

Figure 4b. Inhibitory effect of boiled and unboiled seed extract of *D. edulis* on Fe^{2+}- induced lipid peroxidation in rat liver. Data show means ± SEM values averages from 3 to 4 independent experiments performed in triplicate. 'b' and 'c' indicate a significant difference from the control 'a' at $p < 0.05$.

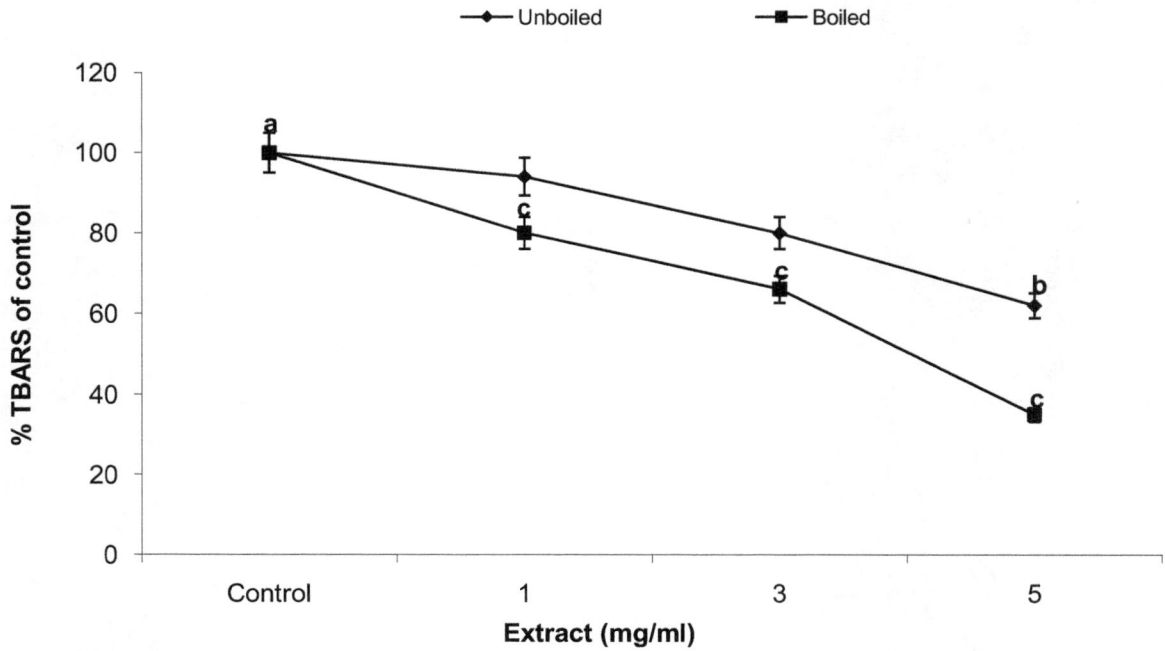

Figure 5a. Inhibitory effect of boiled and unboiled seed extract of *D. edulis* on SNP- induced lipid peroxidation in rat brain. Data show means ± SEM values averages from 3 to 4 independent experiments performed in triplicate. 'b' and 'c' indicate a significant difference from the control 'a' at p < 0.05.

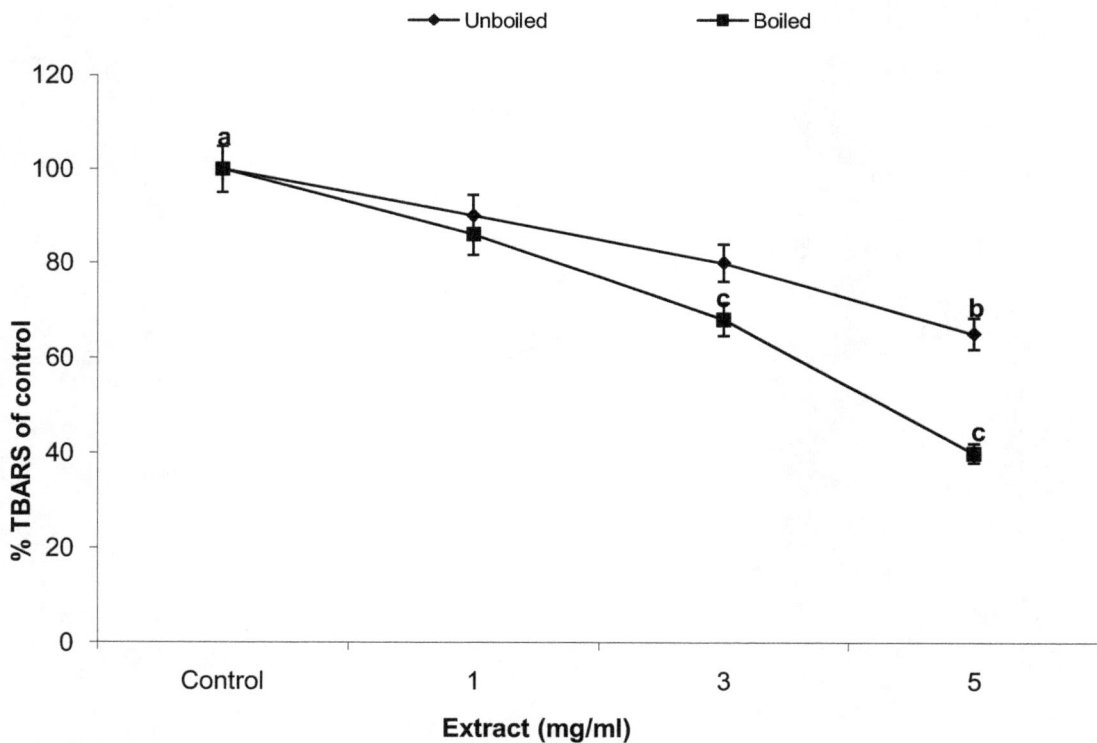

Figure 5b. Inhibitory effect of boiled and unboiled seed extract of *D. edulis* on SNP- induced lipid peroxidation in rat liver. Data show means ± SEM values averages from 3 to 4 independent experiments performed in triplicate. 'b' and 'c' indicate a significant difference from the control 'a' at p < 0.05.

to lipids and other critical macromolecules.

However, pro-oxidants differ in their mechanism of causing oxidative havoc to macromolecules specifically lipids, hence, a good antioxidant should be able to inhibit pro-oxidant-induced lipid peroxidation regardless of the proxidant employed. Hence, two pro-oxidants were employed in this study to investigate the antioxidant potentials of *D. edulis* seed to offer protective benefits to lipids subjected to several oxidative assaults.

Iron (II) sulphate as pro-oxidant

Meanwhile, iron has been reported to cause deleterious effect on biological macromolecules by reacting with superoxide anion (O_2^-) and hydrogen peroxide (H_2O_2) to produce the hydroxyl radical ($OH^•$) via the Fenton chemistry (Graf et al., 1984). These radicals can also lead to the formation of other reactive oxygen species (ROS) (Klebanoff et al., 1992). Interestingly, Figure 4a and b, respectively showed that *D. edulis* extract exhibited marked inhibitory effect against Fe^{2+}- induced cerebral and hepatic lipid peroxidation. Although, both extracts demonstrated marked inhibitory effect against TBARS formation, boiled extract showed a higher inhibitory effect probably due to the its higher phenolics and flavonoid content. Better still, it could be speculated that *D. edulis*, being a good iron chelator must have prevented the oxidation of iron (II), thereby preventing the generation of hydroxyl radical and inhibiting oxidative attack on lipids.

Sodium nitroprusside

Moreover, reports have shown that sodium nitroprusside (SNP) elicits cytotoxic effect through the release of cyanide and/or nitric oxide (NO) (Rauhala et al., 1998). NO has been implicated in the pathophysiology of strokes, traumas, seizures and Alzheimer's, and Parkinson's diseases (Castill et al., 2000; Prast and Philippou, 2001). Besides, light exposure promotes the release of NO from SNP through a photodegradation process (Arnold et al., 1984; Singh et al., 1995), and data from the literature have shown that after the release of NO and SNP, $[NO\text{-}Fe\text{-}(CN)^5]^{2-}$ is converted to iron containing $[(CN)^5\text{-}Fe]^{3-}$ and $[(CN)^4\text{-}Fe]^{2-}$ species (Loiacono and Beart, 1992). After the release of NO, the iron moiety may react with SNP, which could lead to the formation of highly reactive oxygen species, such as hydroxyl radicals via the Fenton reaction (Graf et al., 1984). The fact that *D. edulis* extract inhibited SNP-induced lipid peroxidation (Figure 5a and b) may indicate that extracts possibly prevented the breakdown of SNP to its constituents, thereby offering protective shield to both cerebral and hepatic tissues since the toxic constituents are presumably prevented from being released.

Conclusion

From the foregoing, it is clear that the wide usage of *D. edulis* in traditional medicine is intrinsically linked with its potent and diverse phytochemical constituents. Moreover, boiling, which incidentally is a common practice of traditional medical practitioners, potentiated the antioxidant potency of *D. edulis*. Hence, the act of boiling the seed of *D. edulis* before use as remedy for ailments should be encouraged as this has been found to boost its antioxidant activity *in vitro*.

REFERENCES

Ajayi IA, Adesanwo O (2009). Comparative study of the mineral element and fatty acid composition of *Dacryodes edulis* pulp and seed. World J. Agric. Sci., 5: 279-283.

Ajibesin KK (2011). Dacryodes edulis Lam: A review of its medicinal, phytochemical and economic properties, 51: 32-41.

Ajibesin KK, Rene N, Bala DN, Essiett UA (2008). Antimicrobial activities of the extracts and fractions of *Allanblackia floribunda*. Biotechnology, 7: 129-133.

Anegbeh PO, Ukafor V, Usoro C, Tchoundjeu Z, Leakey RRB, Schreckenberg K (2005). Domestication of *Dacryodes edulis*: Phenotypic variation of fruit traits from 100 trees in south eastern Nigeria. New Forests, 29: 149-160.

Arnold WP, Longnecker DE, Epstein RM (1984). Photodegradation of sodium nitroprusside: biologic activity and cyanide release. Anesthesiology, 61: 254-260.

Ayuk ET, Duguma B, Franzel S, Kengue J, Mollet M, Tiki-Manga T, Zekeng P (1999). Uses, management and economic potentials of *Dacryodes edulis* (Burseraceae) in the humid lowlands of Cameroon. Econ. Bot., 53: 292-301.

Balch JF, Balch PA (2000). Prescription for Nutritional Healing. Avery Penguin Putnam Inc. New York, pp. 267-270.

Bouet C (1980). The traditional and economic importance of several forest trees from Gabon. Serie Sciences Humaines, 17: 269-273.

Castill J, Rama R, Davalos A (2000). Nitric oxide-related brain damage in acute ischemic stroke. Stroke, 31: 852-857.

Dalziel JM (1937). Flora of West Tropical Africa. Crown Agents for Overseas Government, London, p. 296.

Devasagayam TPA, Boloor KK, Ramsarma T (2003). Methods for estimating lipid peroxidation: Analysis of merits and demerits (minireview). Indian J. Biochem. Biophys., 40: 300-308.

Graf E, Mahoney JR, Bryant RG, Eaton JW (1984). Iron catalyzed hydroxyl radical formation: Stringent requirement for free iron coordination site. J. Biol. Chem., 259: 3620-3624.

Gyamfi MA, Yonamine M, Aniya Y (1999). Free-radical scavenging action of medicinal herbs from Ghana: *Thonningia sanguine* on experimentally-induced liver injuries. Gen. Pharmacol., 32: 661-667.

Hemingway RW, Ohara S, Steynberg EV, Brandt D (1992). In: Heminway RH, Laks PE, (eds) Plant Polyphenols: Synthesis, Properties and Significance, Plenum press, New York. pp. 321-338.

Hutchinson J, Dalziel JM, Herpper FN (1963). Flora of West Tropical Africa II. Macmillan Publishers Ltd. Lagos, pp. 252-260.

Igoli JO, Ogaji OG, Tor-Anyiin, Igoli NP (2005). Traditional medical practices among the Igede people of Nigeria. Part II. Afr. J. Tradit. Complement. Altern. Med., 2: 134-152.

Jacques-Silva MC, Nogueira CW, Broch LC, Flores EM, Rocha JBT (2001). Diphenyl diselenide and ascorbic acid changes deposition of selenium and ascorbic acid in liver and brain of mice. Pharmacol. Toxicol., 88: 119-127.

Jiofack T, Fokunang C, Guedje N, Kumeuze V, Fongnzossie E (2010). Ethnobotanical uses of medicinal plants of two ethnoecological regions of Cameroon. Int. J. Med. Sci., 2: 60-79.

Jisika M, Ohigashi H, Nogaka H, Tada T, Hirota M (1992). Bitter steroid glycosides, Vernon sides A1, A2, and A3 and related B1 from the possible medicinal plant *Vernonia amygdalina* used by wild

Chimpanzees. Tetrahedron, 48: 625-630.

Johns T (2001). Dietary, diversity, global change and human health. Proceedings of the Symposium Managing Biodiversity in Agricultural Ecosystems Montreal, Canada, pp. 1-11.

Klebanoff SJ, Gally JI, Goldstein IM, Snyderman R (eds) (1992). Oxygen metabolities from phagocytes. Raven Press, New York, pp. 541-588.

Kolodziej H, Hemingway RH (1991). Plant polyphenols: Synthesis, Properties and Significance, Plenum press, New York, pp. 259-320.

Leakey RRB (1999). Potential for novel food products from Agroforestry trees: A review. Food Chem., 66: 1-14.

Leaky RRB, Atangana AR, Kengnni E, Waruhiu AN, Usoro C, Anegbe PO, Tchoundjeu Z (2002). Domestication of Dacryodes edulis in West and Central Africa: Characterization of genetic variation. Trees Livelihood, 12: 57-71.

Loiacono RE, Beart PM (1992). Hippocampal-lesions induced by microinjection of the nitric-oxide donor nitroprusside. Eur. J. Pharmacol., 216: 331-333.

Missang CE, Guyot S, Renard CMG (2003). Flavonols and anthocyanins of bush butter, Dacryodes edulis (G. Don) HJ. Lam, fruit. Changes in their composition during ripening. J. Agric. Food Chem., 50: 7475-7480.

Nguefack EC (2009). Hypoglycemic, Hypolipidemic and Antioxidant Activity of Some Cameroonian Medicinal Plants. Lyon, France. p. 49.

Ohkawa H, Ohishi H, Yagi K (1979). Assay for lipid peroxide in animal tissues by thiobarbituric acid reaction. Anal. Biochem., 95: 351-358.

Okunomo K, Egho EO (2010). Economic importance of some underexploited tree species in Nigeria: Urgent need for separate research centers. Continental J. Biol. Sci., 3: 16-32.

Okwu DE, Nnamdi FU (2008). Evaluation of the chemical composition of Dacryodes edulis and Raphia hookeri mann and wendl exudates used in herbal medicine in south eastern Nigeria. Afr. J. Tradit. Complement. Altern. Med., 5: 194-200.

Prast H, Philippou A (2001). Nitric oxide as modulator of neuronal function. Neurobiology, 64: 51-68.

Pulido R, Bravo L, Saura-Calixto F (2000). Antioxidant activity of dietary polyphenols as determined by a modified ferric reducing/antioxidant power assay. J. Agric. Food Chem., 48: 3396-3402.

Puntel RL, Nogueira CW, Rocha JBT (2005). Krebs cycle intermediates modulate thiobarbituric acid reactive species (TBARS) production in rat brain in vitro. Neurochem. Res., 30: 225-235.

Rauhala P, Khaldi IA, Mohanakumar KP, Chiueh CC (1998). Apparent role of hydroxyl radicals in oxidative brain injury induced by sodium nitroprusside. Free Radical Biol. Med., 24: 1065-1073.

Singh RJ, Hogg N, Neese F, Joseph J, Kalyanaraman B (1995). Trapping of nitric oxide formed during photolysis of sodium nitroprusside in aqueous and lipid phases: an electron spin resonance study. Photochem. Photobiol., 61: 325-330.

Sofowora LA (2008). Medicinal Plants and Traditional Medicine in Africa. Spectrum Books Ltd., Ibadan, Nigeria. pp. 289.

Sofowora A (1982). Medicinal plants and traditional medicine in Africa. John Wiley and Sons, New York. USA.

Valko M, Izakovic M, Mazur M, Rhodes CJ, Telser J (2004). Role of oxygen radicals in DNA damage and cancer incidence. Mol. Cell. Biochem., 266: 37-56.

Verheij EWM (2002). Dacryodes edulis (G. Don) H.J. Lam. [Internet] Record from Protabase. In: PROTA, Oyen LPA and RHMJ Lemmens (eds) Plant Resources of Tropical Africa, Wageningen, Netherlands.

Waruhiu AN, Kengue J, Atangana AR, Tchoundjeu Z, Leakey RRB (2004). Domestication of Dacryodes edulis. In: Phenotypic variation of fruit traits from 200 trees from four populations in the humid lowlands of Cameroon. J. Food Agric. Environ., 2: 340-346.

8

Alterations in the biochemical components and photosynthetic pigments of mulberry (*Morus* Spp.) attacked by leaf – roller (*Diaphania pulverulentalis*) pest

A. Mahadeva[1]* and V. Nagaveni[2]

[1]Residential Coaching Academy, Babasaheb Bhimrao Ambedkar University, Vidya Vihar, Rae Barely Road, Lucknow - 226 025, India.
[2]Department of Crop Physiology, University of Agricultural Sciences, GKVK, Bangalore – 560 065. India.

Leaf roller (*Diaphania pulverulentalis* Hampson) is a serious pest of mulberry (*Morus alba* L.), which is the sole food for silkworm – *Bombyx mori* L. It attacks the tender leaves of the host causing considerable damage which alters the leaf quality. An attempt was made to know the changes in the biochemical components (free amino acids, total soluble proteins, total reducing sugars, total soluble sugars, starch and total phenols) and photosynthetic pigments (total chlorophyll, chlorophyll – a, chlorophyll – b, chlorophyll – a/b ratio and carotenoids) in six popular indigenous mulberry varieties (M_5, MR_2, Mysore local, S_{36}, S_{54} and V_1) under infestation by leaf – roller pest. The results revealed that the biochemical components were reduced in almost all the varieties chosen. Though, there was no reduction in free amino acids in S_{36}; total soluble proteins in V_1; total reducing sugars in MR_2; total soluble sugars in S_{36} and S_{54}; starch contents in Mysore local and S_{54}; total phenols in MR_2. The MR_2 variety showed no alteration in the free amino acids. Total chlorophyll, chlorophyll – a, and chlorophyll – b were reduced drastically in all the varieties. The chlorophyll – a/b ratio was lowered in Mysore local, S_{36} and S_{54}. However, it was increased in M_5 and MR_2 and V_1 due to pest injury. Except Mysore local, the remaining varieties showed reduction in the carotenoid content. The alterations in biochemical components of mulberry foliage will adversely influence the health, growth and development of silkworm. This inturn results in the production of low quality silk.

Key words: Biochemical components, leaf roller, Mulberry, photosynthetic pigments.

INTRODUCTION

Mulberry (*Morus* spp.) is the exclusive source of feed for the silkworm – *Bombyx mori* L. in commercial sericulture. It is a deep rooted, perennial, found globally in almost all types of agro-climates with high biomass producing, proteins rich foliage plant (Ullal and Narasimhanna, 1981). Mulberry plays an important role in the quantity and quality of silk production, contribute to 38.20% for the success of cocoon crop (Miyashita, 1986). Silkworms feed on mulberry leaves during their entire larval period and utilize the leaf protein for the biosynthesis of silk. It is therefore clear that mulberry plays a dominant role in cocoon production as a source of nutrition to the silkworms. However, Mulberry foliage is prone to depredation by disease causing organism such as pathogens and pests. Among the several pests known to attack mulberry, the lepidopteron leaf roller, *Diaphania pulverulentalis* (Hampson) (family; Pyralidae) has attained a serious status in the Southern India. The leaf roller collected from mulberry gardens and agricultural fields are the potential carriers of microsporidia *Nosema bombycis*, a detrimental pathogen causing a pebrine disease to the silkworm. Thus constitute a potential threat of gaining access to silkworm rearing through contaminated mulberry leaf and perpetuate infection

*Corresponding author. E-mail: amdeva2007@gmail.com.

despite routine care taken in mother moth examination and sanitation thereby may have an adverse impact on the sericulture industry (Ifat et al., 2011).

Leaf roller infestation is high during the rainy and winter season (Geethabai et al., 1997). It attacks the crop throughout the year but, peak during January to October. The pest caused 66% reduction in the leaf yield of M_5, followed by MR_2 (65%), V_1 (52%) and S_{54} (30%) mulberry varieties (Reddy and Narayanaswamy, 2003). The early larval stage inhabits the apical part (unopened leaves) of the mulberry shoot and feed by scraping the tender leaf tissue, resulting in drying of the terminal portion. The infested leaves are brought together and bound through silk web formed by the larva. Sometimes, a single leaf is rolled into folded shape with the web from the larva inside and hence the pest is called as leaf roller. Occasionally, the larva also bores into the soft apical stem resulting in the drying of the shoot. The pest-infested plants show stunted growth. Heavily infested fields show considerable amount of webbing over the leaves. Such leaves are completely eaten up by late-age larva, resulting in considerable decline in the leaf yield (Siddegowda et al., 1995). The damage caused by leaf roller to mulberry is to an extent of 20 to 40% in some traditional regions of Karnataka (Veeranna, 1997). The literature available on the nutritive status of the leaf roller infested mulberry leaves in very scanty (Narayanaswamy, 2003). Therefore, an attempt was made to know the influence of leaf roller infestation on the biochemical components and photosynthetic pigments of mulberry leaves.

MATERIALS AND METHODS

The healthy and leaf roller infested leaves of six popular indigenous mulberry varieties viz., M_5, MR_2, Mysore local, S_{36}, S_{54} and V_1 were collected from plantations in and around Tumkur, Kanakapura taluk and Ramanagara districts (Karnataka, India). The leaves were oven dried and processed to analyse the free amino acid, total soluble proteins, total soluble sugars, starch, reducing sugars and total phenols. Fresh mulberry leaves were utilised to estimated the photosynthetic pigments (total chlorophyll, chlorophyll – a, chlorophyll – b, chlorophyll – a/b ratio and carotenoids).

Estimation of total free amino acids

Total free amino acids were determined by Ninhydrin method (Moore and Stein, 1948). 50 mg of dried mulberry leaf powder was ground in 5 ml of 80% methanol. The methanol layer (2 ml) was taken in a test tube and 1 ml of ninhydrin reagent (4% ninhydrin in methyl cellosolve and 0.2 M acetate buffer in the ratio of (1:1) was added to it. The samples were boiled for 20 min and cooled; the volume was made up to 10 ml with distilled water. Absorbance was noted at 570 nm.

Total soluble proteins

Protein content was determined using the method of Lowry et al. (1951). To 1 ml of sample (leaf extract), 5 ml of alkaline copper reagent (1% $CuSO_4$ + 1% Na-K-tartarate + 2% Na_2CO_3 in 1 N

NaOH) was added and incubated at room temperature for 10 min. For this, 0.5 ml of folin-phenol reagent was added and allowed to stand for 30 min. The OD was measured at 750 nm by UV-spectrophotometer.

Total soluble sugars and starch

The total soluble sugars and starch contents were determined as described by Yemm and Willis (1954) (Anthrone method). 100 mg of leaf powder was ground in 20 ml of 80% ethanol and incubated at 95°C for 10 min from. To 1 ml of the supernatant sample, 4 ml of anthrone reagent was added. The reaction mixture was shook gently and kept over a boiling water bath for 10 min and allowed to cool down. The OD of blue green solution was measured at 625 nm. Glucose was used for plotting a standard curve.

For starch estimation, 5 mL sample was treated with 52% perchloric acid for 30 min. The samples were centrifuged at 6000 rpm for 15 min. 1 mL supernatant was mixed with 2 mL of cold anthrone reagent in ice bath. The samples were then boiled for 10 min at 100°C in water bath, cooled and recorded the absorbance at 630 nm. The starch concentration was calculated by multiplying with 0.9 to the values obtained from the standard curve.

The reducing sugars

The reducing sugars were estimated by Dinitro salicylic acid (DNS) method explained by Miller (1972). 500 mg of dried leaf powder was ground in 10 ml of 80% methanol. To 3 ml aliquot of the extract, 3 ml of di-nitrosalicylic acid (DNS) reagent was added and the mixture was boiled for 5 min in a water bath. 1 ml of 40% sodium potassium tartarate was added and OD was measured at 575 nm. Glucose was used as a standard.

Total phenols

Folin-ciocalteu was used to determine the total phenols (Bray and Thorpe, 1954). 500 mg of dried mulberry leaf powder was ground with 5 ml of methanol. The residue was re-extracted twice with the same volume of the solvent and the pooled supernatants was evaporated to dryness. The residue thus obtained was dissolved in 5 ml of distilled water and used for the estimation of total phenols. To 1 ml of sample, 1 ml of Folin's reagent (1:2) and 2 ml of 20% sodium carbonate was added. The reaction mixture was shook and heated over a boiling water bath for exactly one minute. The OD was measured at 650 nm. Caffeic acid was used for standard graph.

The photosynthetic pigments

100 mg of fresh mulberry leaf tissue was placed in a vial containing 7 ml of dimethyl sulphoxide (DMSO) and chlorophyll was extracted into the fluid without grinding at 65°C, incubated for three hours. Liquid was transformed to graduated tube and made up to a total volume of 10 ml with DMSO and absorption spectra were recorded at 663 and 645 nm using DU-40 spectrophotometer immediately. The content of total chlorophyll, chlorophyll-a and chlorophyll-b were estimated using the method suggested by Arnon (1949). The chlorophyll content was calculated by using the formula:

Total chlorophyll (mg/ml) = $(0.0202) \times (O.D. 645) + (0.00802) \times (O.D. 663)$
Chlorophyll 'a' (mg/ml) = $(0.0127) \times (O.D. 663) - (0.00269) \times (O.D. 645)$

Chlorophyll 'b' (mg/ml) = (0.0229) × (O.D. 645) – (0.00468) × (O.D. 663)

The carotenoid content was estimated using the formula of Kirk and Allen (1965) and expressed in milligrams per gram fresh weight.

Total carotenoids (µg/ml) = O.D. 480 + (0.114 × O.D. 663 – 0.638 × O.D. 645)

Data obtained were analyzed by student's t - test according to the equation of Dixon and Massay (1957). Significant differences were established at $P<0.05$ and $P<0.01$ levels. The information was also subjected to percentage of changes (decrease/increase) in the infested and healthy leaves and was calculated as:

$$\% \text{ decrease/increase} = \frac{(\text{Values of healthy leaves} - \text{values of infested leaves})}{(\text{Values of healthy leaves})} \times 100$$

RESULTS AND DISCUSSION

Biochemical components

Free amino acids

The free amino acid was decreased noticeable in leaf roller infested leaves of M_5, Mysore local, S_{54} and V_1 varieties. But, the free amino acid was increased (17.65%) in S_{36} variety and there was no alteration in MR_2 variety. The reduction was in the range of 8.07 (V_1 variety) to 71.12% (M_5 variety) (Table 1). Raman et al. (1994) observed an increased total free amino acids content in mealy bugs (*Maconellicoccus hirsutus* Green) infested mulberry varieties viz., M_5, MR_2, BC_{259}, Tr4, S_{13} (indigenous), Kosen, Ichinose and Goshoerami (exotic). Mahadeva and Shree (2003) reported a reduction in free amino acid content in M_5, MR_2, S_{36} and it was increased in the V_1 mulberry variety infested by Jassids (*Empoasca flavescenes* F.). Shree and Mahadeva (2005) observed a significant increase in the free amino acids of Mysore local variety. Whereas, it decreased significantly in the S_{54} due to jassids infestation. The amino acid content was increased (Kurangi, Thysong-3, Vietnam and Thailand) as well as decreased (M_5, Kajali, Hakkikalu and *Morus nigra*) in the leaves of mulberry varieties when they were infested by thrips – *Pseudodendrothrips mori* (Niwa) (Latha, 1999).

Total soluble proteins

The total soluble protein was decreased significantly in M_5, Mysore local, S_{36}, $S_{54,}$ non-significantly in MR_2. It was non-significantly increased in the V_1 (0.64%) mulberry variety infested by *D. pulverulentalis*. The reduction was in the range of 0.70 (MR_2) to 13.51% (Mysore local) (Table 1). Narayanaswamy (2003) also observed reduced protein content (15.31%) in the leaf roller infested M_5 variety. Shree et al. (1989) have noticed

10.50% reduction and 40.00% increase in the total soluble proteins of tukra affected leaves in Kajali and Kanva-2 mulberry varieties, respectively. Shree and Umesh (1989) noticed decrease in the protein content of *Morus macroura* and *Morus nigra*; the reduction was, however, negligible in *M. nigra*. But it was increased in *Morus australis* and *Morus cathayana*. Umesh et al. (1989) found an increase in the total soluble proteins of tukra affected mulberry leaves (variety; Vietnam) compared to healthy ones. Bose et al. (1992) investigated a statistically significant reduction in the water soluble proteins in the mealy bugs infested leaves of six mulberry varieties (K_2, S_{30}, S_{36} S_{41} and S_{54}). Raman et al. (1994) recorded the effect of tukra disease on the total soluble proteins in eight mulberry varieties. The total proteins decreased (9 to 26%) in all the mulberry varieties except MR_2. Veeranna (1997) reported an increase in the total proteins due to tukra in M_5 and DD varieties. Narayanaswamy et al. (1999) revealed that the crude protein contents were drastically reduced in spiralling whitefly (*Aleurodicus dispersus* Russell) infested leaves of M_5 mulberry variety. The total soluble proteins were more in the four mulberry cultivars (M_5, MR_2, S_{36} and V_1), when they were infested by jassids (Mahadeva and Shree, 2003).

The total soluble proteins were significantly increased in Mysore local and decreased in S_{54} due to Jassids infestation (Shree and Mahadeva, 2005). Latha (1999) reported enhanced protein content in M_5, Kurangi, Kajali, Vietnam, Thailand and *M. nigra*. It was decreased in Hakkikalu and Thysong mulberry varieties infested by thrips (*P. mori*). The crude protein contents significantly decreased in thrips infested tender leaves in all the varieties (K_2, S_{13}, S_{34}, S_{36} and V_1) considered, maximum decrease was in the variety S_{13} followed by S_{34}, K_2, V_1 and least in S_{36}. Also, a significant decrease in the protein content of medium leaves was recorded in all the varieties except V_1. The highest decrease was observed in S_{13} followed by S_{34}, K_2 and least in S_{36}. The decrease in the crude protein contents may be attributed to the damage caused by the insect, thus altering the metabolic functions leading to either decline in protein synthesis or mobilization of proteins for repair of the damaged tissues in order to develop resistance to insect bite. Increase in the protein content may be due to changes in the protein synthesis pattern to overcome the injury and develop resistance (Sathya et al., 2002a).

Total reducing sugars

The total reducing sugar was reduced in all mulberry varieties, except MR_2 where it showed significant increase (55.88%). The reduction was maximum in the *D. pulverulentalis* infested leaves of M_5 (65.01%) and minimum in Mysore local (1.51%) (Table 1). Shree and Umesh (1989) have noticed a significant increase in the

Table 1. Changes in the biochemical components of leaf roller infested mulberry leaves.

Mulberry varieties	Free amino acids (µg/g)		Total soluble proteins (mg/mg)		Total reducing sugars (mg/g)		Total soluble sugars (mg/g)		Total starch (mg/g)		Total phenols (mg/g)	
	Healthy	Infested	Healthy	Infested	Healthy	Infested	Healthy	Infested	Healthy	Infested	Healthy	Infested
M_5	7.48	2.16** (-71.12)	109.85	101.40* (-7.69)	113.20	39.60** (-65.02)	1.87	1.76 (-5.93)	1.38	1.28** (-7.14)	2.88	1.82** (-36.81)
MR_2	11.16	11.16 (-)	184.60	183.30 (-0.70)	40.80	63.60** (+55.88)	2.14	2.06 (-3.70)	1.53	1.13** (-26.32)	4.44	4.32 (-2.70)
Mysore local	15.36	12.24** (-20.31)	72.15	62.40** (-13.51)	185.20	182.40* (-1.51)	2.52	2.38 (-5.66)	1.05	0.86** (-17.68)	4.04	4.41 (+2.48)
S_{36}	8.16	9.16** (+17.65)	85.15	77.35* (-9.16)	191.60	182.40** (-4.80)	2.31	2.00** (-13.70)	1.31	1.42** (+8.16)	3.66	3.25* (-11.29)
S_{54}	16.68	15.20** (-8.87)	109.85	99.45** (-9.47)	43.20	37.20* (-13.89)	1.90	1.98 (+4.17)	1.50	1.55** (+2.98)	2.50	2.76** (+10.40)
V_1	19.84	18.24** (-8.07)	101.40	102.05 (+0.64)	45.20	43.20** (-4.43)	1.76	1.52** (-13.51)	1.34	1.28** (-4.67)	4.96	3.20** (-35.48)

**Significant at 1% level; * Significant at 5% level.; Values in the parenthesis indicate % difference over healthy (+ = more than; - = less than; ---- = not altered).

tukra affected *M. macroura* and *M. australis* but it was negligible in latter variety. In *M. cathayana* and *M. nigra*, it was unaltered. Similar reductions were observed in the Vietnam mulberry variety infested by mealy bugs (Umesh et al., 1989). Bose et al. (1992) found significant reduction in the non-reducing sugars and total reducing sugars in the leaves of K_2, S_{30}, S_{36}, S_{40} and S_{54} mulberry varieties infested by *M. hirsutus*. There was a drastic decrease in the sugar contents of M_5 mulberry leaves infested by leaf roller (Narayanaswamy, 2003). The reducing sugars were decreased in M_5 and V_1. But, it was increased in MR_2 and S_{36} due to hopper burn (Mahadeva and Shree, 2003). Similar reductions of sugar contents were observed in jassids infested Mysore local and S_{54} mulberry varieties by Shree and Mahadeva (2005). Latha (1999) showed a drastic increase in the total reducing sugars of thrips infested leaves in four indigenous (M_5, Kurangi, Kajali and Hakkikalu) and four exotic (Thysong-3, Vietnam, Thailand and *M. nigra*) mulberry varieties. Sathya et al. (2002b) observed changes in the reducing and non-reducing sugars in the leaves of K_2, S_{13}, S_{34} and S_{36} and V_1 due to thrips infestation. But, it was least in the K_2 and S_{36} varieties. Alteration in the reducing sugars may be due to reduction in leaf lamina and malformation of leaves in pest affected plants resulting in less productivity (Shree and Umesh, 1989).

Total soluble sugars

The total soluble sugar was decreased in leaf roller infested leaves of M_5, MR_2 (3.70%), Mysore local, and it was significant in S_{36} (13.70%) and V_1. But it increased (4.17%) non-significantly in S_{54} mulberry variety (Table 1). Narayanaswamy (2003) observed a 31.34% increase in the sugar contents of M_5 mulberry leaves due to leaf roller infestation. The total soluble sugars were increased by 35% and 36% in the Kajali and Kanva-2 mulberry varieties, respectively. Similar variations were observed in various other cases due to pest attack. The sugar content was marginally decreased in tukra affected *M. australis, M. cathayna* and *M. nigra* varieties. However, there was an increase in *M. macroura* (Shree and Umesh, 1989). Umesh et al. (1989) have noticed a reduction in the leaves of Vietnam variety due to mealy bugs infestation. Umesh et al. (1990) have studied the changes in sugar contents of mealy bugs infested leaves of four indigenous (Berhampore, S_{30}, S_{36} and S_{31}) and six exotic (Kosen, *M. multicaulis*, Philippine, Okinawa-2, Tsukasaguwa and Italian) mulberry varieties. The sugar content showed an increase in Berhampore, Okinawa-2 and Philippine varieties. Whereas, it decreased in Italian, Kosen, *M. multicaulis*, S_{30}, S_{36} and S_{41} varieties. No difference was observed in Tsukasaguwa.

Raman et al. (1994) observed increased soluble sugar content in the tukra affected leaves of M_5, MR_2, BC_{259}, S_{13}, Kosen, Goshoerami. However, in Tr4 and Ichinose, it was decreased. Veeranna (1997) found a high level of soluble carbohydrates in M_5 and DD varieties due to tukra. Sugars were drastically reduced in M_5 due to spiralling whitefly infestation (Narayanaswamy et al., 1999). Chandramohan et al. (2002) have noticed 43.75% reduction in the total sugars of spiralling whitefly infested mulberry leaves. Mahadeva and Shree (2003) noticed a reduction in the soluble sugars of Jassids (*E. flavescenes*) infested leaves of M_5, MR_2, S_{36} and V_1 mulberry varieties. Similar reduction was noticed in Mysore local and S_{54} by Shree and Mahadeva (2005) due to hopper burn. Latha (1999) noticed an increase in the total sugars of thrips infested leaves of M_5, Kurangi, Hakkikalu, Thysong-3, Vietnam, Thailand and *M. nigra*. But, it was decreased in Kajali.

Starch

The starch content was decreased significantly in M_5, MR_2, Mysore local and V_1. It was increased significantly in S_{36} (8.16%) and S_{54} (2.98%) varieties due to *D. pulverulentalis* infestation. The highest reduction was detected in MR_2 (26.32%) and the lowest was in V_1 (4.67%) variety (Table 1). Shree et al. (1989) observed 27 and 36% increase in starch content of mealy bugs affected Kajali and Kanva-2 mulberry varieties. The starch content was increased in the *M. macroura* and *M. nigra*. The increase was significant in *M. macroura*. There was only a negligible reduction in *M. australis* and *M. cathayana* due to *M. hirsutus* infestation (Shree and Umesh, 1989). Umesh et al., (1989) found a significant increase in the starch content in the leaves of Vietnam mulberry variety due to mealy bugs infestation. There was a significant decrease in the starch content of K_2, S_{30}, S_{36}, S_{41} and S_{54} affected by tukra (Bose et al., 1992). Raman et al. (1994) observed variation in starch contents of five indigenous (M_5, MR_2, BC_{259}, Tr4 and S_{13}) and

three exotic mulberry varieties (Kosen, Ichinose and Goshoerami) due to tukra. Shree and Mahadeva (2005) observed significant reduction in the starch content of Jassids infested Mysore local and S_{54} mulberry varieties.

Total phenols

The total phenol was decreased significantly in M_5, S_{36}, V_1 and non-significantly in MR_2; but it increased significantly in S_{54} and non-significantly in Mysore local. The leaf roller infested M_5 variety showed maximum reduction (36.81%) and it was minimum in MR_2 (2.70%). The increase was 10.40 and 2.48% in S_{54} and Mysore local varieties, respectively (Table 1). Shree and Umesh (1989) reported a decreased phenolic level in tukra affected varieties. However, there was a significant increase in *M. australis* and *M. cathayana*. Umesh et al. (1989) found that increased total phenols in the leaves of Vietnam variety due to *M. hirsutus* infestation. Muthegowda et al. (1990) noticed that the pattern of phenol accumulation varied depending upon the positional status of leaves in the tukra affected C_{15} (Conoor series) variety. There was an initial increase followed by a sudden decrease from the leaf of the 2nd to 10th order. The tukra affected leaves had more phenolics as a result of insect bite. This clearly shows that the pests certainly altered phenolic metabolism in the host leading to biochemical changes.

Raman et al. (1994) observed a lower content of total phenol in M_5, Kosen, Ichinose, Goshoerami, BC_{259} and S_{13} varieties due to mealy bugs attack. However, it was increased in MR_2 and Tr4. Mahadeva and Shree (2003) showed a decreased phenolic content in the Jassids infested leaves of M_5, MR_2, S_{36} and V_1 varieties. Similar results were also noticed by Shree and Mahadeva (2005) in Mysore local and S_{54} varieties. Latha (1999) studied the changes in the biochemical composition of leaves of four indigenous and four exotic mulberry varieties infested by thrips. M_5, Hakkikalu and *M. nigra* varieties showed a drastic increase in the total phenolic content; while in the other varieties, it was significantly reduced. The accumulation of phenolics in the host may inactivate the enzyme which inhibits the further advance of the pathogenic organism by limiting its source of nutrients (Uritani, 1961).

The most important phenolic compounds implicated in the defense mechanism of plants against pathogens are coumaric acid, phloretin, umberlliferons, caffeic acid, chlorogenic acid and ferulic acid (Agrios, 1969). When such substances are ingested by the phytophagous insects along with the food, they get access to the natural defense mechanism.

Photosynthetic pigments

The leaf roller infested mulberry leaves of six indigenous

Table 2. Changes in the photosynthetic pigments (mg/g. fresh weight) of leaf roller infested mulberry leaves.

Mulberry varieties	Total chlorophyll		Chlorophyll – a		Chlorophyll – b		Chlorophyll – a/b		Carotenoids	
	Healthy	Infested	Healthy	Infested	Healthy	Infested	Healthy	Infested	Healthy	Infested
M_5	1.31	1.13** (-13.40)	1.12	0.97** (-13.37)	0.19	0.16* (-14.13)	5.90	5.93 (+0.49)	0.90	0.67** (-26.02)
MR_2	2.52	2.11** (-16.29)	1.99	1.68** (-15.46)	0.53	0.42** (-19.24)	3.80	3.98* (+4.66)	1.26	0.93** (-26.01)
Mysore local	2.36	1.52** (-35.53)	1.92	1.22** (-36.40)	0.44	0.30** (-31.97)	4.35	4.06* (-6.60)	1.05	1.64** (+39.01)
S_{36}	1.34	0.82** (-39.03)	1.06	0.63** (-40.63)	0.28	0.19** (-33.57)	3.77	3.38 (-10.48)	0.72	0.38** (-47.50)
S_{54}	1.95	1.21** (-38.07)	1.56	0.94** (-39.41)	0.40	026** (-33.84)	3.94	3.61* (-8.43)	0.90	0.57** (-36.99)
V_1	3.79	3.31** (-12.74)	2.98	2.64** (-11.70)	0.81	0.67** (-16.71)	3.69	3.91** (+5.93)	1.43	1.08 (-24.58)

** Significant at 1% level; * Significant at 5% level; , Values in parenthesis indicate % difference over healthy (+ = more than; - = less than; ---- = not altered).

(M_5, MR_2, Mysore local S_{36}, S_{54} and V_1) varieties showed almost significant changes in the photosynthetic pigments (total chlorophyll, chlorophyll – a, chlorophyll – b, chlorophyll – a/b ratio and carotenoids). There was a significant decrease in the total chlorophyll, chlorophyll – a and chlorophyll – b in the leaves of all six varieties infested by *D. pulverulentalis*. S_{36} variety showed maximum reduction in the total chlorophyll, chlorophyll – a and chlorophyll – b by 39.03, 40.63 and 33.57%, respectively. Total chlorophyll and chlorophyll – a were reduced by 12.74 and 11.70% in V_1 variety respectively; chlorophyll – b by 14.13% in M_5 mulberry variety (Table 2).

Narayanaswamy (2003) observed a reduced amount of total chlorophyll, chlorophyll – a and chlorophyll – b by 32.40, 31.98, 32.87%, respectively in the leaf roller infested mulberry plants (Variety; M_5). In the present study, the chlorophyll – a/b ratio was lowered significantly in Mysore local, S_{54} and non-significantly in S_{36}. It was increased significantly in MR_2, V_1 and non-significantly in M_5. The reduction was least (6.60%) in Mysore local but maximum (10.46%) in S_{36} variety. Similarly, the least (0.49%) increase was in M_5 and maximum reduction (5.93%) was in V_1 variety. The carotenoids were decreased significantly in M_5, MR_2, S_{36}, S_{54} and

non-significantly in V_1 mulberry varieties attacked by leaf roller pest. But, there was a drastic increase in the Mysore local variety. The decrease in carotenoids was found to be maximum (47.50%) in S_{36} and minimum (24.58%) in V_1 variety. The Mysore local showed an increase of 39.01% compared to healthy ones (Table 2). Mahadeva and Shree (2003) showed a decrease in the photosynthetic pigments (total chlorophyll, chlorophyll – a, chlorophyll – b, chlorophyll – a/b ratio and carotenoids) due to *E. flavescenes* infestation (M_5, MR_2, S_{36} and V_1), except the chlorophyll – a/b ratio in M_5, which showed an increase. Similar trend was noticed in Mysore local and S_{54} varieties by Shree and Mahadeva (2005). The photosynthetic pigments (total chlorophyll, chlorophyll – a, chlorophyll – b, chlorophyll – a/b ratio and carotenoids) were reduced because of pest (Jassids) injury.

However, in the Mysore local variety, the chlorophyll – a/b ratio was increased. Similar observations were made in other cases when mulberry leaves were infested by various pests such as mealy bugs (Umesh et al., 1989), thrips (Das et al., 1994) and giant African snails (Ravi, 1997). Thus, the content of photosynthetic pigments varied depending upon the intensity of pest-infestation, extent of damage and mulberry varieties. The altered chlorophyll content adversely affected the photosynthetic activity (Heldt, 1997), productivity which ultimately leads to reduce protein synthesis (Burd and Elliot, 1996). Consequently, the mulberry foliage suffers from nutritional inferiority.

Pest attacks generally set in motion or accelerate a complicated series of metabolic disturbances in the host, rather than effecting a simple change in a unique process. Malformation of leaves due to pests will affect the photosynthesis by the crop in 3 ways: by lowering light interception, by reducing photosynthetic efficiency, or by altering normal distribution of assimilates within the plant. Ultimately results in the variation of net availability of plant productivity (Hewett, 1977). Obviously, the foliage suffers from nutritional inferiority. Therefore, the pest attacked leaves when fed to silkworms will have an adverse impact of their growth and development, leading to cocoon crop failures (Pradeep et al., 1992; Ravi and Shree, 1998; Doureswamy and Chandramohan, 1999; Mahadeva and Shree, 2004a). The pest attacked and diseased mulberry leaves should not be used for silkworm feeding as they are known to affect the commercial characters of cocoons (Ravi and Shree 1998; Chandramohan et al., 2000; Mahadeva and Shree, 2004b). Necessary arrangements must be made to manage the pests and diseases of mulberry plant as it is the only source of food for silkworms.

REFERENCES

Agrios GN (1969). Plant defense against pathogen. Biochemical defense. In Plant Pathology. (eds). G.N. Agrios, Academy Press. New York.

Arnon DJ (1949). Copper enzymes in isolated chloroplasts. Polyphenol-oxidase in Beta vulgaris. Plant Physiol., 24: 1-15.

Bray HG, Thorpe WV (1954). Analysis of phenolic compounds of interest in metabolism. Meth. Biochem. Anal., 1: 27-52.

Bose PC, Majumdar SK, Sengupta K (1992). Effect of tukra disease on the nutritional composition of mulberry (Morus alba L.). Sericologia, 32: 311-316.

Burd JD, Elliot NC (1996). Changes in chlorophyll a flourescence induction kinetics in cereals infested with Russian wheat aphid (Homoptera: Aphidiae). J. Econ. Entomol., 89: 1332-1337.

Chandramohan N, Doureswamy S, Asia MM (2002). Status and management of spiralling whitefly, Aleurodicus dispersus (Russel). Proceedings of National Seminar on "Advances in Tropical Sericulture". University of Agricultural Sciences, Bangalore – 560

065, pp. 76-80.

Das C, Shivanath Rao KN, Ghosh PL, Sengupta D, Sengupta K (1994). Biochemical changes due to thrips infestation in mulberry. National Symposium on "Plant Science in Nineties". University of Kalyani, APSR., 5: 4.

Doureswamy S, Chandramohan N (1999). Investigation on the economic characters of silkworm Bombyx mori L. influenced by the spiralling whitefly damage on mulberry. National Seminar on Tropical Sericulture, December, 28-30, University of Agricultural Sciences, Bangalore, 65: 82.

Dixon WJ, Massey FJJR (1957). Introduction to Statistical Analysis, 2nd edn. McGraw-Hill Book. Company, New York.

Geethabai M, Marimadaiah B, Narayanaswamy KC, Rajagopal D (1997). An outbreak of leaf roller pest, Diaphania (=margaronia) pulverulentalis (Hampson) on mulberry in Karnataka. Geobios New Reports, 16(2): 73-79.

Heldt HW (1997). The use of energy from sunlight by photosynthesis. In plant Biochemistry and Molecular Biology. Institute of Plant Biochemistry, Oxford University Press. New York, pp. 39-59.

Hewett EW (1977). Some effects of infestation on plants: A physiological viewpoint. The New Zealand Entomologis, 6(3): 235-243.

Ifat B, Sharma SD, Shabir AB (2011). Screening of different insect pests of mulberry and other agricultural crops for microsporidian infection. Int. J. Biotechnol. Mol. Biol. Res., 2(8): 138-142.

Kirk JTO, Allen RL (1965). Dependence of chloroplast pigment synthesis on protein synthesis: Effect of acs-idione. Biochem. Biophys. Res. Commun., 21: 530-532.

Latha MK (1999). Biochemical analysis of thrips infested leaves of different mulberry varieties. M.Sc., (Seric.) dissertation, Bangalore University, Bangalore. 12-16 pp.

Lowry OH, Rosebrough NJ, Fan AL, Randall RJ (1951). Protein measurement with Folin-phenol reagent. J. Biol. Chem., 193: 265-275.

Mahadeva A, Shree MP (2003). Biochemical studies in the mulberry (Morus spp.) leaves infected by jassids (Empoasca spp.) Proceedings, National Conference on "Tropical Sericulture for Global Competitiveness". 5th – 7th November Central Sericultural Research and Training Institute. Mysore.

Mahadeva A, Shree MP (2004a). Impact of feeding mealy bug (Maconellicoccus hirsutus) infested mulberry (Morus spp.) leaves (tukra) on the nutritional efficiency and economic characters of silkworms (Bombyx mori L.). Proceedings of the 91st Session of Indian Science Congress, Chandigarh, 14: 22-23.

Mahadeva A, Shree MP (2004b). Effect of feeding spiralling whitefly (Aleurodicus dispersus Russel.) infested mulberry leaves on the nutritional efficiency and economic parameters of silkworm (Bombyx mori L.). Proceedings. National Seminar on "Disease and Pest Management in Sericulture". 12th and 13th September. Sericulture College, University of Agriculture Sciences, Chintamani, pp. 86-90 .

Miller GL (1972). Use of Dinitro-salicylic acid reagent for determination of reducing sugar. Anal. Chem., 31: 426-428.

Miyashita Y (1986). A report on mulberry cultivation and training methods suitable to bivoltine rearing in Karnataka. pp. 7-30.

Muthegowda M, Santa Kumari, Umesh KNN, Boraiah G (1990). Changes in total phenolics of tukra-affected mulberry plants. Sericologia, 30(2): 263-266.

Moore S, Stein WH (1948). Photometric method for use in chromatography of amino acid. J. Biol. Chem., 176: 367-388.

Narayanaswamy KC, Ramegowda T, Raghuraman R, Manjunatha MS (1999). Biochemical changes in spiralling whitefly (Aleurodicus dispersus Russell) infested mulberry leaf and their influence on some economic parameters of silkworm (Bombyx mori L.). Entomon, 24(3): 215-220.

Narayanaswamy KC (2003). Biochemical composition of leaf roller infested mulberry leaf. Insect Environ., 8(4): 166-167.

Pradeep K, Kishore R, Noamani MKR, Sengupta K (1992). Effect of feeding tukra affected mulberry leaves on silkworm rearing performance. Indian J. Seric., 31(1): 27-29.

Raman SB, David D, Vivekanandan M (1994). Changes in morpho-physiology, water relations and nutrients in tukra diseased leaves of a few mulberry varieties. J. Seric. Sci. Jpn., 63(3): 183-184.

Ravi KK (1997). Biochemical changes in the leaves of snail infested mulberry plants and their effect on the growth and development of silkworms and cocoon characters. M.Phil., (Seric.,) thesis. Bangalore University, Bangalore, pp. 34-40.

Ravi KK, Shree MP (1998). Impact of feeding leaves from snail infested mulberry plants on the growth and development of silkworms and cocoon characters. Him. J. Env. Zool., 12: 1-10.

Reddy DNR, Narayanaswamy KC (2003). Pest of Mulberry. Zen Publishers. Bangalore, pp. 56-60.

Sathya PK, Sujatha CR, Manjunatha D, Datta RK (2000a). Screening of popular mulberry varieties for tukra infestation. National Conference on "Strategies for Sericulture Research and Development", Central Sericultural Research and Training Institute. Mysore.

Sathya PK, Sreedhar S, Singhvi NR, Kodandaramaiah J, Sen AK (2002b). Post thrips infestation biochemical changes in leaves of mulberry (Morus spp.). Plant Arch., 2(1): 85-88.

Siddegowda DK, Gupta VK, Sen AK, Benchamin KV, Manjuanath D, Prasad KS, Magdum SB, Datta RK (1995). Diaphania species infests mulberry in South India. Indian Silk, 34(12): 6-8.

Shree MP, Sreedhara VM, Umesh Kumar NN, Boraiah G (1989). Some biochemical changes in "tukra" affected leaves of mulberry varieties. Sericologia, 29(1): 141-144.

Shree MP, Umesh KNN (1989). Biochemical changes in tukra affected exotic mulberry plant. Curr. Sci., 58(22): 1251-1253.

Shree MP, Mahadeva A (2005). Impact of jassids (Empoasca flavescens F.) infestation on the biochemical constituents and photosynthetic pigments of mulberry (Morus spp.) foliage. National seminar on "Scenario of Sericulture in India". 25th - 26th March. Sri Padmavathi Mahila Visvavidyalayam, Thirupathi, Andhra Pradesh, p. 13.

Veeranna G (1997). Biochemical changes in tukra leaves of mulberry and its effects on economic characters of mulberry silkworm, Bombyx mori L. Entomon, 22(2): 129-133.

Ullal SR, Narasimhanna MN (1981). Hand Book of Practical Sericulture. Central Silk Board. Bombay.

Umesh KNN, Shree MP, Ravi Kumar R (1989). Effect of "Tukra" on the biochemical contents of mulberry variety Vietnam. Geobios, 16(6): 283-284.

Umesh KNN, Shree MP, Muthegowda, Boraiah G (1990). Changes in proteins, sugars, phenols and total chlorophyll content of mulberry plants affected by "tukra". Indian J. Seric., 29(1): 93-100.

Uritani I (1961). Changes in phenolic compounds in rice varieties as influenced by Xanthomonos oryzae infection. Il Riso., 25: 55-89.

Yemm EW, Willis AJ (1954). The estimation of carbohydrates in plant extracts by anthrone. Biochem. J., 57: 508-514.

9

Effects of fungal (*Lachnocladium spp.*) pretreatment on nutrient and antinutrient composition of corn cobs

A. Olagunju , E. Onyike, A. Muhammad, S. Aliyu and A. S. Abdullahi

Department of Biochemistry Ahmadu Bello University, Zaria, Kaduna, Nigeria.

The nutritive value of corn cob following pretreatment by fermentation with fungal species of *Lachnocladium spp.* was investigated. Corn cob was milled and subjected to incubation with fungal species for a period of one week. Significant increase ($p < 0.05$) in protein, ash and some mineral elements were observed. Calcium, magnesium, zinc and sodium content were observed to be higher in fermented cobs while significant decrease ($p < 0.05$) were observed in the fiber and antinutrient composition. Phytate, saponin, and oxalate levels were particularly lower in the fermented cobs. Biological pretreatment of corn cobs by fermentation with *Lachnocladium* species was found to significantly improve the nutritive value of corn cob, thus its potential usage in balanced feed for animal product was greatly improved.

Key words: Corn cob, nutrients, *Lachnocladium spp.*, fermentation.

INTRODUCTION

Huge quantities of agro-industrial biomass are produced worldwide annually, although, these materials are potential feed resources for ruminant livestock, their use is limited due to high fiber components (Jarommi et al., 2011). Vast quantities of crop residues are generated as a result of agricultural practices. These residues pose both disposal and environmental pollution problem. These results in the loss of nutritionally valuable materials which when processed could yield various valuable products like biofuels, chemicals and cheap energy sources for fermentation, improved animal feeds and human nutrients (Soliman et al., 2013). There is tremendous potential of agro- industrial by-products and crop residues in upholding the aims of livestock production. However the high amount of lignin content of the residues underscores optimal utilization. Lignin interferes by acting as a physical barrier that prevents the contact of cellulase to cellulose and other nutrients (Umamaheswari et al., 2010). Currently, about

eight million metric tonnes of corn are produced annually (Nwanma, 2009) and a production forecast for 2010-2015 envisaged a 23% growth. The maize plant comprises of the stalks, husks, shanks, silks, leaf blades, leaf sheaths, tassels and cobs. The corn cob carries the grain and together with associating husks, shanks and silks are harvested from the farm. The other parts are left on the farm to rot (Kludze et al., 2010). Corn cob has high percentage of lignin (45% cellulose, 35% hemicellulose and 15% lignin) and has low nutritive value and degradation rate (Sun and Cheng, 2002). This is because rumen micro flora lack enzymes for degradation of lignin and cellulose and hemicellulose are embedded within the lignin structures. Thus, the nutritive values of corn cob not only depend on the availability of nutrient but on such attributes as lignifications and crystallinity of cellulose.

The degree of utilization of crop residues or wastes by livestock is affected by pretreatment (Wong et al., 1991;

Bolanle et al., 2012). Treatment of crop residues for improving their nutritional value has been undertaken since the beginning of the 20th century (Doyle et al., 1990). Since then, tremendous efforts has been directed towards treatment via physical, physiochemical, chemical and biological means. Physical treatment for example milling and pilleting, soaking, boiling, and steaming leads to increase in surface area and density and an increase in the metabolizable energy (Beadsley, 1993), but have high-energy cost and ineffectiveness in improving feeding value of crop residue. Chemical treatment methods involve steeping by use of chemicals basically acids and alkali like sodium hydroxide and calcium hydroxide. The chemical methods are also more effective compared to the physical methods but their limitations are numerous. Environmental concerns are associated with disposal of spent acids and alkali, and unspecific side reactions which occur to yield non-specific by products. Also, there is requirement of extreme corrosive conditions of high temperature and pH thus necessitating highly trained personnel and expensive equipment (Grethlein and Converse, 1991; Zhu et al., 2009). Biological treatments with microorganisms such as fungi for example white rot fungi have several advantages when compared with physical and chemical methods. In this case, hydrolysis of polysaccharides occurs via microbial enzymes though fermentation under much milder conditions, do not produce undesirable products and are environmentally friendly (Smith et al., 1997; Rubin, 2008; Palmqvist et al., 2000).

White rot fungi grow well and produce lignocellulosic enzymes under solid state fermentation (SSF) because the medium conditions are closer to their natural habitats. (Salvachúa, 2011; Davinia, 2013). Thus, considering the substantial amount of cobs available for free or sold at very low prices by agro-industries, the upgrading of residues from fermented cobs for use in balanced feed for animal production is of potential advantage (Stamford et al., 2004). This research work has thus been designed to investigate the ability of fungal *Lachnocladiun Spp.* pretreatment to upgrade the nutrient quality of corn cobs and enhance its potential usage.

MATERIALS AND METHODS

Collection and preparation of sample

Air dried corncob residues or agricultural waste were collected in clean bags from Samaru market opposite Ahmadu Bello University main Campus Samaru Zaria. The residues were milled in a mortar and subsequently sieved with a 40 mm mesh size.

Organism and fermenting conditions

The test organism *Lachnocladium* spp. (white rot fungi), was cultivated on potato dextrose agar (PDA) slants until sporulation. The spores were harvested using 0.1% between 80 and spore number was estimated by direct microscopic counting using a haemocytometer. The white rot fungi was cultivated in mineral salt agroresidue media as described by Ali et al. (1991), in a 250 ml conical flask containing 30 ml in g/l of 10.0, KH_4PO_4; 10.5, $(NH_4)_2SO_4$; 0.33, $MgSO_4.7H_2O$; 0.5, $CaCl_2$; 0.013, $FeSO_4.7H_2O$; 0.004, $MnSO_4.H_2O$; 0.5% yeast extract and 10 g of corn cobs. The agroresidue media was inoculated with spore suspension of 7.5 ± 10^5 spores/ml and incubated at $28 \pm 3°C$. Another set of fermenting conditions was set up as above but without inoculation with the fungi to serve as the unfermented control, while the dried milled residues served as pure control for both media.

Chemical analysis

The proximate composition of the fermented, unfermented and control samples was carried out. Samples were analyzed for moisture, dry matter, crude protein, lipid, total carbohydrate, fiber, organic matter and mineral matter (ash) using AOAC (1990) methods. Antinutrients components determined include tannin, using Trease and Evans (1978), Saponin (Hudson and El-Difrawi, 1979), Phytate and oxalate (AOAC, 1990) and cyanide (Ikediobi et al., 1980). Sodium and potassium analysis were carried out using flame photometer. Atomic absorption spectroscopy (AAS) was used to determine Ca, Fe, Mn, Zn, and Mg.

Statistical analysis

Data was subjected to one-way analysis of variance (ANOVA) and least of significant difference (LSD) at 0.05 probability level. All statistical analyses of data were performed using SPSS 17.0 software and the data were reported as mean values± standard deviation (SD).

RESULTS

The effects of fungal fermentation on proximate composition of corn cobs are shown in Table 1. Proximate analysis of all the three group of samples showed significant difference (p<0.05) in the amount of crude protein, ash content and crude fiber in the fermented cobs compared to the unfermented and control corn cobs. Significant difference at p<0.05 was not observed in the fat, moisture, total carbohydrate, dry matter and organic matter in all the three group of samples. Table 2 shows the results of mineral analysis on the three groups of samples. Significant difference (p<0.05) was observed in sodium, potassium, and Zinc between the fermented and control samples.

The mineral values for sodium, potassium and zinc in the unfermented samples did not show significant difference when compared to the control and fermented samples except sodium which did not vary significantly with either the fermented or the control samples. Other minerals; calcium, iron, manganese, and magnesium determined in the three samples did not show significant variability. The results of antinutrients analysis on the fermented, unfermented and control samples are shown in Table 3. Significant difference was observed in levels of phytate, saponin, and oxalate as the levels decreased in fermented samples as compared to the unfermented and control samples. Other antinutrients; cyanide and

Table 1. Effects of *Lachnocladium spp.* fermentation on proximate composition of corn cobs.

Parameter	Fermented cobs	Unfermented cobs	Control
Crude protein	4.79 ± 0.017^a	4.10 ± 0.0057^a	3.42 ± 00.010^b
Fat	9.69 ± 0.005^a	9.96 ± 0.0059^a	9.55 ± 0.005^a
Moisture	3.98 ± 0.010^a	5.42 ± 0.021^a	5.43 ± 0.017^a
Ash	5.69 ± 0.010^a	4.46 ± 0.340^b	4.41 ± 0.004^b
Carbohydrate	74.51 ± 0.006^a	74.51 ± 0.0055^a	74.51 ± 0.0061^a
Fibre	6.83 ± 0.023^a	7.72 ± 0.011^b	$7.3\text{-}\pm 0.001^b$
Dry matter	94.55 ± 0.0058^a	94.56 ± 0.0056^a	94.52 ± 0.035^a
Organic matter	95.59 ± 0.0058^a	94.31 ± 0.011^a	95.34 ± 0.011^a

Values are means of triplicate determination ± standard deviations (n= 3). Values not having the same superscript on the row are significantly different (p<0.05).

Table 2. Effect of *Lachnocladium spp.* fermentation on mineral composition of corn cobs.

Parameter	Fermented cobs	Unfermented cobs	Control
Calcium	0.024 ± 0.0057^a	0.021 ± 0.0001^b	0.025 ± 0.010^a
Sodium	0.056 ± 0.0057^a	0.049 ± 0.005^{ab}	0.041 ± 0.005^b
Potassium	0.311 ± 0.50^a	0.241 ± 0.0057^b	0.241 ± 0.047^b
Magnesium	0.044 ± 1.00^{ab}	0.047 ± 0.010^a	0.031 ± 0.013^b
Iron	0.0025 ± 0.001^a	0.0024 ± 0.010^a	0.0024 ± 0.057^a
Manganese	0.0033 ± 0.0011^a	0.033 ± 0.011^a	0.0031 ± 0.019^a
Zinc	0.0063 ± 0.0058^a	0.0018 ± 0.015^b	0.0018 ± 0.120^b

Values are means of triplicate determination ± standard deviations (n= 3). Values not having the same superscript on the row are significantly different (p<0.05).

Table 3. Effect of *Lachnocladium* spp. fermentation on antinutrients composition in corn cobs.

Parameter	Fermented cobs	Unfermented cobs	Control
Phytate	0.803 ± 0.0057^a	1.32 ± 0.0061^b	1.32 ± 0.010^b
Cyanide	0.75 ± 0.005^a	0.75 ± 0.01^a	0.75 ± 0.005^a
Saponin	2.21 ± 0.0043^a	3.84 ± 0.030^b	3.86 ± 0.032^b
Tannin	0.022 ± 0.002^a	0.024 ± 0.0033^a	0.024 ± 0.0083^a
Oxalate	0.024 ± 0.001^a	0.087 ± 0.005^b	0.088 ± 0.0055^b

Values are means of triplicate determination ± standard deviations (n= 3). Values not having the same superscript on the row are significantly different (p<0.05).

tannin were not found to be significantly different in the three samples analyzed.

DISCUSSION

The proximate analysis of the fermented cobs showed a significant increase in crude protein. The increase in the protein content of the corn cobs fermented with *Lachnocladium* spp. could be attributed to the possible secretion of laccase and manganese peroxidase and some extracellular enzymes (proteins) such as amylases, and cellulases (Oboh and Akinwumi, 2003) into the fermenting media by the fermenting organism (Sidharth et al., 2013) as well as increase in the growth and proliferation of the fungi in the form of single cell proteins (Omer et al., 2012). Most agricultural wastes are known to support the growth of microorganism as single cell protein and thus enhance feed quality (Sharma and Arora, 2013). For example, fermented cassava has been reported for similar potentials by Obuekwe (1993) using Rhizopus as fermenting organism. Fermentation was also seen to generally improve the protein value of the same seed flour (Aderonke and Beatrice, 2013). The reduction of fiber content in fermented corn cobs is of significant importance, as fiber is often used as a negative index of nutritive value in the prediction of total digestible nutrients (TDN) and net energy. It is assume that higher fiber means lower digestibility.

The physical characteristics of fiber (particularly particle size) are also important in regulating rate of passage, rumination, insalivations and the pH of the rumen (Mahesh and Madhu, 2013). The total fiber or cell wall fraction of plants comprises cellulose, hemicellulose, lignin, cutin, silca and a variety of minor substances. The proportions of these components vary among parts of the same plant and also change as plants mature. The proximate analysis of the fermented, unfermented and control samples did not show any significant difference at $p < 0.05$ in the percenttage dry matter, organic matter, moisture, moisture content and crude fat. The dry matter and organic matter are common denominators for comparing nutrient contents of feeds; other determinations in proximate analysis are expressed on dry matter basis. The results obtained have shown that fermentation does not have effect on these compositions.

Another factor limiting the wider feed use of many crop residues is the ubiquitous occurrence of a diverse range of natural compounds which act to reduce nutrient utilization and low food intake which are referred to as antinutritional factors (Osagie, 1998; Sarwar, 2012). Fermentation brings about numerous biochemical and nutritional changes in the raw materials, including the breakdown of certain constituents, the reduction of antinutritional factors and the synthesis of B vitamins (Egounlety and Aworh, 2000). Analysis of fermented corn cobs showed a significant reduction ($p < 0.05$) in the saponin, oxalate, and phytate, while cyanide and tannin did not show significant changes. Saponins are steroid or triterpinoid glycosides which are characterized by their bitter or stringent foaming properties and their hydrolytic effects on red blood cells. But the effectiveness of saponins and other antinutrients depend on the amount present in the feed and effect of pretreatment process on these feeds.

Thus, a significant reduction in the level of the saponin following fermentation is quite advantageous. (Adeniran, 2013). Phytates are hexaphoshate derivative of inositol and storage form of phosphorus in plants. Phytate are insoluble and form salt with metals such as calcium, iron, zinc and magnesium, rendering these metals unavailable for absorption (Osagie, 1998). A reduction in phytate observed in fermented corn cobs may be due to secretion of phytates bond or due to change in the pH of the medium which affects the attachment of water and thus configuration of phytic acid by altering the strong water molecules attached (Onigbinde, 2005; Adeniran, 2013). The ability of fermentation process to reduce

phytate levels have been reported by Mulimani et al. (2003) and Marfo et al. (1990). Oxalate is a dicarboxilic acid anion present as insoluble salts of potassium, sodium and ammonium or as calcium oxalate. It can be toxic when consumed in large quantities. Thus, a reduction in its level through fermentation is quite beneficial to the feeding value of corn cobs. The ash content which is defined as the total mineral was seen to have increased in value. Specific analysis of the mineral elements showed changes in calcium, potassium and zinc. This could be attributed to the effect of fermentation on reducing the levels of antinutrients; phytate and oxalate which increase the bioavailability of mineral elements. (Onyango, 2013)

Conclusion

Fermentation with white rot fungi could serve as a good means of pre-treating corn cobs to improve its nutritional value, as it has been demonstrated to improve the protein content, fiber level, ash and some mineral elements like calcium, sodium, potassium and zinc. It has also been seen to reduce the level of some antinutrients like saponin, phytate and oxalate. With the combined effects of fungal fermentation on the nutritive value of corn cobs, its usage as an animal feed has been greatly improved. With this alternative use, environmental pollution concerns by maize cobs available for free or sold at very low prices by agro industries and the upgrading of residues from fermented cobs for use in balanced feed for animal production is of potential advantage.

REFERENCES

Adeniran, HA, Farinde, EO, Obatolu, VA (2013). Effect of heat treatment and fermentation on anti-nutrients content of lima bean (Phaseolus lunatus) during production of Daddawa analogue. Ann. Rev. Res Biol. 3(3):256-266.

Aderonke IO, Beatrice OT, Ifesa (2013). Changes in nutrient and antinutritional contents of sesame seeds during fermentation. J Microbiol. Biotechnol. Food Sci. 2(6)2407-2410.

Ali SS, Saker ART, Akin R (1991). Factors affecting cellulase production by Aspergillus terrus using water hyacinth. World J. Microbiol. Biotechnol. 62-66.

Association of analytical chemist (AOAC 1990) methods of analysis. 15th edition Washington D C.

Beadsley DW (1993). Symposium on forage utilization: Nutritive value of forage as affected by physical form. Part II Beef cattle and sheep studies. J. Anim. Sci. 23:239-245.

Bolanle Kudirat Saliu, Alhassan Sani (2012). Boethanol potentials of corn cob hydrolysed using cellulases of Aspergillus niger and Penicillium decumbens. EXCLI J. 11:468-479.

Davinia S, Angel TM, Ming T, María F L, Francisco G, Vivian R, María J M, and Alicia Pr; (2013) Differential proteomic analysis of the secretome of Irpex lacteus and other white-rot fungi during wheat straw pretreatment. Biotechnol. Biofuels 6:115.

Doyle PT, Pearce GR, Denvendra C (1986). Rice straw as a feed for ruminants. International development programme of Australian Universities and colleges, Canberra Australia. p. 177.

Egounlety, M, Aworh OC. (2000). Biochemical changes in soyabean (Glycine max Merr.), cowpea (Vigna unguiculanta L. Walp), and groundbean (*Macrotyloma geocarpa* Harms) during fermentation with *Rhizopus oligosporus*. (Paper presented at the International Seminar on Traditional African Fermented Foods. Accra, Ghana). 4-6, 25.

Grethlien HE and Converse AO (1991). Common aspects of acid prehydrolysis and steam explosion for pretreting wood. Biores. Technol. 38:156-168.

Hudson BJ, El-Difwari KB (1979). The sapogens of the seeds of four *Lupin species*. J. Plant Foods Human Nutr. 3:181-186.

Ikediobi CO, Onia GO, Eluwa CE (1980). A rapid and intensive enzymatic assay for total cyanide in cassava products. Agric. Biol. Chem. 44(12):2803-2809.

Jaromi MF, Liang M, Rosiarizan YM, Goh P, Shokryazdan YWH (2011). Efficiency of rice straw lignocellulosises degradability by *Aspergillus terrus* ATCC 74135 in solid state fermentation. Afr. J. Biotech. 10(21):4428-4435.

Kludze H, Deen B, Weersink A, van Acker R, Janovicek K, De Laporte A. (2011). Assessment of the availability of agricultural biomass for heat and energy production in Ontario. 01.27.

Mahesh MS, Madhu M (2013). Biological treatment of crop residues for ruminant feeding. Afric. J. Biotechnol. 12(27):4221-4231.

Marfo EK, Sampson BK, Idowu JS, Oke OL (1990). Effect of local food processing on phytate levels in cassava, cocoyam, yam, sorghum, rice, cowpea and soybeans. J. Agric. Food Chem. 1580-1583.

Mulimani VH, Kadi MS, Thippes Wamy S. (2003). Effect of processing and phytic acid content in different red gram (*Cajanus cajan L.*) varieties. J. Food Sci. Technol. 40:371-373.

Nwanma V. (2009) Nigerian corn study shows production could more than double: A report presented to government by IITA. Bloomberg.

Oboh G, Akindahunsi AA (2003). Biochemical changes in cassava products (flour and gari) subjected to saccharomyces solid media fermentation. Food Chem. 82(4):599-602.

Obuekwe CO, Okunbgowa JO (1986). Assesment of biomass production of potential of some fungal isolates. Nig. J. Microbiol. 6:120-130.

Omer HAA, Ali FAF, Gad SM (2012). Replacement of clover hay by biologically treated corn stalk in growing sheep rations. J. Agric. Sci. 4:257-268.

Onigbinde AO (2005). Food and human nutrition (Biochemical integration) Revised Ed Alva Corperate org. Benin City 287-334.

Onyango CA, Ochanda SO, Mwasaru MA, Ochieng JK, Mathooko FM, Kinyuru, JN (2013) Effects of malting and fermentation on anti-nutrient reduction and protein digestibility of red sorghum, white sorghum and pearl millet. J. Food Res. 2(1):41-49.

Osage AU, Offiong UE (1998). Nutritional quality of plat foods, Post Harvest Research Unit. Dept. of Biochemistry University of Benin, Benin City Nigeria. pp. 221-244.

Palmqvist E, Hahn-Hägerdal B (2000): Fermentation of Lignocellulosic hydrolysates. I: inhibition and detoxification. Biores. Technol. 74:17-24.

Rubin EM (2008). Genomics of cellulosic biofuels. Nature 454:841-845.

Salvachúa D, Prieto A, López-Abelairas M, Lu-Chau T, Martínez AT, Martínez MJ (2011). Fungal pretreatment: an alternative in second-generation ethanol from wheat straw. Biores. Technol. 6:7500-7506.

Sharma RK, Arora DS (2013). Fungal degradation of lignocellulosic residues: An aspect of improved nutritive quality. Crit. Rev. Microbiol. 16:223.

Shereen A. Soliman, Yahia A. El-Zawahry and Abdou A. El-Mougith (2013). Fungal biodegradation of agro-industrial waste licensee InTech. Open access article distributed under the terms of the Creative Commons. pp. 1-28.

Sidharth V, Devendra PM, Mohmed S, Ayushi A and Sangeeta N (2013); Development of microbial consortium for production of blend of enzymes for hydrolysis of agricultural wastes into sugars. J. Sci. Ind. Res. 72:585-590.

Smith JE, Anderson JG, Senior EK (1987). Bioprocessing of lignocelluloses Phill Trans R. Soc. Lond Ser. A 321:507-521

Stanford RB, Guerra NY, Medeiros CP, Defreitas MSR and Cavalcante, M I (2004) Protein enrichment of pineapple waste for animal feeds. In Bioconversion of vegetable, animal and industrial wastes by means of fungi mycelia in an artificial rumen. Laboratory of food sciences Department of nutrition Health Science centre, federal University of Pernambuco Recife, Brazil. pp. 1-5

Sun y and Cheng j (2002) Hydrolysis of lignocellulosic material from ethanol production. A review. Bioresour. Technol. 83:1-11

Trease GE, Evans WC (1978). A textbook of pharmacology 11th Ed MaCraw Hill publishers United Kingdom.

Van Soest PJ (1982) Nutritional ecology of the ruminant Cornell University Press, Ithacaa NY USA.

Wong KKY, Dellerrell KF Keith LM Clerk TA, Donnalson LA (1991). The relationship between fiber porosity and cellulose digestibility in steam exploded pinups. Biotech. Bioeng. 31:447-456.

Zhu JY, Pan XJ, Wang GS, Gleisner R. (2009). Sulfite pretreatment (SPORL) for robust enzymatic saccharification of spruce and red pine. Bioresour. Technol. 100:2411-2418.

10

Production of yeast using acid-hydrolyzed cassava and poultry manure extract

L. I. Osumah and N. J. Tonukari*

Department of Biochemistry, Faculty of Science, Delta State University, P. M. B. 1, Abraka, Delta State, Nigeria.

Cassava is made up of starch as its major nutritive reserve. Starch which is one of the most important products synthesized by plants is consumed as food and can be used in industrial processes. This investigation seeks to explore the availability of cassava as a source of glucose as well as poultry manure as a source of nitrogen in the production of yeast. Cassava flour was hydrolyzed with 0.5% (v/v) concentrated H$_2$SO$_4$ as carbon source for the production of yeast. It was found that pH 6.5 gave optimum yeast growth. Increased concentrations of acid-hydrolyzed cassava and poultry manure extracts led to significant (P < 0.05) increase in yeast biomass after 36 h culture. The residual glucose concentration was also determined and was found to be significantly (P < 0.05) increased with increase in the concentration of poultry manure extract. Therefore, yeast can be produced using acid hydrolyzed cassava flour as carbon source with poultry manure extract as nitrogen source. The methods described in this work can be used in the development of a rapid method of producing glucose and simple sugars from cassava through acid hydrolysis and combining this with poultry manure for yeast production.

Key words: Yeast, poultry manure extract, acid-hydrolyzed cassava.

INTRODUCTION

Yeasts (*Saccharomyces cerevisae*) have been known to humans for thousands of years as they have been used in fermentation processes like in the production of alcoholic beverages and bread leavening (Broach and John, 1991). The industrial production and commercial use of yeast started at the end of the 19th century after their identification and isolation by Pasteur (Bekatorov et al., 2006). During commercial production, yeast is grown under carefully controlled conditions on a sugar containing media typically composed of beet and cane molasses. Under ideal growth conditions, a yeast cell reproduces every two to three hours (Bekatorov et al., 2006). Studies show that organic nitrogen sources, such as yeast extracts support rapid growth and high cell yields of microorganisms because they contain amino acids and peptide, water soluble vitamins and carbohydrates (Peppler, 1982). The basic carbon and energy source for yeast culture are sugars (Dubai and

Muhammad, 2005). Starch cannot be used because yeast does not contain the appropriate enzymes to hydrolyze this substrate to fermentable sugars. Beet and cane molasses are commonly used as raw material because the sugars present in molasses, a mixture of sucrose, fructose and glucose, are readily fermentable. In addition to sugar, yeast also requires certain minerals, vitamins and salts for growth. The number of yeast cells increases about five to eight-fold during fermentation (Glen and Dilworth, 2002).

Cassava (*Manihot esculenta*), also known as *manioc*, *tapioca* or *yucca*, is one of the most important food crops in the humid tropics, being particularly suited to conditions of low nutrients availability and is able to survive drought (Burelli, 2003). The major harvested organ is the tuber, which is actually a swollen root. Cassava is a source of calories for both human and animal feeding (Tonukari, 2004). The nutrient reserve of cassava is made up of starch which is consumed as food and used in industrial processes (Tonukari, 2004); although, cassava leaves are sometimes consumed. The acid catalyzed hydrolysis of starch is a complex

---

*Corresponding author. E-mail: tonukari@gmail.com.

heterogenous reaction. It involves physical factors as well as hydrolytic chemical reaction. The molecular mechanism of acid catalyzed hydrolysis of starch involves cleavage of glucan bonds. This is mainly on the $\alpha(1,4)$ bonds of amylose and $\alpha(1,6)$ bonds of amylopectin to produce several thousands of glucose residues. The hydrolysis is therefore controlled by both the reaction conditions (which are acid concentration and temperature) and the physical state of starch (Oboh and Akindahunsi, 2003). The treatment of starch as a mixture of several glucans with concentrated H_2SO_4 showed a reaction rate, two orders of magnitude higher than that of untreated starch about the same magnitude. However, the starch when treated with a varying level of H_2SO_4 undergoes an abrupt change in the physical structure of the glycosidic bond linking. $\propto$-amylose and amylopectin glycosidic bonds are broken to produce glucose and oligosaccharide residues (Burelli, 2003). The major hydrolytic product of cassava starch (glucose) can serve as a suitable carbon source for the production of S. cerevisae. The production of yeast is an important step to the commercial use of the product in baking, brewing and other applications (Broach and John, 1991).

The millions of kilograms of nitrogen collected and disposed off each day in animal manure and municipal waste represent a valuable reservoir of nitrogen potentially available for conversion to protein for livestock and poultry feeds as well as other valuable products. Traditionally, this nitrogen has been applied to land, where some of it is recycled into plant protein. However, the so called waste materials are now coming under close scrutiny because of increases in demand for plant proteins for human food. As a result, new and old systems are now being considered for their potential, in converting animal and municipal wastes into acceptable animal products (Anthony, 1971; Kyoung et al., 2007).

Various works have been carried out on the utilization of poultry manure on various applications (Obeta et al., 2009; Kargi and Shuler, 2005; Orhan et al., 2009). However, the present experimental design involves the use of acid-hydrolyzed cassava as carbon source and poultry manure extract as nitrogen source for the production of yeast. The purpose of this research is to develop a method for producing glucose and simple sugars from cassava (a locally available raw material) through acid hydrolysis and combining this with poultry manure for yeast production. The significance of this work is to use locally abundant agricultural products, cassava as well as agricultural waste (poultry manure) for the production of an industrial raw material, yeast.

MATERIALS AND METHODS

Cassava and poultry manure

Cassava (M. esculenta) tubers were purchased from Abraka Market in Delta State, Nigeria. It was peeled and thoroughly washed (to reduce the cyanide content). It was then sliced into small pieces, dried under the sun (open), after which, it was ground to flour and passed through a sieve of 0.25 mm before it was stored at room temperature in a dry plastic container, ready for use.

Acid hydrolysis of cassava

To get the best hydrolysis of cassava flour, varying weights (5 - 25 g) of cassava flour used at constant acid volume was weighed into six (6) conical flasks. To each of the conical flask, was added 0.5 ml of 0.5% (v/v) H_2SO_4 acid. The conical flask was then made up to 100 ml with distilled water and shaken thoroughly to have an even mix. It was autoclaved at about 115°C for 20 min, cooled and filtered into different sterilized test tubes as hydrolyzed glucose (sample glucose) and stored in a cool dry place.

Glucose estimation

The sample glucose estimation was carried out to determine the optimum hydrolysis of cassava flour. This was done using glucose estimation kit (Randox Laboratories Ltd, Diamond Road, Crumlin, Co. Antrim United Kingdom, BT29 4QY). Five test tubes were labeled test tubes 1 - 5 with tube 1 serving as the control. To each of the test tubes, 100 µl of hydrolyzed glucose filtrate, 2.0 ml of distilled water and 2.0 ml glucose buffer was added. This was incubated for 5 min at room temperature and optical density (O.D) taken using the spectrophotometer at 500 nm to estimate the amount of glucose as well as the optimum hydrolysis of cassava flour (Barham and Trended, 1972).

Poultry manure extract

Poultry manure (chicken droppings) was gotten from one of the poultries in Abraka, Delta State, Nigeria. It was dried under the sun (open air) and mashed to fine particles using mortar and pestle. It was then passed through a sieve of 0.25 mm and then stored in a dry plastic container, ready for use. To get a pure filtrate (extract) from the collected sample of poultry manure, 10 g of poultry manure was weighed into different conical flasks (at desired number). To each of the conical flask was added 100 ml of water. It was then filtered to remove the debris before autoclaving for 30 min. After autoclaving, it was filtered again using filter paper into different test tubes and stored at room temperature in a cool dry place to be used as nitrogen source for yeast culture.

Yeast tablets

Yeast tablets were purchased from a pharmaceutical store in Abraka, Delta State, Nigeria.

Preparation of yeast culture (YPD media) for inoculation

To prepare an uncontaminated yeast culture which is used for inoculation of different reaction media for yeast production, 2 g peptone, 2 g D-glucose and 1 g yeast extract was weighed into a 250 ml autoclaved conical flask. The content was mixed with a little amount of distilled water and the solution was made up to 100 ml with distilled water. The content was autoclaved for 30 min; it was brought out, cocked with cotton wool to make it air tight and then autoclaved again for another 30 min. After which, it was brought out and allowed to cool at room temperature. After cooling, 100 µl of 20% (w/v) antibiotics (ampiclox) was added to the solution to

Table 1. Glucose estimation after acid-hydrolysis of cassava.

Acid-hydrolyzed cassava	5%	10%	15%	20%	25%
O.D (500 nm)	1.757	1.723	1.707	1.785	1.268
Glucose concentration (mmol/L)	9.800	9.611	9.521	9.956	7.072
Amount of glucose (g)/100 ml	1.764	1.730	1.714	1.792	1.273

prevent bacteria growth before adding two tablets of yeast. Its content was thoroughly shaken to allow the tablet dissolve and the culture was incubated for 6 - 12 h before using the culture for inoculation of other substrate for yeast culture.

Yeast culture with varying poultry manure extract

To determine the yeast biomass using acid hydrolyzed cassava as carbon source and poultry manure as nitrogen source, 20 ml of the 20% acid-hydrolyzed cassava was measured into six (6) conical flasks having the glucose control flask; the poultry manure extract control flask containing 10 ml of poultry manure extract and four other flasks containing 2.5, 5.0, 7.5 and 10.0 ml of poultry manure extract, respectively. To the poultry manure extract control flask, there was no addition of 20 ml acid hydrolyzed cassava and no poultry manure extract was added to the glucose control flask. The pH in the conical flasks was adjusted to 6.5. This was done using the pH meter and NaOH solution (1.0 M) as alkaline medium. The total volume in each conical flask was made up to 100 ml with distilled H_2O. The flasks were then autoclaved for 30 min, allowed to cool, corked with cotton wool and re-autoclaved for another 30 min. After autoclaving the second time, it was allowed to cool and 100 µl of 20% (w/v) antibiotics (ampiclox) was added to each flask to prevent bacteria growth. 2 ml of the incubated yeast culture was then used to inoculate each of the conical flasks. The flask was thoroughly shaken and allowed to grow for 36 h. After the growth period, the OD was taken at 600 nm to determine the yeast biomass.

Yeast culture with varying acid-hydrolyzed cassava

To determine the yeast biomass using varying amount of hydrolyzed cassava as carbon source with poultry manure extract as nitrogen source, a varied amount of the 20% acid-hydrolyzed cassava was added from 2.5 - 20 ml into different conical flasks (flasks 2 - 9). To flask one, there was no addition, serving as control. 5% of poultry manure extract was then added to the different conical flasks. After these additions, subsequent procedures following the yeast determination is as previously described. In estimating the glucose after yeast culture using varied acid-hydrolyzed cassava, nine test tubes were used and labeled test tubes 1 - 9. All other procedures followed procedures previously described above. After the yeast biomass has been determined, the pH readings were also determined as previously described.

Yeast culture with varying pH

To determine the yeast biomass with varying pH using acid hydrolyzed cassava as carbon source and poultry manure extracts as nitrogen source, 20 ml of the 20% acid-hydrolyzed cassava was measured into six (6) conical flasks followed by the addition of 5 ml of poultry manure extract to the conical flasks. The pH of contents

of the conical flasks labeled 2 - 6 was adjusted to 2.5, 3.5, 4.5, 5.5, and 6.5, respectively, with the exception of conical flask 1, which served as the control. This was done using the pH meter and NaOH (1.0 M) solution as alkaline medium. The total volume in each conical flask was made up to 100 ml with distilled H_2O after adjusting the pH. It was then autoclaved for 30 min, allowed to cool, corked with cotton wool and re-autoclaved for another 30 min. After autoclaving the second time, it was allowed to cool and 100 µl of 20% (w/v) antibiotics (ampiclox) was added to each flask to prevent bacteria growth. 2 ml of the incubated yeast culture was used to inoculate each of the conical flasks. The flask was shaken very well and allowed to grow for 36 h. After the growth period, the optical density (OD) was taken at 600 nm to determine the yeast biomass.

RESULTS

Acid hydrolysis of cassava flour

Acid hydrolysis of the cassava flour was carried out to determine the best hydrolysis of cassava flour varying the weights of cassava flour from 5 - 25 g at constant acid volume. At the end of the hydrolysis, the hydrolyzed glucose (sample glucose) was filtered into different sterilized test tubes and stored in a cool dry place. Glucose estimation of the varied hydrolyzed cassava was carried out as well as standard glucose estimation in comparison. This was to test for the optimal hydrolysis of cassava. This sample glucose estimation was carried out using the glucose kit and the result shows that 20% of the cassava flour gave the optimum hydrolysis as shown in Table 1. This gave the highest amount of glucose on hydrolysis which was significant at 5% level using F-test for statistical analysis. Thus, subsequent analysis for yeast production was carried out using 20% cassava flour for acid hydrolysis and the glucose produced was used for yeast production.

Yeast biomass estimation with varying poultry manure extract

The yeast biomass estimation analysis was aimed at determining the amount of yeast as well as determining the amount of residual glucose and the pH level (whether acidic or alkaline) after the yeast culture. This was done using the acid-hydrolyzed cassava (glucose) as carbon source and the poultry manure extract as nitrogen source. The results obtained using poultry manure extract is shown in Table 2. Yeast grows very well in the

Table 2. Yeast biomass estimation after 36 h culture using acid-hydrolyzed cassava (glucose) as carbon source and varying poultry manure extract as nitrogen source.

Acid-hydrolyzed cassava (%, v/v)	Poultry manure extract (%, v/v)	Yeast biomass ($OD_{600\ nm}$)**	Residual glucose (mmol/L)**	pH after yeast culture
20	0	0.507 ± 0.002	5.438 ± 0.002	4.39
20	2.5	0.550 ± 0.004	3.581 ± 0.002	4.45
20	5.0	0.568 ± 0.001	1.774 ± 0.001	4.44
20	7.5	0.629 ± 0.002	2.806 ± 0.002	4.46
20	10.0	0.677 ± 0.003	1.841 ± 0.002	4.48
0	10	0.306 ± 0.002	0.011 ± 0.000	4.94
	YPD	0.992 ± 0.002	7.351 ± 0.045	4.79

**Values are mean ± standard deviation of triplicate experiments. YPD = Yeast peptone dextrose culture medium containing 2% glucose.

presence of carbon and nitrogen. From Table 2, it is observed that, under the normal yeast culture (YPD), the yeast grew optimally compared to when another carbon source (acid-hydrolyzed cassava filtrate - glucose) and nitrogen source (poultry manure extracts) was used.

Using the constant percentage of acid-hydrolyzed cassava filtrate - glucose as carbon source together with varied percentage of poultry manure extract as nitrogen source, increased yeast biomass was observed as the percentage of the poultry manure extract increases which was significant at 5% level (Table 2). This indicates that, yeast grows very well in the presence of high amount of nitrogen and carbon. Table 2 also depicts glucose filtrate (acid-hydrolyzed cassava) control and poultry manure extract control. It was observed from the table that, in the presence of high amount of glucose filtrate (acid-hydrolyzed cassava) and no amount of poultry manure extract (poultry manure extract control), there was higher amount of yeast biomass compared to biomass in the presence of high amount of poultry manure extract and no amount of glucose filtrate (glucose filtrate control) at $P < 0.05$.

After the growth of the yeast at the specified incubation period, the residual glucose as well as the pH of the media after growth was measured. From Table 2, it was observed that, with increase in yeast biomass, there was significant decrease in the residual glucose concentration, which indicates that, much of the glucose have been used during the yeast growing process. It was also observered that the pH was acidic after yeast culture. Yeast grows optimally at pH 6.5. The reduced pH observed indicates that, after the optimal growth of yeast, the media in which the growth occurred became more acidic, because of the production of organic acids like lactic acid and malic acid.

Yeast biomass estimation with varying acid-hydrolyzed cassava

The yeast biomass estimation with varied acid-hydrolyzed

cassava (glucose) was aimed at determining the yeast growth using various percentage of acid-hydrolyzed cassava (glucose). The amount of residual glucose and the pH after the yeast culture was also determined. The results obtained using poultry manure extract as nitrogen source is shown in Table 3. From the table, it was observed that, as the percentage of acid-hydrolyzed cassava (glucose) increases from 0 - 20%, there was also significant ($P < 0.05$) increase in the yeast biomass which indicates that, yeast biomass is dependent on carbon source in the presence of nitrogen (poultry manure extract).

After the growth of the yeast at the specified incubation period, the residual glucose concentration as well as the pH of the media after culture was measured. From Table 3, it was observed that, there was lesser yeast biomass, with high concentration of residual glucose. This indicates that, the lesser the yeast biomass, the more the residual glucose after yeast culture.

Also measured was the pH after yeast culture, which was observed to be acidic (reduced pH). The low pH observed indicates that, after the optimal growth of yeast, the media in which the growth occurred becomes more acidic because of the production of organic acids, like lactic and malic acid, as yeast grows optimally at pH 6.5 (Table 3).

Yeast biomass estimation with varying pH

The yeast biomass estimation with varying pH was aimed at determining the pH at which optimal yeast growth is observed. This experiment is necessary because, after acid hydrolysis, the medium became very acidic (about pH 1.3); thus, varying the pH helps to ascertain the minimum amount of NaOH needed to adjust the pH. The amount of residual glucose and the pH level (whether acidic or alkaline) after the yeast biomass was also determined. The results obtained using poultry manure extract as nitrogen source is shown in Table 4.

Table 3. Yeast biomass estimation after 36 h culture with varying amount of acid-hydrolyzed cassava (glucose) as carbon source and poultry manure extract as nitrogen source.

Acid-hydrolyzed cassava (%, v/v)	Poultry manure extract (%, v/v)	Yeast biomass (O.D$_{600\,nm}$)**	Residual glucose (mmol/L)**	pH after yeast growth
0	5	0.383 ± 0.008	1.902 ± 0.004	5.17
2.5	5	0.486 ± 0.008	2.214 ± 0.002	5.88
5.0	5	0.761 ± 0.008	2.194 ± 0.001	5.61
7.5	5	0.945 ± 0.013	2.181 ± 0.001	5.03
10.0	5	0.979 ± 0.008	2.298 ± 0.001	4.98
12.5	5	0.995 ± 0.019	2.479 ± 0.004	4.95
15.0	5	1.109 ± 0.010	3.561 ± 0.000	4.71
17.5	5	1.135 ± 0.005	3.583 ± 0.017	4.67
20.0	5	1.192 ± 0.003	3.587 ± 0.001	4.4

**Values are mean ± standard deviation of triplicate experiments.

Table 4. Yeast biomass estimation after 36 h culture with varying pH using acid-hydrolyzed cassava (glucose) as carbon source and poultry manure extract as nitrogen source

Acid-hydrolyzed cassava (%, v/v)	Poultry manure extract (%, v/v)	pH variation	Yeast biomass (O.D$_{600\,nm}$)**	Residual glucose (mmoles/L)**	pH after yeast growth
20	5	1.47***	0.220 ± 0.003	11.557 ± 0.011	2.35
20	5	2.5	0.269 ± 0.002	10.994 ± 0.016	3.36
20	5	3.5	0.454 ± 0.000	10.904 ± 0.004	3.96
20	5	4.5	0.456 ± 0.001	9.851 ± 0.002	4.78
20	5	5.5	0.743 ± 0.002	5.254± 0.006	5.24
20	5	6.5	0.796 ± 0.002	5.126 ± 0.008	5.37

**Values are mean ± standard deviation of triplicate experiments; ***pH after acid hydrolysis and before adjusting with NaOH (1 M).

Yeast grows best (optimally) at pH 6.5. From Table 4, the pH of the yeast media was adjusted to varying pH range, from pH 2.5 - 6.5 using 1 M sodium hydroxide (NaOH) solution as the alkaline medium while the acid-hydrolyzed cassava (glucose) was used as the acidic medium as well as carbon source; the poultry manure extract was, however, used as the nitrogen source. The pH of the medium before adjusting with 1 M NaOH (serving as the control for the analysis) was 1.47 for poultry manure extract.

Using constant percentage of acid-hydrolyzed cassava filtrate - glucose as carbon source and constant percentage of poultry manure extract as nitrogen source, high yeast biomass was observed as the pH variation increases from pH 2.5 - 6.5 and pH 6.5 gave the best (optimum) yeast growth (Table 4).

After the growth of the yeast at the specified incubation period, the residual glucose concentration as well as the pH of the media after growth was measured. From Table 4, it was observed that, as the pH increases, the residual glucose concentration after yeast growth decreases, which indicates that, as the pH variation increases from

2.5 - 6.5, there was high yeast biomass which resulted in low residual glucose concentration due to the high utilization of the carbon content in the acid-hydrolyzed cassava (glucose filtrate). This was significant at $P < 0.05$ using F-test for statistical analysis.

The results showed that the pH after yeast culture was acidic. However, it was also observed that, at other adjusted values of pH ranging from 2.5 - 4.5 with the controls inclusive, the pH values after the yeast culture was higher than the adjusted values, which indicates that, yeast does not grow very well in a highly acidic medium. Fair yeast growth was observed for adjusted pH 5.5, as the pH value after the yeast culture was seen to be reduced (acidic), but not as good as pH 6.5 (Table 4).

DISCUSSION

More than two-third of the total production of cassava is used as food by humans, with lesser amounts being used for animal feeds and industrial purposes (Nwokoro et al., 2002). Starch is one of the most important plant product to man. Cassava starch, which is very bland in flavour, is used in processed baby foods as a filler material, and bonding agent in confectionary and biscuit industries (Tonukari, 2004).

Various industries have exploited the use of cassava in the production of many items such as textiles, cosmetics, glue and adhesive, pharmaceutical and cement (Tonukari, 2004). This experimental design seeks to explore the use of cassava flour as a source of glucose and poultry manure as source of nitrogen in the production of yeast. The results of the present study, shows that, acid hydrolyzed cassava flour (glucose) can serve potentially as a cheap carbon source for the production of yeast. This is due to the composition of starch in cassava and also the relative high availability of cassava (Ihekonronye and Ngoddy, 1985; Alais and Linden, 1999; Magnolia et al., 2006). Acid hydrolysis of cassava flour was used because of the advantages it has over saccharification by enzymes. The processes involved in acid hydrolysis of cassava flour is fast, cheap, high yielding and not affected by contamination (Ipsita and Munishwar, 2003).

The results obtained showed that, the hydrolysis of cassava flour gave a high yield of glucose which serves as good and cheap carbon source for yeast production. This is in agreement with the findings of Oboh and Akindahunsi (2003), who observed that, the hydrolytic product of cassava starch - glucose, serves as a suitable substrate in providing carbon source for the growth of yeast (*S. cerevisiae*). It also confirms Dubai and Muhammad (2005) observation that the basic carbon and energy source for yeast culture are sugars.

This research also shows that yeast culture can be enhanced with the use of poultry manure which serves as an alternative to peptone. Peptone is commonly used as nitrogen source for the growth of yeast. Poultry manure as a way of recycling environmental waste has been seen to be efficient as peptone, because of their similar level of yeast growth, and it can also be considered as a better and preferred source of nitrogen because of its abundant availability and the presence of mineral salts like phosphorus, potassium, calcium and magnesium which can as well aid in the growth and development of yeast (Albers et al., 1996; Yao et al., 2006).

The optimal growth of yeast as shown by studies is best in the presence of carbon and nitrogen sources; hence, glucose stands as an important carbon source for the growth of yeast with a dual role in biosynthesis, and energy generation and for microbial fermentation processes (Stanbury et al., 1995; Dubai and Muhammad, 2005) which was also confirmed in this experimental design. Poultry manure was used as nitrogen source for the production of yeast and CO_2 as by-products by micro-organisms as a result of the fermentation of sugar (Table 2).

In the course of the research, yeast growth was determined at varied pH values of 2.5 - 6.5 using poultry manure extract as nitrogen source and acid-hydrolyzed cassava as carbon source. The results showed that, yeast grows closer to neutral pH (Table 4). At pH 5.5 – 6.5, there was a sharp increase in yeast growth. This result is in agreement with the findings of Glen and Dilworth (2002), who studying the effect of nitrogen and carbon sources, showed that, the growth of the yeast reached a peak at a pH of about 7 (neutral pH) and was not able to grow when the pH of the medium was lower than 4.

Yeast biomass was also determined at varied values of the acid-hydrolyzed cassava using poultry manure extract as nitrogen source. The results showed that, yeast biomass increases with increasing amount of acid-hydrolyzed cassava as carbon source. Thus, confirming Dubai and Muhammad (2005), who found that, the basic carbon and energy source for yeast culture are sugars.

Conclusion

The methods described in this work can be used in the development of a rapid method for producing glucose and simple sugars from cassava through acid hydrolysis and combining this with poultry manure for yeast production. This research has taken the advantage of cassava flour in its rich content of starch to assess its ability to serve as a carbon source for the production of yeast using poultry manure extract as an alternative nitrogen source to peptone because of its availability. Yeast is currently imported into Nigeria for various uses. Local industries should take advantage of the results presented here to start yeast processing plants using readily available materials- cassava and poultry manure. These can yield an incredible savings in operation cost. This research can be extended using laboratory and pilot fermentors to optimize the various parameters for yeast production using these readily available local raw materials.

REFERENCES

Alais C, Linden G (1999). Food Biochemistry. Aspen publishers Inc. Maryland.
Albers E, Larsson C, Lidén G, Niklasson C, Gustafsson L (1996). Influence of the nitrogen source on Saccharomyces cerevisiae anaerobic growth and product formation. Appl. Environ. Microbiol. 62(9): 3187-3195.
Anthony WB (1971) Cattle manure as feed for cattle. In livestock waste management and pollution abatement. Livestock wastes, Columbus, Ohio. pp. 293-296.
Barham D, Trinder P (1972). An improved colour reagent for the determination of blood glucose by the oxidase system. Analyst 97(151): 142-145.
Bekatorov A, Psarianos C, Athanasios A (2006) Production of food grade yeasts. Food Technol. 44(3): 407-415.
Broach JR, John R (1991). Pringle and biology of the yeast Saccharomyces cerevisiae: Genome dynamics protein synthesis and

energetics, cold spring harbour laboratory press, New York.

Burelli MM (2003). Starch: the need for improved quality or quantity-- an overview. J. Exp. Bot. 54(382): 451-456.

Dubai YU, Muhammad S (2005). Cassava starch as an alternative to agar-agar in microbiological media. Afr. J. Biotechnol. 4(6): 573-574.

Glen ST, Dilworth EA (2002). Growth and Survival of Yeast in dairy product. Food Res. Int. 34: 791-796.

Ihekonronye AI, Ngoddy PO (1985) Integrated food science and technology for the tropic. Macmillian Education Ltd. Oxford.

Kargi F, Shuler ML (2005). A mixed yeast-bacteria process for the aerobic conversion of poultry waste into single-cell protein. Biotechnol. Lett. 3(8): 409-414.

Ipsita Roy, Munishwar Nath Gupta (2003). Hydrolysis of starch by a mixture of glucoamylase and pullulanase entrapped individually in calcium alginate beads.Enzyme Microb. Technol. 34(1): 26-23.

Obeta Ugwuanyi J, Brian McNeil, Linda Harvey M (2009). Production of Protein-Enriched Feed Using Agro-Industrial Residues as Substrates.Utilisation, DOI 10,1007/978-1-1-4020-99942-7-5. Springer science + Business media B.V.

Kyoung SR, Keri C, Douglas E, Patrick GH (2007). Catalytic Wet Gasification of Municipal and Animal Waste. Ind. Eng. Chem. Res. 46(26): 8839–8845.

Magnolia A-N, Maria TS, Larry IH, Ying H, Ross MW, Raymond PG (2006). Cassava (Manihot esculenta) has high potential for iron biofortification. FASEB J. pp. 20-24.

Nwokoro SO, Orheruata AM, Ordiah PI (2002). Replacement of maize with cassava sieviates in cockerel starter diets: effects on performance and carcass characteristics. Trop. Anim Health Prod. 34(2): 103-107.

Oboh G, Akindahunsi AA (2003). Biochemical changes in cassava products (flour and garri) subjected to Saccharomyces cerevisiae solid media fermentation. Food Chem. 52(4): 599-602.

Orhan O, Bulent I, Gulay O (2009). Pretreatment of poultry litter improves Bacillus thuringiensis-based biopesticides production. Bioresour. Technol. 101(7): 2401-2404.

Peppler HJ (1982) Yeast extract. In fermented foods ed. Rose, A.H. London: Academic press. pp. 93-312.

Stanbury PF, Whitaker A, Hall SJ (1995) Media for industrial fermentations. In principles of fermentation technology. Oxford; pergamon press pp. 93-121.

Tonukari NJ (2004) Cassava and the future of starch. Electron. J. Biotechnol. 7(1). www.ejbiotechnology.info//content/vol7/issue1/issues/2/.

Yao L, Li G, Dang Z (2006). Major chemical components of poultry and livestock manures under intensive breeding. J. Appl. Ecol. 17(10): 1989-1992.

11

Antinociceptive properties of *Trigonella foenumgreacum* seeds extracts

A. Laroubi[1]*, L. Farouk[1], R. Aboufatima, A. Benharref[2], A. Bagri[3] and A. Chait[1]

[1]Laboratory of Animal Physiology unit of Ecophysiology, Cadi-Ayyad University, Faculty of Science Semlalia Marrakech, Morocco.
[2]Chemistry Laboratory of the Natural Substances and the Heterocycles, Cadi-Ayyad University, Faculty of Science Semlalia Marrakech, Morocco.
[3]Laboratory of Biochimestry and Neurosciences, Faculty of Sciences and Technology, University Hassan 1[er], B.P.: 577, Settat 26000, Morocco.

Trigonella foenum-graecum L. (Leguminosae), known in Morocco as "Helba", is used in folk medicine for its anti-ulcer, anti-inflammatory, cicatrizing activities and to treat various pain-related physiological conditions. In the present study, we attempted to verify the possible antinociceptive action of different extracts obtained from the seeds of this plant. Three experimental models were used (acetic acid, formalin, and hot-plate tests) in order to characterize the analgesic effect. The extracts significantly, and in a dose-dependent manner, reduced the pain induced by intraperitoneal injection of acetic acid. In the formalin test, the extracts, except ethyl acetate extract (Tfge), significantly reduced the painful stimulus but only in the early phase of the test. On the contrary, these extracts, except Tfge, were ineffective to increase the latency of licking or jumping in the hot plate test. These results suggest that the compounds present in the extracts activated both central and peripheral mechanisms to elicit the analgesic effect.

Key words: *Trigonella foenum-graecum* seeds, writhing test, formalin test, hot-plate test, nociception, mice, rats.

INTRODUCTION

Trigonella foenum-graecum [(Tfg), Fenugreek] (leguminosae), locally known by its Arabic name "Helba", is one of the oldest medicinal plants, originating in India and Northern Africa. It is extensively cultivated in most regions of the world (Bellakhdar, 1997). The applications of Tfg were documented in ancient Egypt, where it was used in incense and to embalm mummies (Basch et al., 2003). In 2003). In Chinese traditional medicine, the seeds of this plant have been prescribed as a tonic for stomach disorders, and the whole aerial part of the plant is used as a folk medicine for the treatment of renal diseases in the Northern-east region of China and Morocco (Laroubi et al., 2007). The seeds of Tfg which are commonly used as a condiment in Moroccan eating are reported to have nutritive properties and to stimulate digestive process. Its leaves are used internally and externally to reduce swelling, prevent falling of hair and in the treatment of burns (Bellakhdar, 1997). Tfg is known to have several pharmacological effects such as hypoglycaemia (Abdel-Barry et al., 2000), hypocholestrolemia (Kholsa et al., 1995), anti-oxidation (Dixit et al., 2005), and laxation (Dirk et al., 1999).

Most protocols for the control of pain rely on using non-steroidal anti-inflammatory drugs (NSAIDs) and opioid analgesics. However, both of them produce several side effects. NSAIDs produce gastrointestinal disturbances and ulceration, renal damage and hypersensitivity reac-

---

*Corresponding authors. E-mail : lar_amine@yahoo.fr.

Non-standard Abbreviations

ASA, acetylsalicylic acid; Tfga, aqueous extract; Tfgb, butanolic extract; COX I, cycloxygenase I; COX II, cycloxygenase II; Tfgd, dichloromethane extract; Tfge, ethyl acetate extract; Tfgh, hexane extract; I.P, intraperitoneal; I.C.V, intra-cerebro-ventricular; NSAIDs, nonsteroidal anti-inflammatory drugs; Tfg, *Trigonella foenum-graecum*; s.c, subcutaneously; S.S., saline solution; SN, nervous system.

tion resulting from a non selective inhibition of cycloxygenase I (COX I) and cycloxygenase II (COX II) (Tjolsen et al., 1992; Vane and Botting, 2003). Opioids induce nausea, constipation, confusion, respiratory depression, and possibly dependence (Dray and Urban, 1996). Therefore, searching for less harmful compounds is still an out-standing domain of investigation. Some research focused on plant medicines used in traditional medicine as they could be good sources for natural analgesic agents. There are several reports concerning the antinociceptive and anti-inflammatory effects of Tfg seeds in Moroccan traditional medicine (Bellakhdar, 1997). This plant is known to contain alkaloids, saponins, flavonoides, Sali-cylate, and nicotinic acid (Saxena and Shalem, 2004; Yingmei et al., 2001). The present study was designed to investigate if the Tfg seeds extract has antinociceptive effect.

MATERIAL AND METHODS

Animals

Male Swiss mice weighing 20 - 30 g and male Sprague-Dawley rats weighing 180 - 280 g were used. The animals were kept in a room maintained on a 12h/12h light/dark cycle, on 25 °C constant temperature and on 55% relative humidity. They had free access to food and water. Before testing, they were allowed to adapt in the test room for at least 12 h. Each rat was used in a single experiment. All experiments were carried out in accordance with the European community guidelines (EEC Directive of 24 November 1986; 86/609/EEC). All efforts were made to minimise animal suffering and to reduce the number of animals.

Plant materials

T. foenum-graecum seeds were collected in the Chaouia region of Morocco, it was identified and stored by Pr A.Ouhamou in the Herbarium of Faculty of Science Semlalia Marrekech (voucher number 4228).

Preparation of extracts

The seeds were dried and coarsely powdered. A 210 g powder was extracted (24 h) in a Soxhlet apparatus using methanol and concentrated on a rotaevaporator. The methanolic extract (51.71 g) was successively separated with water, hexane, dichloromethane, ethyl acetate and n-butanol according to the method of Shaheen et al. (2000). The extraction has given 18.06 g of aqueous extract (Tfga), 7.51 g of hexane extract (Tfgh), 12.57 g of dichloromethane extract (Tfgd), 4.34 g of ethyl acetate extract (Tfge) and 8.52 g of butanolic extract (Tfgb).

The extracts were prepared just before use. A preliminary experiment was made to check effective doses. Three doses (200, 350 and 500 mg/kg) of each extract were selected for intraperitoneal (I.P) injections and two doses (50 and 90 µg/3 µl/rat) were selected for intra-cerebro-ventricular (I.C.V) injections. Control animals were treated with saline solution (S.S.)

The dose employed in present research is based on that used in the traditional medicine (Bellakhdar, 1997), and the precedents researches (Laroubi et al., 2007).

Writhing test

The anti-nociceptive effect was evaluated in mice by the writhing test induced by 0.6% acetic acid (0.1 ml/10 g; I.P). Each dose of the extracts was administered 30 min before the acetic acid injection. 5 min after the administration of the acid, the number of writhes and stretching movements (contraction of the abdominal musculature and extension of hind limbs) was counted over a 5 min for a period of 30 min. The strength of the elicited analgesic effect was compared to that of an effective dose of acetylsalicylic acid (ASA, 200 mg/kg) (De Miranda et al., 2001).

Formalin test

Each mouse was placed 5 min before formalin injection in a transparent plastic cage for habituation to the new environment. A dose of 20 µl of 2% formalin was injected subcutaneously (s.c) to the plantar region of its right hind paw. The doses of the extracts and ASA were injected I.P 30 min before the formalin injection. The time spent licking the injected paw was recorded every 5 min using a chronometer. Observations were carried out for 30 min (Tjolsen et al., 1992).

Hot plate test

The heated surface of a hot plate analgesia meter (Ugo Basil,Italy; Socrel DS-37) was maintained at 55 ± 0.2 °C. Each animal was placed into a glass cylinder (diameter 20 cm) on the heated surface of the plate. The latency to exhibit nociceptive reaction was determined before and 30, 45 and 60 min after IP injections and also before and 10 and 30 min after ICV injection. Licking of paws and jumping were the parameters evaluated as the thermal reactions. In order to minimise damage to the animal paw the cut-off time for latency of response was taken as 20 s (Shaheen et al., 2000).

Surgical preparation and technique of intra-cerebro-ventricular injection

The rats were anaesthetized with Ketamine (60 to 80 mg/kg, I.P) and were implanted stereotaxically with a cannula that descended into the lateral ventricle (coordinates: 1.3 mm posterior to the Bregma, lateral 1.6 mm from midline, deep 3.2 mm from the dura). The cannula was fixed to the skull by mean of dental cement. Animals were allowed to recover for 7 days during which they were handled daily.

On the day of the experiment, an injection cannula, connected by a polyethylene tube type PE-10 to an inhalation syringe of 10 µl, was introduced into the fixed cannula. A volume of 50 and 90 µg/rat of every extract of Tfg seeds, or S.S were injected into the lateral ventricle (volume of injection: 3 µl) through the injection cannulae (0.15 mm inner diameter).

At the end of the experiments, the rats were anaesthetized and perfused intracardially with 0.9 saline followed by a 10% formalin solution. The brain were extracted, fixed in 10% formalin for 2 days, and cut at 80 µm. Localization of the cannulae tips was determined according to the Atlas of Paxinos and Watson (1986).

Phytochemical screening

Phytochemical screening of the tested extracts was performed to detect the eventual presence of different classes of constituents, such as: alkaloids with H_2SO_4 and Dragendorff's reagents, flavonoids with the use of Mg and HCl, tannin with Fecl$_3$ solution, anthocyanes with HCl, sterols and/or terpenes with acetic anhy-

Table 1. The effect of IP administration of *T. foenum-graecum* seeds extracts on abdominal constriction test of mice.

Extracts	Dose (mg/kg)	Number of abdominal constriction	Inhibition %
Control	....	80.50 ± 14,74	....
ASA	200	40.75 ± 10,43***	49.38
	200	71.50 ± 17.27	11.18
Dichloromethane	350	59.13 ± 14.66*	26.55
	500	52.88 ± 12.56**	34.31
	200	74.38 ± 11.58	7.6
Ethyl Acetate	350	57.75 ± 15.16**	28.26
	500	48.13 ± 8.04***	40.21
	200	73.12 ± 20.61	9.16
Hexane	350	76.00 ± 18.75	5.59
	500	67.62 ± 9.55	16
	200	76.62 ± 9.66	4.81
Aqueous	350	69.25 ± 7.92	13.97
	500	64.63 ± 9.41*	19.71
	200	75.62 ± 14.90	6.05
Butanolic	350	72.37 ± 13.54	10.09
	500	63.75 ± 13.69*	20.81

*Denotes significant difference from the corresponding values obtained from control rats.
*$P < 0.05$; ** $P < 0.01$; *** $P < 0.001$.

dride and H_2SO_4, quinons with HCl and ammoniac, and saponin with ability to produce suds (Farouk et al., 2008).

Drugs

Drug solutions were prepared just before the start of the experiments. Intra-peritoneal (I.P) injections were performed using a volume of 10 ml/kg body weigh whereas intra cerebro-ventricular (ICV) injections were performed using a volume of 3 µl/rat. Each drug was dissolved in appropriate solvents as follows: Acetic acid (0.6 %) and formalin (2 and 10%) in water, extracts of plant and acetylsalicylic acid in saline solution. The chemicals used in the extractions were: methanol, hexane, dichloromethane, ethyl acetate and butanol.

Statistical analysis

The results were presented as means ± S.E.M and the comparesons between the experimental groups were made using Student's t-test and ANOVA. *: $P < 0.05$; **: $P < 0.01$; ***: $P < 0.001$) were considered as indicative of significance. The inhibition percents were calculated by the following formula:

Inhibition percent = (1-Vt/Vc) x 100,

Where Vt and Vc represent the number of writhes or the licking paw time of the treated and control groups respectively.

RESULTS

Tfg seeds extracts, dichloromethane (Tfgd) and ethyl acetate (Tfge), significantly (p<0.01 for most doses and p<0.001 for 500 mg/kg of Tfge) reduced the writhing and the stretching reactions induced by 0.6% acetic acid. As shown in Table 1, there was a dose dependent effect. The percent of reduction were 26.55 and 28.26% for 350 mg/kg, whereas it was 34.31 and 40.21% for 500 mg/kg for Tfgd and Tfge respectively. The aqueous (Tfga) and the butanolic (Tfgb) extracts at 500 mg/kg induced a percent of reduction of the writhing response of 19.71 and 20.81% respectively (Table 1). ASA 200 mg/kg was effective in reducing the writhing response by 49.38%.

Intraplantar injection of 2% formalin evoked a characteristic biphasic licking response. The duration of licking for the early phase (0 - 5 min) was 84.25 ± 9.52 s, whereas for the late phase (15 - 30 min) it was 49.98 ± 24.54 (control group, Table 2). The doses of 500 mg/kg Tfgd produced a marked reduction of 18.86 and 50.26% of the licking time in the early and late phase, respectively (Table 2) but weaker doses have no significant effect. The Tfge inhibited significantly the two phases of the formalin response but higher inhibition (58.48% at 500 mg/kg) was seen in the second phase (Table 2). ASA was significantly more active in the second phase (61.1%; P< 0.01). As shown in Table 2, a pre-treatment with different doses of hexane extract (Tfgh) has no significant effect on the duration of licking in both phases.

In the hot-plate test, I.P administration of 500 mg/kg of Tfgh or of Tfga produced a significant (P<0.05) increase in the latency 45 min after the administration of extract (Figure 1A and E). However, Tfgh (200, 350 and 500 mg/kg) and Tfga (200, 350 and 500 mg/kg) showed no significant anti-nociceptive effect in this test (Figure 1C

Table 2. Effects of *T. foenum graecum* seeds extracts on the nociceptive responses in the formalin test.

Extracts	Dose mg/kg	Licking response (Sec)		Inhibition %	
		Early phase	Late phase	Early phase	Late phase
Control	....	84.25 ± 9.52	49.98 ± 24.54	....	....
ASA	200	75.28 ± 7.87*	19.44 ± 12.03**	10.65	61.1
	200	81.38 ± 12.90	40.89 ± 25.10	3.4	18.19
Dichloromethane	350	78.34 ± 9.00	28.47 ± 14.19*	7.01	43.04
	500	68.36 ± 14.89*	24.86 ± 10.45*	18.86	50.26
	200	87.50 ± 8.93	46.14 ± 25.40	....	7.68
Ethyl acetate	350	73.16 ± 7.38*	29.00 ± 12.71*	13.16	41.98
	500	71.53 ± 7.99**	20.75 ± 14.19**	15.1	58.48
	200	86.32 ± 7.37	42.89 ± 15.58	....	14.18
Hexane	350	81.96 ± 5.76	46.93 ± 15.03	2.72	6.1
	500	83.63 ± 5.11	45.00 ± 21.91	0.73	9.96

*denote the significance levels as compared with control groups (Saline solution).
*$P < 0.05$; ** $P < 0.01$; *** $P < 0.001$.

and D). Figure 1A and E also shows that, neither the Tfgd doses 200 and 350 mg/kg nor the Tfgb doses 200 and 350 mg/kg exerted a significant analgesic effect. Tfge (200, 350 and 500 mg/kg) produced an analgesic effect that was most pronounced with the dose 500 mg/kg ($P<0.05$) (Figure 1B). Acetylsalicylic acid (200 mg/kg) induced a weak protection against heat-induced pain (Figure 1).

Intra-ventricular injection of Tfge (50 and 90 µg/rat) significantly increased the pain reaction latency (Figure 2B). Injection of the other Tfg extracts was ineffective (Figure 2).

Phytochemical screening indicated the presence of tannins at high concentration in the aqueous extract while high flavonoids content were found in the ethyl acetate, aqueous and butanolic extracts. Quinons were not detected and Alkaloids were found in low concentrations in the dichloromethane extract. Positive reactions to Saponins were found in the butanolic and aqueous extracts.

DISCUSSION

The results of this study indicate that Tfg seeds extract has potent analgesic effect. The extracts were shown to possess anti-nociceptive effects evident in three pain models thus indicating that the observed effects may involve both central and peripheral mechanisms. Indeed, the acetic acid-induced abdominal constriction is believed to show the involvement of peripheral mechanisms, whereas the hot plate test is believed to show that of central mechanisms (Paulino et al., 2003). The formalin test is used to investigate both peripheral and central mechanisms (Tjolsen et al., 1992). Besides, our results bring scientific evidence for the use, in Moroccan traditional medicine, of Tfg as antinociceptive (Bellakhdar, 1997).

The analgesic effects observed were dose dependants which indicate that the compounds presents in the extracts exert their effects by activation of specific recaptors. Tfg seeds contain saponins, alkaloids and flavonoids that have been shown to possess analgesic activity in other plant extracts (Golshani et al., 2004). The different degree of effectiveness between the extracts may probably depend on their compounds concentrations and on some physical factors such as the polarity which is related to the nature of the solvent used. Besides, the effectiveness may also depend on the nature of the recaptors that could be activated. Some hypothesis on this will be suggested as the results obtained in the three pain models are discussed.

Assessment of the abdominal constrictions elicited by acetic acid revealed that the extract of Tfg seeds, when given IP, produces significant dose-related analgesic effect. It has been suggested that acetic acid acts by releasing endogenous mediators that stimulate the nociceptive neurons (Collier et al., 1968). It was postulated that the abdominal constriction response is induced by local peritoneal receptors activation (Bentley et al., 1983) and involved prostanoids mediators. As a matter of fact, increased levels of PGE2 and PGF2 in peritoneal fluids (Derardt et al., 1980) as well as lipooxygenase production were reported (Dahara et al., 2000). The results of the present study showed that ASA, which inhibit cycloxygenase, cause significant inhibition of acetic acid-induced pain. This is in accordance with previous reports indicating that this test is sensitive to non-steroidal anti-inflammatory drugs (NSAIDs) (Biswal et al., 2003; Vane and Botting, 2003). Therefore, the analgesic and anti-inflammatory actions of Tfg extracts seems to be mediated by inhibition of lipoxygenase and/or cyclo-oxygenase activity or by release of cytokines such as TNF-α, interleukin-1β and interleukin-8; by resident peritoneal macrophages

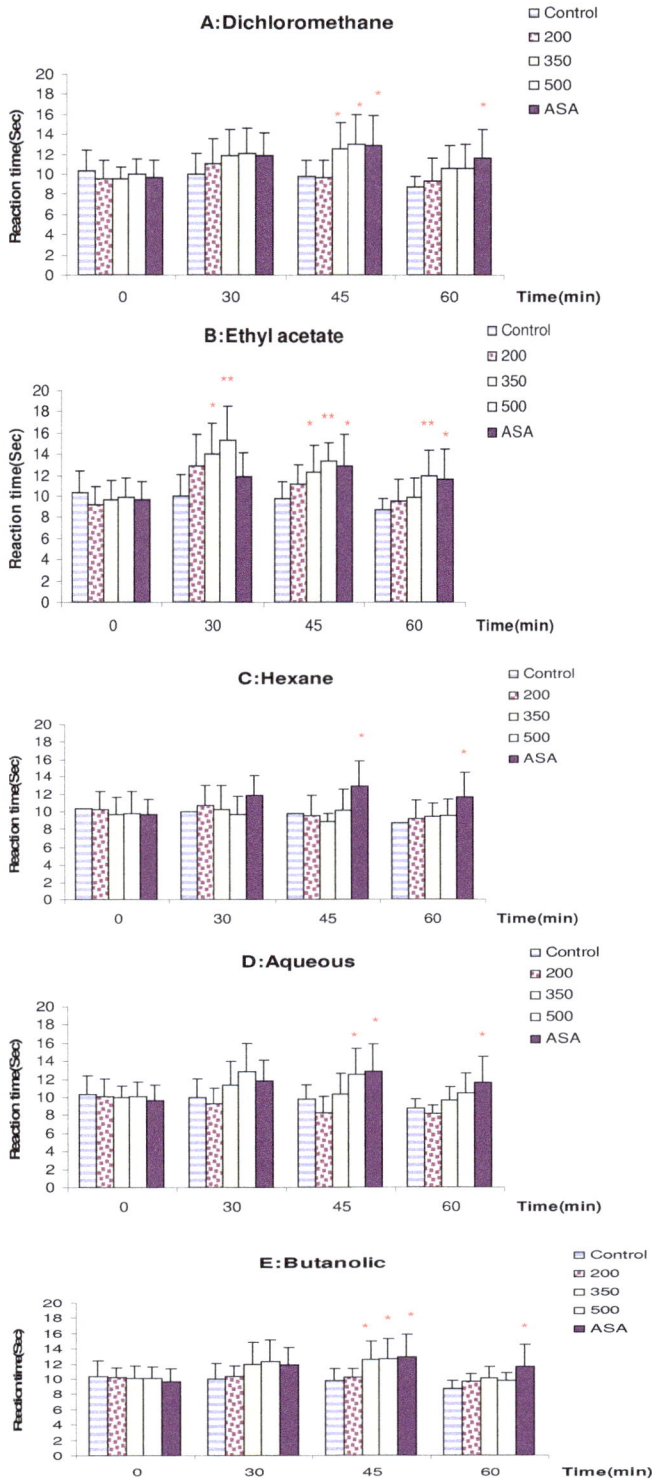

Figure 1. The effect of dichloromethane (A), ethyl acetate (B), hexane (C), aqueous (D) and butanolic (E) extracts of *T. foenum graecum* seeds on the hot plate test. Each column and vertical bar represents mean ± S.E.M. of six to eight mice. The extracts were administered intraperitoneally at doses of 200, 350 and 500 mg/kg.
"*" Denotes significant differences (P< 0.05) from the corresponding values from control.

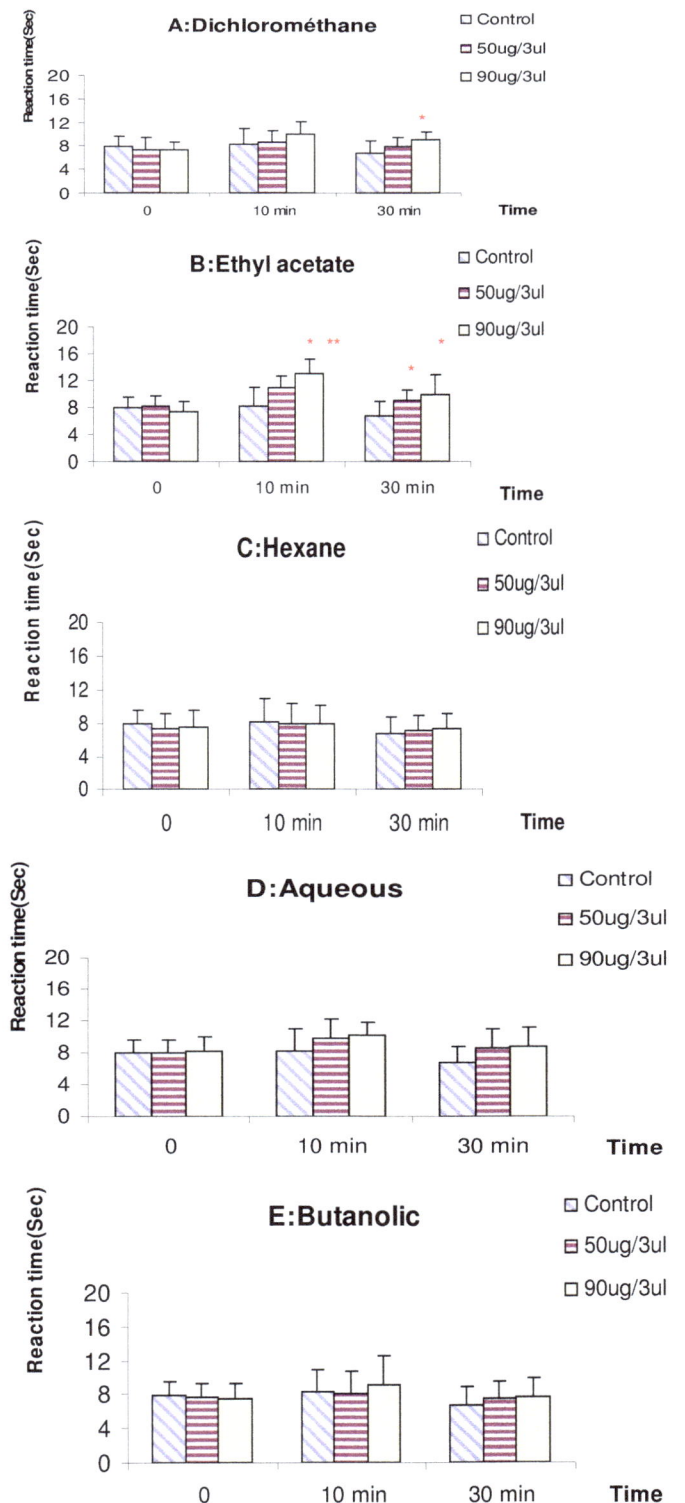

Figure 2. The effect of dichloromethane (A), ethyl acetate (B), hexane (C), aqueous (D) and butanolic (E) extracts of *T. foenum graecum* seeds on the hot plate test. Each column and vertical bar represents mean ± S.E.M. of six rats. The extracts were administered intra-cerebro-ventriculary at doses of 50 and 90 ug/3ul.
"*" Denotes significant differences (P< 0.05) from the corresponding values from control.

and mast cells, as shown by Reibero et al. (2000), or by both mechanisms.

In our experiment using the formalin test, the ethyl acetate and dichloromethane extracts suppressed both phases suggesting that both extracts contains molecular products active on the SN centrally and peripherally. Indeed, in this test there is a distinctive biphasic nociceptive response termed early and late phases (Hunskaar and Hole, 1987). Drugs that act primarily on the central nervous system inhibit both phases equally, whereas peripherally acting drugs inhibit the late phase. The early phase is probably a direct result of stimulation of nociceptors in the paw. The late phase is due to the release of serotonin, histamine, bradykinin and prostaglandins during the inflammatory process (Tjolsen et al., 1992). But also could be due, to a lesser degree, to the activation of central nociceptive neurons (Tjolsen et al., 1992; Parvizpur et al., 2006). The Tfg extracts anti-nociceptive and anti-inflammatory properties reported in our study resemble the NSAIDs properties, specifically the salicylates and their derivatives.

In the hot plate test, only the ethyl acetate extract significantly increase the latency. It could be suggested that ethyl acetate Tfg extract contains products that may exert analgesic effect through activation of central mechanisms. Indeed, the hot-plate test is commonly used to assess opioidergic analgesic mechanisms (Araujo et al., 2005) and narcotic analgesia (Asongalem et al., 2004). Our hypothesis is further confirmed by the observed analgesic effect elicited by ICV injections of this extract. The nature of the neurochemical substrate of such effect is not known but it could be suggested that probably an activation of the opioidergic system may occur. However, pharmacological experiment using naloxones to reverse such analgesic effects are needed to support this assumption (Parvizpur et al., 2004). The remaining Tfg extracts were ineffective in the hot plate test suggesting that the compounds they contains have no central action. They have a similar profile as ASA which exerted little or no influence on the response in tests with phasic stimuli such as the hot-plate and early phase of formalin test. This suggests that the compounds of these extracts may have similar properties as NSAIDs as it was suggested from the results obtained in the formalin test.

In conclusion, our results support the traditional use of Tfg in some painful conditions. However, further investigations are needed to elucidate the mechanisms related to the actions of the Tfg seeds extracts. As a next step, studies in our laboratory are currently under way to isolate and characterize the active principles of each extracts.

ACKNOWLEDGEMENT

We gratefully acknowledge Mr. Regragui Abderazzak for giving us animals.

REFERENCES

Abdel-Barry JA, Abdel-Hassan IA, Jawad AM, Al-Hakiem MHH (2000). Hypoglycaemic effect of aqueous extract of the leaves of Trigonella foenum-graecum in healthy volunteers. East .Mediterr. Health. J. 1: 83-88.

Asongalen EA, Foyet HS, Ngogang J, Folefoc GN, Dimo T, Kantchouing P (2004). Analgesic and antiinflammatory activities of Erigeron floribundus. J. Ethnopharmacol. 91: 301-308.

Basch E, Ulbricht C, Kuo G, Szapary P, Smith M (2003). Therapeutic application of fenugreek. Alternative Medicine Review. 1: 20-27.

Bellakhdar J (1997). La pharmacopée marocaine traditionnelle, médicine arabe ancienne et savoir populaires. Ibis press, Paris. pp 320-321.

Bentley GA, Newton SH, Starr J (1981). Evidence for an action of morphine and the enkephalins on sensory nerve endings in the mouse peritoneum. Br. J. Pharmacol. 73: 325-332.

Bentley GA, Newton SH, Starr J (1983). Studies on the anti-nociceptive action of a agonist drugs and their interaction with opioid mechanisms. Br. J. Pharmacol. 79: 125-134.

Biswal S, Das MC, Nayak P (2003). Antinociceptive activity of seeds of Trigonella foenum graecum in rats. Indian J. Physiol. Pharmacol. 47: 479-480.

Collier HO, Dinneen LC, Johnson CA, Schneider C (1968). The abdominal constriction response and its suppression by analgesic drugs in the Mouse. Br. J. Pharmacol. Chemother. 32: 295-310.

De Araujo Pinho FVS, Coelho–de Sanza AN, Marais SM, Ferreira Santos C, Leal-Cardoso JH (2005). Antinociceptive effects of the essential oil of Alpina zerumbet on mice. Phytomedicine 12: 482-486.

De Miranda FGG, Vilar JC, Alves IAN, Cavalcanti SCH, Antoniolli AR (2001). Antinociceptive and antiedematogenic properties and acute toxicity of Tabebuia avellaneadae lor.ex Griseb. Inner bark aqueous extract. BMC Pharmacol. 1: 6.

Deraedt R, Jouquey S, Delevallée F, Flahaut M (1980). Release of prostaglandins E and F in algogenic reaction and its inhibition. Eur. J. Pharmacol. 51: 17-24.

Dhara AK, Suba V, Sen T, Pal S, Nagchaudhuri AK (2000). Preliminary studies on the anti-inflammatory and analgesic activity of the methanolic fraction of the root extract of Tragia involucrate. J. Ethnopharmacol. 72: 265-268.

Dirk LMA, Vander Krol AR, Vregdenhil D, Hilhorst HWM, Bewley JD (1999). Galactomanan, soluble sugar and starch mobiliziation following germination of Trigonella foenum graecum seeds. Plant Physiol. Biochem. 37: 41-50.

Dixit P, Ghaskadbi S, Mohan H, Devasagayan TP (2005). Antioxidant properties of germinated fenugreek seeds. Phytother Res. 19: 977-983.

Dray A and Urban L (1996). New pharmacology strategies for pain relief. Annu Rev. Pharmacol Toxicol. 32: 34-36.

Farouk L, Laroubi A, Aboufatima R, Benharref A, Chait A (2008) Evaluation of the analgesic effect of alkaloid extract of Peganum harmala L.: possible mechanisms involved. J. Ethnopharmacol. 115(3): 449-54.

Golshani S, Karamkhani F, Monsef-Esfehani HR, Abdallahi M (2004). Antinociceptive effects of the essential oil Dracocephalum kotschyi in the mouse writhing test. J. Pharm .Pharmaceut. Sci. 7: 76-79.

Hunskaar S and Hole K (1987). The formalin test in mice: dissociation between inflammatory and non-inflammatory pain. Pain 30: 103-104.

Khosla P, Gupta DD, Nagpal RK (1995). Effect of Trigonella foenum graecum (fenugreek) on serum lipids in normal and Diabetic rats. Indian. J. Pharmacol. 27: 89-93.

Laroubi A, Touhami A, Farouk L, Zrara I, Aboufatima R, Benharref A, Chait A (2007).

Prophylaxis effect of Trigonella foenum graecum L. seeds on renal stone formation in rats. Phytother. Res. 21: 921-5.

Le Bars D, Gozariu M, Cadden S (2001). Animal models of nociception. Pharmacol.Rev. 53: 597-652.

Paulino N, Dantas AP, Bankova v, Longhi DT, Scremin A, Decastro SL, Calixto JB (2003). Bulgarian propolis induces analgesics and anti-inflammatory effects in mice and inhibits in-vitro contraction of airway smooth muscle. J. Pharmacol. Sci. 93: 307-313.

Paxinos G, Watson C (1986). The rat brain in stereotaxic coordinates. 2nd ed. Acedemis Press. San Diego (CA), USA.

Parvizpur A, Ahmadiani A, Kamalinejad M (2004). Spinal serotonergic system is partially involved in antinociception induced by Trigonella foenum graecum (TFG) leaf extract. J. Ethnopharmacol. 95: 13-17.

Parvizpur A, Ahmadiani A, Kamalinejad M (2006). Probable role of spinal purinoceptors in the analgesic effect of Trigonella foenum (TFG) leaves extract. J. Ethnopharmacol. 104: 108-112.

Reichert JA, Daughters RS, Rivard R, Simone DA (2001). Peripheral and preemptive opiod antinociception in a Mouse visceral pain model. Pain 89: 221-227.

Reibero RA, Vale ML, Thomazzi SM, Pascholato ABP, Poole S, Ferreira SH, Cunha FQ (2000). Involvement of resident macrophages and mast cells in the writhing nociceptive response induced by zymosan and acetic acid in mice. Eur. J. Pharmacol. 387: 111-118.

Santos ARS, Vedama EMA, Freitas GAG (1998). Antinociceptive effect of meloxican, in neurogenic and inflammatory nociceptive models in mice. Inflamm. Res. 47: 302-307.

Saxena VK and Shalem A (2004). Yamogenin 3-O-β-D-glucopyanosyl(1→4)-0-α-D-Xylopyranoside from the seeds of Trigonella foenum-graecum. J. Chem. Sci. 116: 79-82.

Shalheen HM, Badreldin HA, Alquarawi AA, Bashir AK (2000). Effect of Psidium guajava leaves ons ome aspects of the central nerveus system in mice. Phytother. Res. 14: 107-111.

Tjolsen A, Berge OG, Hunskaar S, Rosland JH, Hole K (1992). The formalin test: an evaluation of the method. Pain. 51: 5-17.

Vane JR and Botting RM (2003). The mechanism of action of aspirin. Thromb. Res. Rev. 110: 255-258.

Yingmei H, Sansei N, Yakari N, Zhexiong J (2001). Flavonol glycosides from the stems of Trigonella foenum graecum. Phytochemistry 58: 577-580.

12

Ferric reducing antioxidant power and total phenols in *Cordia africana* fruit

Tewolde-Berhan Sarah[1,2], **Remberg Siv Fagertun**[2], **Abegaz Kebede**[3], **Narvhus Judith**[2], **Abay Fetien**[1] **and Wicklund Trude**[2]

[1]Department of Land Resources and Environmental Protection/Department of Crop and Horticultural Sciences, Mekelle University, P.O. Box 231, Mekelle, Ethiopia.
[2]Department of Chemistry, Biotechnology and Food Science/Department of Plant and Environmental Sciences, University of Life Sciences, P.O. Box 5003, NO-1432 Aas, Norway.
[3]School of Human Nutrition, Food Science and Technology, Hawassa University, P.O. Box 5, Hawassa, Ethiopia.

Antioxidants are beneficial compounds found in a lot of foods. *Cordia africana* (Lam.) is a small fruit eaten all over Tigray and other parts of Ethiopia. The fruit was tested for its antioxidant content using the ferric reducing antioxidant power (FRAP) assay and total phenols (TP) measured with Folin Ciocalteu`s reagent, across four different agroecological zones and three land use classes in Tigray. The average FRAP value on dry weight basis was 30.8 ± 1.45 mg Trolox equivalent 100 g^{-1} fruit, and the average TP value on dry weight basis 2317.0 ± 104.0 mg gallic acid equivalent 100 g^{-1} fruit. Both FRAP and TP values were found to be significantly ($p < 0.05$) different across the agroecology with the lower altitude agroecology giving the highest value and the dry mid altitude agroecology giving the least value. The difference in land use showed no effect on the FRAP value; however the TP values were significantly ($p < 0.01$) different across the different land use. The highest value of TP was found in the wild and the lowest was found in the backyard land uses. *C. africana* is a fruit with good quantities of TP, and small amounts of antioxidants measured with FRAP. Both FRAP and TP values showed variation across agroecology, while only the TP content vary across land use. The fruit was also found to have 9.07 mg 100 g $^{-1}$ fruit, which makes it a good source of the vitamin to meet part of the daily requirement. As antioxidants and vitamin C are highly beneficial to general health, the consumption of this fruit should thus be recommended and promoted.

Key words: *Cordia africana* fruit, ferric reducing antioxidant power, total phenol, agroecology, land use.

INTRODUCTION

The benefits of antioxidants have been studied and discussed by many scientists in food science, medical science and general health areas (Baumann, 2009; Cadenas and Packer, 2002; Packer et al., 2000; Sen et al., 2000; Tardif and Bourassa, 2006). These show the multi-disciplinary nature of the studies and the multiple use and application of antioxidants. When it comes to *Cordia*, several species have been investigated for their antioxidant properties of the fruits, roots, barks and leaves. For example the leaves of *Cordia wallichii* and

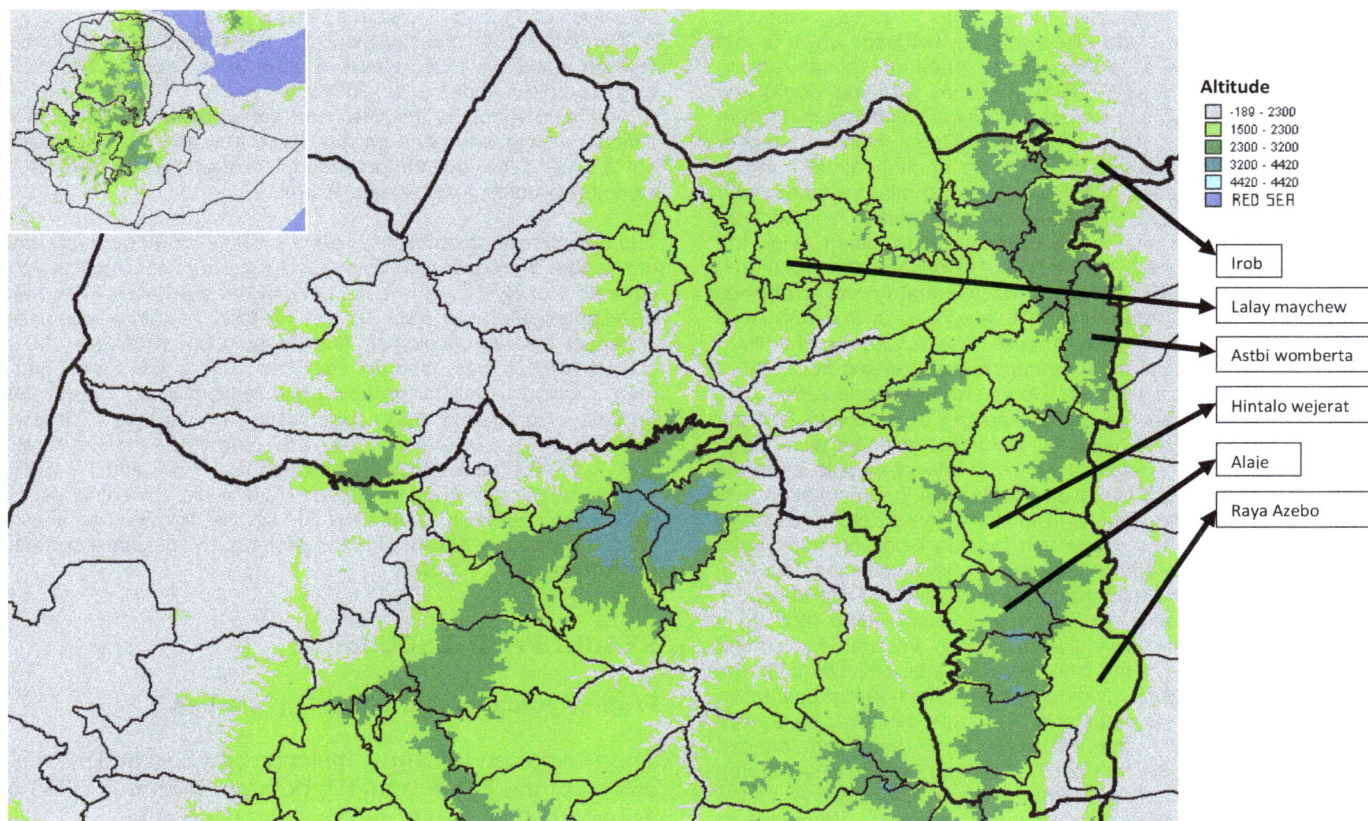

Figure 1. Map of selected woredas in Tigray, Ethiopia, East Africa, showing their relative location and altitude (Generated on DIVA-GIS software). Altitude: Irob; Lalay maychew; Astbi womberta; Hintalo wejerat; Raya Azebo; Alaje.

Cordia verbenacea were looked into by two different studies (Makari et al., 2008; Michielin et al., 2011).

The antioxidant content of fruits vary due to cultivar (genetic variance, provenance) (Cordenunsi et al., 2002; Howard et al., 2003; Kotíková et al., 2011; Wicklund et al., 2005), ripening stages (Gull et al., 2012; Kotíková et al., 2011; Vendramini and Trugo, 2000), various climate conditions such as season and production location (Howard et al., 2003; Iqbal and Bhanger, 2006), temperature (Howard et al., 2003; Wang and Zheng, 2001), altitude and ultraviolet radiation (Bhattacharya and Sen-Mandi, 2011) in addition to overall environmental conditions (Oh et al., 2009; Wang, 2006; Yuri et al., 2009).

Cordia africana is a fruit found wide spread in the Middle East, West, East, and Southern Africa. It is known by the name Sudan teak, East African Cordia, large-leafed Cordia, and Sebastian fruit (ICRAF, 2008). In Tigray, *C. africana* fruit is eaten by the local community during its fruiting season of April to June. It is collected and eaten by shepherds and children when found in the wild, and collected, and eaten or sold by women and children when grown in farms or backyards. Generally, the fruit is eaten fresh, however traditionally the fruit is also dried and kept for use during off season. The objective of this study was to determine the antioxidant levels of *C. africana* within

the context of the typical Tigrian diet when consumed fresh.

MATERIALS AND METHODS

C. africana fruit sampling strategy

In Tigray, it is found that *C. africana* grows in the wild (natural forests, community afforestation sites, church forests), farm lands, grazing lands and people's backyards (home gardens) within the altitude range of 1500 to 2950 m.a.s.l. Thus, the present experimental design took into account the different agroecological and land use patterns. In two studies, the diversity of *C. africana* was observed by looking at genetic markers (Derero et al., 2011), seed physical characteristics and germination time (Loha et al., 2006; Loha et al., 2009). The results of these studies found that the populations of *C. africana* investigated had more genetic diversity within the populations rather than between populations. Thus, the present experimental design took into account both inter- and intra-population diversities, looking at variations across the different populations at the different agroecology and land uses.

The existing study areas where *C. africana* grows were divided into four agroecological zones. A woreda (second smallest level administrative body in Ethiopia) was randomly selected from each of the agroecological zone so as to represent the agroecological area. Figure 1 shows an altitudinal map of Tigray Regional State showing the three different agroecological zones based on altitude. Estimated rainfall data were added to this map in order to determine the selection of the four woredas. The four randomly selected woredas were: Irob (Weyna dega, midland) which is mid altitude 1500-2300 m.a.s.l., moist with mean annual rainfall ranging from 316 to 823 ml year[-1]; Laelay Maychew (Weyna dega) mid altitude 1500-2300 m.a.s.l., dry mean annual rainfall ranging from 639 to 673 ml year[-1]; Atsbi Womberta (Dega, highland) higher altitude 2300- 3200 m.a.s.l., moist mean annual rainfall ranging from 577 to 608 ml year[-1]; and Raya Azebo (Kolla, lowland) lower altitude 500-1500 m.a.s.l., dry mean annual rainfall ranging from 633 to 770 ml year[-1]. These agroecological classifications follow the standard set for Ethiopia by the Soil Conservation Research Programme (Hurni, 1986). However, no adequate number of trees could be found in the three land use categories in Irob and Atsbi Womberta. As a result, Hintalo Wajerat (Weyna dega) mid altitude 1500-2300 m.a.s.l., moist mean annual rainfall ranging from 516 to 716 mL year[-1] and Alaje woredas (Dega) higher altitude 2300- 3200 m.a.s.l., moist mean annual rainfall ranging from 624 to 839 mL year[-1] were selected as substitutes. Within the selected woredas, a village was purposively selected where C. africana could be found growing in the wild, farm lands and backyards in consultation with the woreda level forestry experts. Ten trees were selected randomly from each site of the wild, farm and grazing land and backyards. From each tree, 250 to 450 g mature fruits were collected, labelled, and placed in a cooler (which had an average of 4°C) for transport. The transport from Laelay Maychew and Raya Azebo took 24 h from collection to placement in the laboratory, and that of Alaje and Hintalo Wajerat arrived in the laboratory 5 h after collection. On the same day of arrival, the size, colour and weight of 10 representative fruits from each tree was measured. Two representative fruits were taken in triplicate for the moisture level and ash content determination. The whole fruit and fruit stones were also separated out and ashed for ash content determination. The reminder of the fruit was placed in a refrigerator (4°C) until processed. Within two to three days (stored at 4°C), fruits from each tree were homogenized. For homogenization, initially the fruit cap was removed; then the fruit skin was removed. Following this, the sticky flesh was dissolved into a specified amount of water (50-150 mL depending on number of fruits collected and fruit flesh size) and by blending it with an egg whisk. When the stone and flesh are separated, the skin is placed into the dissolved fruit flesh and blended into a homogenized fruit pulp paste.

Analytical methods

The size, colour, weight, moisture and ash were determined on individual fruits, while TP, FRAP, and vitamin C values were determined from the homogenized samples. As the homogenisation process involved dilution, TP, FRAP and vitamin C values were calculated back to discount the dilution. The principles followed in the analytical measurements were the following:

1. Antioxidants: 3 g of the homogenate was extracted in 30 mL of methanol and the antioxidant levels were determined by ferric reducing activity power (FRAP), and total phenols (TP) using Folin Ciocalteu's reagent. For both analysis, the Konelab 30i outline and method was followed (Volden et al., 2008; Zargar et al., 2011).
2. Size: the size was measured on both diagonal and vertical directions of the fruit (Bertin et al., 2009). This was done using a micro-calliper.
3. Colour: the colour was initially measured using a colour chart from the Natural Colour Systems (Hård and Sivik, 1981). The Natural Colour Systems colours were then converted to the CIE L*a*b* colour (Osorio and Vorobyev, 1996; Özkan et al., 2003) reading using a Minolta colour meter.
4. Weight: the weight of a selected representative 10 fruits was measured in grams using a portable digital balance with a

sensitivity of 0.001 g (Ercisli and Orhan, 2007).
5. For vitamin C measurement 2,6-dichloroindophenol titrimetric method using oxalic acid as extractant was used, AOAC 967.21 was used (Hernández et al., 2006) .
6. Ash: As separating the flesh and the stone without dilution was difficult, ashing was done on the whole fruit and on the separated out stones. The overall procedure followed the AOAC 940.26 standard (Horwitz and Latimer, 2005).

The overall experimental setup gave a nested or hierarchical design of the four weredas, with three land uses and ten replicas. Each laboratory analysis was measured with three parallels. The results were aggregated per tree. To test for land use (inter population), variances and agroecological (intra population) variances, the fully nested ANOVA was used (Minitab 16.1, USA). The FRAP and TP means where tested for grouping and ranking using Tukey's tests. Further investigation was done using a Principal Component Analysis (PCA, Unsrambler 10X, USA and PAST (Hammer et al., 2001)), to test the relationship between the FRAP and TP laodings and agroecology. Another set of PCA was applied to test the relationship between FRAP and TP across agroecology and land use. This relationship was further explored by making a correlation matix (Minitab 16.1, USA).

RESULTS AND DISCUSSION

Fruit physical and chemical properties

The mean, standard error, minimum, median and maximum values of the FRAP and TP, both on fresh fruit (FW) and dry weight (DW) basis, are presented in Table 1.

The antioxidant levels in Table 1 show the measured values. There is a difference between the fresh and dry weight FRAP and TP values because the fresh fruit contained a lot of moisture with a mean of 56.89%. The FRAP average is comparable to that found from the bark extract in a similar species Cordia dichotoma bark, with 22.8 mg mL[-1] TE on a dry weight basis (Ganjare et al., 2011). The average total phenol values (DW) are at least 5.7 times higher than fruit extract found in a similar species, Cordia myxa, with a variation found in the literature of 373.9-400 mg 100 g[-1] GAE (Aberoumand and Deokule, 2009b; Aberoumand, 2011b; Souri et al., 2008). TP values in fresh fruits are comparable with that found in Cordia exaltata fruit with 190 mg 100 g[-1] on a FW basis (Silva et al., 2007). To date, there is no daily requirement set for the consumption of antioxidants and total phenols. The American average daily intake of total phenols has been set to 450 mg GAE (Chun et al., 2005), and the Mexican average daily intake of antioxidants is 170.2 - 240.2 mg trolox equivalents day[-1], which is stated to be similar to 222-1004 mg GAE per day as found in Spanish Mediterranean diets, which is used as a reference for European countries (Hervert-Hernández et al., 2011). Taking these figures as bench marks, on the one hand, these antioxidant levels are too low for daily intake levels to be reached as 5222 g would be needed to reach the American recommended values. On the other hand, the total phenol content is high enough to meet the American and Mexican/Spanish Mediterranean daily intake levels

Table 1. Basic statistical results of the FRAP and TP measurements on both fresh (FW) and dry weight (DM) basis, and other basic fruit parameters measured.

Variable	Mean	SE Mean	Minimum	Median	Maximum
FRAP FW (mg 100 g^{-1}) TE	4.6	0.2	1.1	4.2	10.4
FRAP DW (mg 100 g^{-1})	30.8	1.45	8.7	27.8	78.5
TP FW (mg GAE 100 g^{-1})	264.1	12.2	70.5	248.5	687.9
TP DW (mg GAE 100 g^{-1})	2317.0	104.0	578.0	2215.0	4980.0
Ash whole fruit (%)	2.00	0.02	1.62	1.97	2.38
Average size (cm)	1.33	0.02	0.90	1.30	1.76
Moisture (%)	56.89	0.64	41.94	58.15	74.97
L (L scale)	31.48	0.22	25.08	31.42	43.71
a (a scale)	3.37	0.27	-5.97	3.43	9.87
b (b scale)	31.20	0.32	20.36	31.12	49.81
Weight (g)	15.34	0.50	6.60	14.00	33.70
Vitamin C (mg/100 g FW)	9.07	0.25	4.96	8.25	18.93
Vitamin C (mg/100 g DW)	20.20	0.31	14.15	19.77	30.38

FRAP = Ferric reducing activity power, TP = total phenols, TE = TROLOX equivalent, GAE = gallic acid equivalent.

Table 2. Nested ANOVA analysis of the FRAP and TP values of *C. africana* fruits.

Variable	Tested parameter	DF	Variance explained	F	P	Significance level
FRAP (DW)	Agroecological	3	10.83	4.54	0.04	*
	Land use	8	0.29	1.03	0.42	ns
TP (DW)	Agroecological	3	21.62	4.17	0.05	*
	Land use	8	13.99	3.17	0.00	**

FRAP = Ferric reducing activity power, TP = total phenols, dry weight (DW), * = significant at 95 %, ** = significant at 99% and ns = not significant.

by consuming 170.4 g. The fruiting season lasts on average for three months, as each tree has a different time of fruit maturation, and per tree fruits have different times of maturation. Assuming people consume the fruit during the fruiting season, it helps them to meet the daily recommended rates of total phenol need. On average people will eat about 100 g of the fruit at any given time, unless they are using it for treating gastro-intestinal illnesses, for which they would consume about 750 g at one time.

The ash content was 2 ± 0.02%, lower than that reported by Murray et al. (2001) (5.1 to 7.8% for *Cordia sinensis*) Aberoumand and Deokule (2009a) (6.7and 6.7 + 0.80%) and Aberoumand (2011a) for *Cordia myxa*. The average fruit diameter was 1.33 cm with an average moisture content of 56.89%. The average colour is light yellow (L 31.48, a 3.37 and b 31.2). The weight of the 10 representative fruits sampled was on average of 15.34 g. The fresh weight vitamin C was found to be 9.07 mg 100 g $^{-1}$ fruit. This value is similar to that found in banana and apples (Planchon et al., 2004; Wall, 2006). According to the FDA guidelines, this value meets 15.12% of the daily required vitamin C levels for an average adult and according to FAO/WHO guidelines, this value meets 30.23% of the daily requirements

(FAO/WHO, 2002; Food and Drug Administration, 2011).

Looking at the relatively good content of total phenols and vitamin C, it is obvious that this fruit is a nutritionally important fruit, with a limited range of use. Taking into account the benefits of antioxidants and vitamin C, it is clear that work is needed in the promotion and popularisation of this fruit. It is also clear that there is need for further study on its other nutritional benefits, processing and marketing potential to aid its promotion. The high variation in its size and weight also shows that there is great variation, which has implication on selection of the fruit source for promotion. As it grows in most parts of Africa and the Middle East its use and promotion has a wider significance than that of the studied area.

Fully nested ANOVA, principal component analysis (PCA) and correlation results

As the experimental setup had four agroecological zones and three land use systems, from which 10 trees were selected randomly, the experimental setup fits best to the nested or hierarchical design. ANOVA for a fully nested design was run and the results are presented in Table 2.

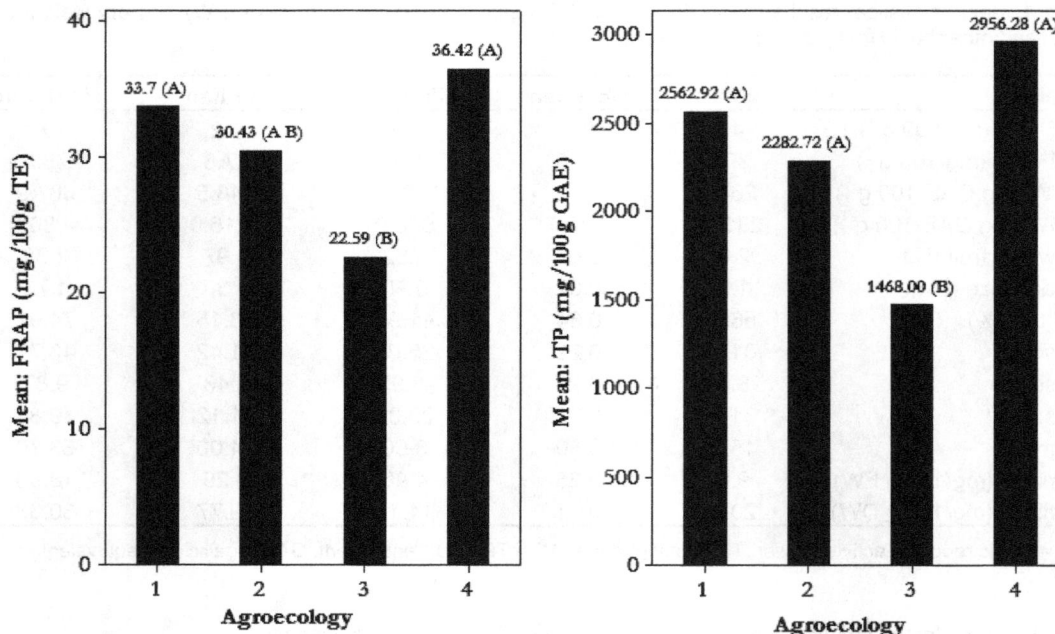

Figure 2. Tukey's ranking and grouping of mean values for FRAP and TP (based on dry weight (DW) across agroecology. 1 = highland, 2 = moist mid altitude, 3 = dry mid altitude, 4 = lowland.

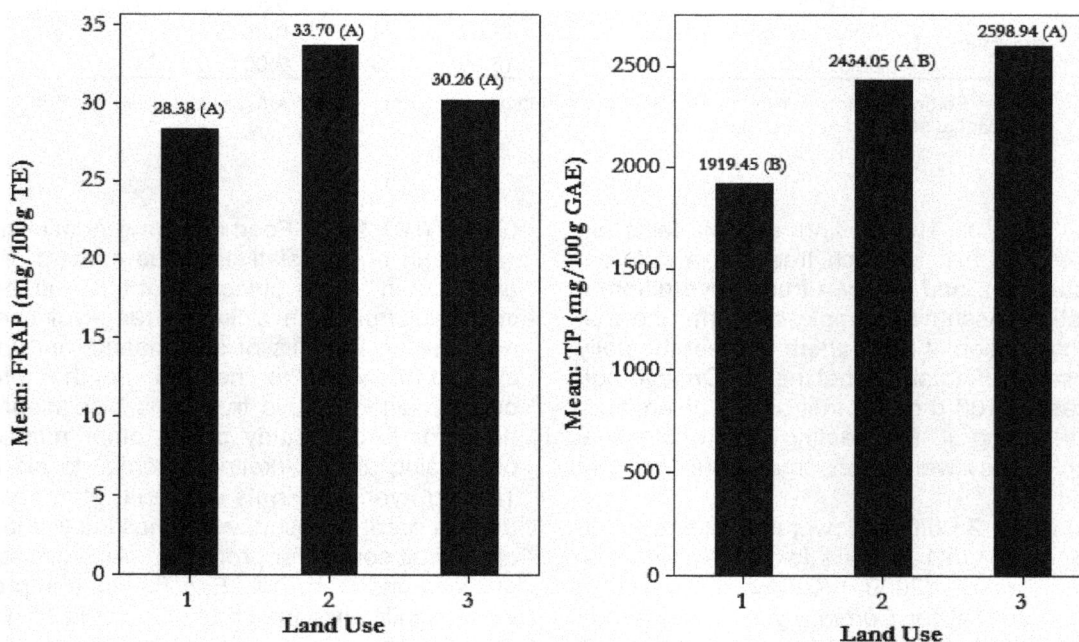

Figure 3. Tukey's ranking and grouping of mean values for FRAP and TP (based on dry weight (DW)) across land use. 1 = backyard, 2 = farm land, 3 = wild.

As can be seen in Table 2, both the FRAP and TP values were significantly different for the different agroecologies tested. The FRAP values were not significantly different from each other across the different land uses, while the TP values were significantly different across the different land uses. All the parameters were also tested for ranking and grouping using Tukey's test. The results are summarised in Figures 2 and 3. As can be seen in Figure 2,

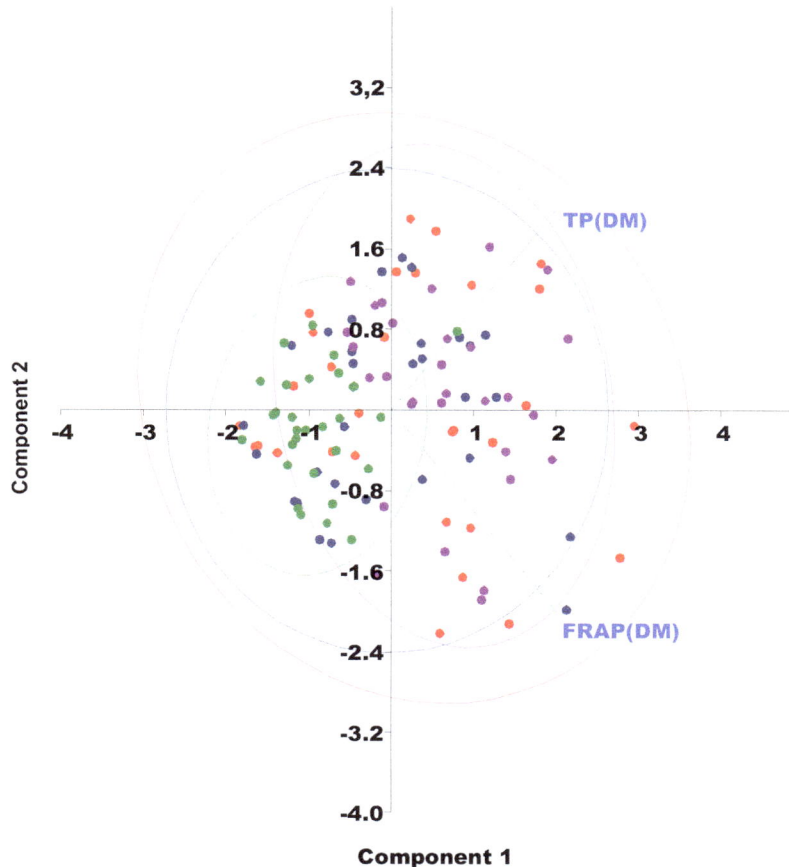

Figure 4. PCA of FRAP and TP values based on agroecology. Red = higher altitude, dark blue = moist mid altitude, green = dry mid altitude, purple = lower altitude.

the lower altitude site had the highest values for both FRAP and TP, while the mid altitude dry site had the lowest values with significant differences between these highest and lowest values. The lower altitude area has higher temperatures as compared to the other sites, and these results are in agreement with a strawberry study, where it was noted that both phenolic content and antioxidant content of the fruits increased with increasing growing temperature (Wang and Zheng, 2001).

The FRAP values in Figure 3 were not significantly different across land use in both the nested ANOVA and Tukey's tests. The TP values showed a significant difference with highest values measured for fruits from the wild and lowest values for fruits from the backyard, while that in the farm land was in between. As the trees in the backyards are purposefully selected, and those from the farm land are semi purposefuly selected, one possible explanation for this difference could be a factor of selection resulting with trees with special traits. Breeding and cultivar (genetic variety, provenace) development starts with this, and several studies have shown that cultivars have a significant effect on the total

phenol values (Cordenunsi et al., 2002; Howard et al., 2003; Kotíková et al., 2011; Wicklund et al., 2005). Though there have not been studies showing that *C. arifcana* has specific cultivars, studies on its genetic variation within provenances (specific geographical location) have shown that there is a high variation (Derero et al., 2011; Loha et al., 2006, 2009). Another reason can be difference in the micro climate of these land uses, with the wild predominantly being marginal where environmental stress is highest, and the backyard being the most conducive with watering, organic matter and ash application creating a difference in the stress levels in the trees. The farm lands are flat and more fertile than the wild areas. In relation to this, several studies have shown that fruits grown under stressful conditions produce higher levels of phenolic compounds (Oh et al., 2009; Tomás-Barberán and Espín, 2001; Yuri et al., 2009).

A principal component analysis of the agroecological groups was run for both FRAP and TP values as presented in Figure 4. As can be seen with respect to the principal components (PC) 1 and 2, the score of the

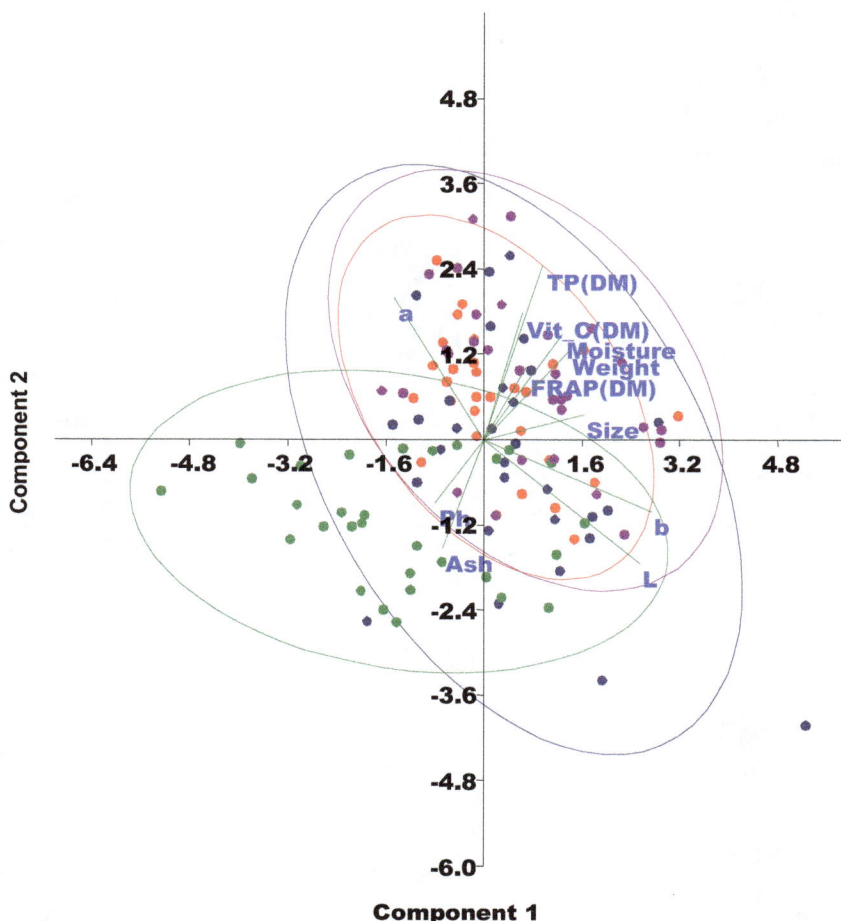

Figure 5. PCA of FRAP and TP and other related fruit parameters based on agroecology. Red = higher altitude, dark blue = moist mid altitude, green = dry mid altitude, purple = lower altitude.

higher altitude agroecology was more widely scattered, while they were narrowly scattered for the dry mid altitude agroecology. The loadings for FRAP and TP best explain the score of the lower altitude agroecology and least explain the score range of the dry mid altitude agroecology concurring with the ANOVA and Tukey's grouping results (Table 2, Figures 2 and 3). Another principal component analysis with agroecological and land use groups was run for both FRAP, TP and other related fruit parameters (Figures 5 and 6). The loadings for the first two principal components show that FRAP, TP, vitamin C, moisture content, weight and average size have a positive relationship, while whole fruit ash content have a negative relationship. The colour parameters L*a*b showed minimal relationship. With respect to the agroecological grouping (Figure 5), the scores of the higher altitude agroecology was the narrowest in scatter, yet the relationship it had with the loadings was similar with that of the moist mid altitude and lower altitude agroecologies. The scores of the dry mid altitude

agroecology were least explained once again by the loadings for FRAP, TP, vitamin C, moisture, weight and colour axis a. This result shows a similar pattern as that of ANOVA and Tukey's grouping; where the dry mid altitude agroecology is separate (Figures 2 and 3).

Looking at the land use groupings of the scores (Figure 6), the backyard and farm land showed similar patterns. On the other hand, the wild land use was slightly different in that it had more response to the colour axis of L*a*b values and was less explained by the other loading directions. This also shows a similar pattern as that of ANOVA and Tukey's grouping, with the wild score being separate (Table 2, Figures 2 and 3).

Following the PCA analysis, a correlation relationship with FRAP and TP of a few related fruit parameters was run to further investigate the observed relationships. The correlation matrix showed a similar picture as that of principal component analysis in Figures 5 and 6, with both positive and negative relationships with the selected fruit parameters except for that of L*a*b*. The FRAP

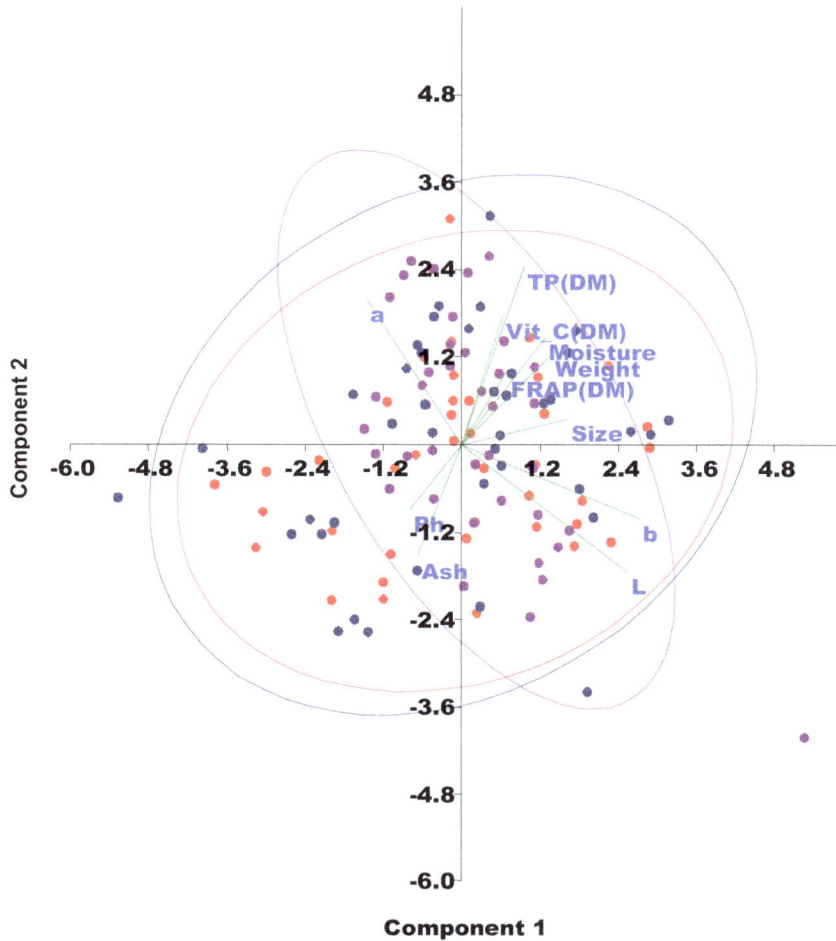

Figure 6. PCA of FRAP, TP and other related fruit parameters based on land use. Red = backyard, dark blue = farm land, purple = wild.

values were only correlated significantly with total phenol and fruit moisture levels, while the TP values are significantly correlated with the moisture, weight, vitamin C and fruit ash content. However, it needs to be noted that though there was a significant relationship considering the p values, the Pearson's correlation coefficient was very low. The correlation relationship of TP with fruit ash content was negative as indicated by the PCA in Figures 5 and 6. The positive relationships could be explained as vitamin C is an antioxidant and the weight and moisture content are indicators of better fruit growth. The negative relationship can be explained by the fact that phenols are acidic, and acidic compounds have lower pH. With respect to the relationship with the fruit ash content, the highest content of total phenols was found in the wild which are marginal lands. The soil fertility in these areas is lower, and previous studies have shown that soil fertility and mineral contents are related (Havlin et al., 2009; Lucas et al., 1942).

These results have implications for the use and promotion of this fruit. Currently, the fruits is consumed locally and sold in local markets (Demel and Abeje,

2005). Currently, in most parts of Ethiopia and Tigray there are massive afforestation efforts underway, that if managed properly have a great potential to contribute to improved food security, poverty alleviation and environmental rehabilitation (Egziabher, 2006; Fentahun and Hager, 2009; Gebrehiwot and Headquarters, 2004; Mengistu et al., 2005; Tewolde-Berhan et al., 2000; Yami et al., 2006). This study has shown that the fruit is useful and beneficial for health, and the TP values were highest in the wild. The promotion and wider use of the fruit can be achieved through its incorporation in the existing enclosure, communal forestry, and hillside distribution efforts. In addition to this, efforts need to be put in place to promote and market the fruit.

Conclusion

Cordia africana is a fruit with good quantities of TP-total phenol antioxidants. The antioxidant content tested with the FRAP, was not high. Both FRAP and TP values showed variation across agroecology, and only the phenolic content with variation across land use. The phenolic content is also strongly related to the FRAP, moisture content, weight, vitamin C and fruit ash contents.

The significance of these results is in its implication for use. These fruits are consumed by the local community and are sold in local markets. As these antioxidants and vitamin C tested are known to have beneficial effects towards health, the continued consumption of these fruits in Tigray is highly recommended. As the fruit is known to grow in most parts of Africa and the Middle East, its use needs to made known and the fruit needs to be promoted. Further study on its overall nutritional value, processing potential and marketability will help in the promotion of the fruit. As the size and weight was also found to have great variability, care needs to be taken to select appropriate seed sources for promotion. With the growing afforestation efforts in the region, this species should be given focus as its TP values where the highest in the wild, which represents the afforestation sites. Understanding their use will help in the up scaling of their use and marketing.

ACKNOWLEDGEMENTS

The authors would like to thank NUFU (The Norwegian Programme for Development, Research and Education), Mekelle University, University of Life Science, and Sokoine University for the collaborative project that made this study possible. We would also like to thank the laboratory staff of the Mekelle University and the University of Life Science.

REFERENCES

Aberoumand A (2011a). Studies on nutritional and functional properties of some plant foods. World J. Sci. Technol. 1(3):26-30.

Aberoumand A (2011b). Survey on some food plants as sources of antioxidants. Innov. Rom. Food Biotechnol. 8: 22-25.

Aberoumand A, Deokule SS (2009a). Determination of elements profile of some wild edible plants. Food Anal. Methods 2(2):116-119.

Aberoumand A, Deokule SS (2009b). Studies on Nutritional Values of Some Wild Edible Plants from Iran and India. Pakistan J. Nutr. 8(1):26-31.

Baumann L (2009). Cosmetic Dermatology: Principles and practices. Second Edition ed. New York. 366 p.

Bertin N, Causse M, Brunel B, Tricon D, Génard M (2009). Identification of growth processes involved in QTLs for tomato fruit size and composition. J. Exp. Bot. 60(1):237-248.

Bhattacharya S, Sen-Mandi S (2011). Variation in antioxidant and aroma compounds at different altitude: A study on tea (Camellia sinensis L. Kuntze) clones of Darjeeling and Assam, India. Afr. J. Biochem. Res. 5(5):148-159.

Cadenas E, Packer L (2002). Handbook of antioxidants. Second edition revised and expanded ed. New York: Marcel Dekker Inc. 602 p.

Chun OK, Kim DO, Smith N, Schroeder D, Han JT, Lee CY (2005). Daily consumption of phenolics and total antioxidant capacity from fruit and vegetables in the American diet. J. Sci. Food Agric. 85 (10):1715-1724.

Cordenunsi BR, Oliveira do Nascimento JR, Genovese MI, Lajolo FM (2002). Influence of cultivar on quality parameters and chemical composition of strawberry fruits grown in Brazil. J. Agric. Food Chem. 50(9):2581-2586.

Demel Teketay, Abeje Eshete (2005). Status of indigenous fruits in Ethiopia. In IPGRI-SAFORGEN (ed.). Review and appraisal on the status of indigenous fruits in Eastern Africa: AFREA/FORNESSA.

Derero A, Gailing O, Finkeldey R (2011). Maintenance of genetic diversity in Cordia africana Lam., a declining forest tree species in Ethiopia. Tree Genetics Genomes 7(1):1-9.

Egziabher TBG (2006). The role of forest rehabilitation for poverty alleviation in drylands. J. Drylands 1 (1): 3-7.

Ercisli S, Orhan E (2007). Chemical composition of white (Morus alba), red (Morus rubra) and black (Morus nigra) mulberry fruits. Food Chem. 103(4):1380-1384.

FAO/WHO (2002). Human Vitamin and Mineral Requirements. In FAO/WHO (ed.). Training materials for agricultural planning. FAO Rome: FAO. pp. 1-286.

Fentahun M, Hager H (2009). Exploiting locally available resources for food and nutritional security enhancement: wild fruits diversity, potential and state of exploitation in the Amhara region of Ethiopia. Food Sec. 1(2):207-219.

Food and Drug Administration (2011). Guidance for Industry: A Food Labeling Guide. (14. Appendix F: Calculate the Percent Daily Value for the Appropriate Nutrients) Silver Spring, US: U.S. Food and Drug Administration. Available at: http://www.fda.gov/Food/GuidanceComplianceRegulatoryInformation/GuidanceDocuments/FoodLabelingNutrition/FoodLabelingGuide/ucm064928.htm (accessed on 05/23/2011).

Ganjare AB, Nirmal SA, Rub RA, Patil AN, Pattan SR (2011). Use of Cordia dichotoma bark in the treatment of ulcerative colitis. Pharm. Biol. 49(8):850-855.

Gebrehiwot K, Headquarters NK (2004). Dryland agro-forestry strategy for Ethiopia. Mekelle, Tigray, Ethiopia. pp. 1-20.

Gull J, Sultana B, Anwar F, Naseer R, Ashraf M, Ashrafuzzaman M (2012). Variation in antioxidant attributes at three ripening stages of guava (Psidium guajava L.) Fruit from different geographical regions of Pakistan. Molecules 17(3):3165-3180.

Hammer Ø, Harper DAT, Ryan PD (2001). PAST: Paleontological statistics software package for education and data analysis. Palaeontol. Electronica. 4(1):1-9.

Hård A, Sivik L (1981). NCS-Natural Color System: A Swedish Standard for Coloer Notation. Color. Res. Appl. 6(3):129-138.

Havlin JL, Tisdale SL, Nelson WL, Beaton JD (2009). Soil fertility and fertilizers: an introduction to nutrient management. New Delhi: PHI Learning. 528 p.

Hernández Y, Lobo MG, González M (2006). Determination of vitamin C in tropical fruits: A comparative evaluation of methods. Food Chem. 96(4):654-664.

Hervert-Hernández D, García OP, Rosado JL, Goñi I (2011). The contribution of fruits and vegetables to dietary intake of polyphenols and antioxidant capacity in a Mexican rural diet: Importance of fruit and vegetable variety. Food Res. Int. 44 (5):1182-1189.

Horwitz W, Latimer GW (2005). Official methods of analysis of AOAC International. Gaithersburg, Md.: AOAC International. No 37.1.18, method 940.26.

Howard LR, Clark JR, Brownmiller C (2003). Antioxidant capacity and phenolic content in blueberries as affected by genotype and growing season. J Sci. Food Agric. 83(12):1238-1247.

Hurni H (1986). Guidelines for development agents on soil conservation in Ethiopia. [Addis Ababa], Ethiopia: Ministry of Agriculture, Natural Resources Conservation and Development Main Dept., Community Forests and Soil Conservation Development Dept. Ethiopia. 100 p.

ICRAF (2008). Cordia africana. A tree species reference and selection guide World Agroforesrty Centre database. Nairobi: ICRAF. Available at: http://www.worldagroforestry.org/treedb2/AFTPDFS/Cordia_africana.pdf (accessed: 4/10//2008).

Iqbal S, Bhanger MI (2006). Effect of season and production location on antioxidant activity of Moringa oleifera leaves grown in Pakistan. J. Food Compost. Anal. 19(6-7):544-551.

Kotíková Z, Lachman J, Hejtmánková A, Hejtmánková K (2011). Determination of antioxidant activity and antioxidant content in tomato varieties and evaluation of mutual interactions between antioxidants. LWT - Food Sci. Technol. 44 (8):1703-1710.

Lucas RE, Scarseth GD, Sieling DH (1942:468). Soil fertility level as it influences plant nutrient composition and consumption. Lafayette, Ind.: Purdue University Agricultural Experiment Station. p. 43.

Loha A, Tigabu M, Fries A (2009). Genetic variation among and within populations of Cordia africana in seed size and germination responses to constant temperatures. Euphytica 165(1):189-196.

Loha A, Tigabu M, Teketay D, Lundkvist K, Fries A (2006). Provenance Variation in seed morphometric traits, germination, and seedling growth of Cordia africana Lam. New For. 32(1):71-86.

Lucas RE, Scarseth GD, Sieling DH (1942). Soil fertility level as it influences plant nutrient composition and consumption. Lafayette, Ind.: Purdue University Agricultural Experiment Station. 43 p.

Makari HK, Haraprasad N, Ravikumar PH (2008). In vitro antioxidant activity of the hexane and methanolic extracts of Cordia wallichii and Celastrus paniculata. Internet J. Aesthet. Anti-aging Med. 1(1).

Mengistu T, Teketay D, Hulten H, Yemshaw Y (2005). The role of enclosures in the recovery of woody vegetation in degraded dryland hillsides of central and Northern Ethiopia. J. Arid Environ. 60(2):259-281.

Michielin EMZ, de Lemos Wiese LP, Ferreira EA, Pedrosa RC, Ferreira SRS (2011). Radical-scavenging activity of extracts from Cordia verbenacea DC obtained by different methods. J Supercrit. Fluids 56(1):89-96.

Murray SS, Schoeninger MJ, Bunn HT, Pickering TR, Marlett JA (2001). Nutritional composition of some wild plant foods and honey used by Hadza foragers of Tanzania. J. Food Compost. Anal. 14 (1):3-13.

Oh MM, Trick HN, Rajashekar CB (2009). Secondary metabolism and antioxidants are involved in environmental adaptation and stress tolerance in lettuce. J. Plant Physiol. 166(2):180-191.

Osorio D, Vorobyev M (1996). Colour vision as an adaptation to frugivory in primates. Proceedings of the Royal Society of London. Series B: Biol. Sci. 263(1370):593-599.

Özkan M, Kirca A, Cemeroğlu B (2003). Effect of moisture content on CIE color values in dried apricots. Eur. Food Res. Technol. 216(3):217-219.

Packer L, Rösen P, Tritschler HJ, King GL, Azzi A (2000). Antioxidants in diabetes managment. New York: Marcel Dekker Inc. 376 p.

Planchon V, Lateur M, Dupont P, Lognay G (2004). Ascorbic acid level of Belgian apple genetic resources. Sci. Hortic. 100(1-4):1-61.

Sen CK, Packer L, Hänninen O (2000). Handbook of Oxidants and Anti-oxidants in Exercise. Amsterdam: Elsevier. 1220 p.

Silva EM, Souza JNS, Rogez H, Rees JF, Larondelle Y (2007). Antioxidant activities and polyphenolic contents of fifteen selected plant species from the Amazonian region. Food Chem.101(3):1012-1018.

Souri E, Amin G, Farsam H, Barazandeh Tehrani M (2008). Screening of antioxidant activity and phenolic content of 24 medicinal plant extracts. DARU J. Pharm. Sci. 16(2):83-87.

Tardif JC, Bourassa MG. (2006). Antioxidants and Cardiovascular Disease. Second edition ed. New York: Springer. p. 494.

Tewolde-Berhan S, Mitlöhner R, Muys B, Haile M (2000). Comparison of Flora development of area enclosures and 'undisturbed'forest in Tigray, Ethiopia. Deutscher Tropentag 2002: Challenges to Organic Farming and Sustainable Land Use in the Tropics and Subtropics, October 9 - 11, 2002, University of Kassel, Witzenhausen: 8 p.

Tomás-Barberán FA, Espín JC (2001). Phenolic compounds and related enzymes as determinants of quality in fruits and vegetables. J. Sci. Food Agric. 81(9):853-876.

Vendramini AL, Trugo LC (2000). Chemical composition of acerola fruit (Malpighia punicifolia L.) at three stages of maturity. Food Chem. 71(2):195-198.

Volden J, Borge GIA, Bengtsson GB, Hansen M, Thygesen IE, Wicklund T (2008). Effect of thermal treatment on glucosinolates and antioxidant-related parameters in red cabbage (Brassica oleracea L. ssp. capitata f. rubra). Food Chem. 109(3):595-605.

Wall MM (2006). Ascorbic acid, vitamin A, and mineral composition of banana (Musa sp.) and papaya (Carica papaya) cultivars grown in Hawaii. J. Food Compost. Anal. 19(5):434-445.

Wang SY (2006). Effect of pre-harvest conditions on antioxidant capacity in fruits. Acta Hort. (ISHS) 712:299-306.

Wang SY, Zheng W (2001). Effect of plant growth temperature on antioxidant capacity in strawberry. J. Agric. Food Chem. 49 (10):4977-4982.

Wicklund T, Rosenfeld HJ, Martinsen BK, Sundfør MW, Lea P, Bruun T, Blomhoff R, Haffner K (2005). Antioxidant capacity and colour of strawberry jam as influenced by cultivar and storage conditions. LWT - Food Sci. Technol. 38(4):387-391.

Yami M, Gebrehiwot K, Moe S, Mekuria W (2006). Impact of area enclosures on density, diversity and population structure of woody species: the case of May Ba'ati-Douga Tembien, Tigray, Ethiopia. Ethiop. J. Nat. Resour. 8(1):99-121.

Yuri JA, Neira A, Quilodran A, Motomura Y, Palomo I (2009). Antioxidant activity and total phenolics concentration in apple peel and flesh is determined by cultivar and agroclimatic growing regions in Chile. International J. Food Agric. Environ. 7(3-4):513-517.

Zargar M, Azizah AH, Roheeyati AM, Fatimah B, Jahanshiri F, Pak-Dek MS (2011). Bioactive compounds and antioxidant activity of different extracts from Vitex negundo leaf. J. Med. Plants Res. 5(12):2525-2532.

3

Effects of water stress on physiological processes and yield attributes of different mungbean (L.) varieties

R.K. Naresh[1], Purushottam[2], S.P.Singh[3], Ashish Dwivedi[1] and Vineet Kumar[3]

[1]Department of Agronomy, Sardar Vallabhbhai Patel University of Agriculture and Technology, Meerut (U. P.), India.
[2]Department of Plant pathology & Microbiology, Sardar Vallabhbhai Patel University of Agriculture and Technology, Meerut (U. P.), India.
[3]Department of Soil Science, Sardar Vallabhbhai Patel University of Agriculture and Technology, Meerut (U. P.), India.

In order to investigate the effect of water stress on Mungbean varieties and its physiological responses to yield, a field experiment was carried out according to split plot design with twenty treatments combination and three replications during 2010 and 2011 at Crop Research Centre of Sardar Vallabhbhai Patel University of Agriculture and Technology, Meerut (Uttar Pradesh). The results indicated that drought tolerant varieties maintained highest xylem water potential (XWP), transpiration resistance (TR) and lowest leaf diffusive resistance (LDR) and canopy temperature minus air temperature (Tc – Ta) while drought susceptible varieties maintained lowest XWP,TR and highest LDR and Tc – Ta during the 1300 -1330 h. The rate of net photosynthesis decreased in all the varieties in treatment samples at various stages with the decrease in external CO_2 concentrations. Increase in free proline was observed from vegetative to active pod filling stages in all the varieties. The resistant varieties accumulated higher level of proline under water stress. SML-668 recorded the highest free proline content than all the other varieties. Variety SML-668 produced significantly higher grain yield.

Key words: Mungbean, water stress, yield and yield components.

INTRODUCTION

Among the pulse crops, Mungbean (L.) has a special importance of intensive crop production due to its short growth period (Ahmed et al., 1978). Climatic conditions of western Uttar Pradesh is suitable for Mungbean cultivation throughout the year (Ali and Kumar, 2004). The crop grown under non-irrigated condition, encounters drought stress at different growth stages. The crop is potentially useful for improving cropping pattern as it can be grown as a cash crop due to its rapid growth and early maturing characteristics. In a symbiotic relationship with the soil bacteria, Mungbean roots can fix atmospheric nitrogen

and thus improve soil fertility (Nabizade et al., 2011). Further, Mungbean plays an important role in protein supplement in the cereal-based low-protein diet of the people in western Uttar Pradesh, but the acreage and production of Mungbean is steadily declining (Ali and Kumar, 2004).

Plant performance in Mungbean under conditions of drought stress has been extensively studied (Mahmoodian et al. 2012). Water stress affects various physiological processes associated with growth, development, and economic yield of a crop (Allahmoradi et al., 2011). Water deficit disturbs normal turgor pressure, and the loss of cell turgidity may stop cell enlargement that causes reduced plant growth (Srivalli et al., 2003). Water stress increases root shoot ratio, thickness of cell walls and amount of cutinization and lignifications (Srivalli et al., 2003). Further, water stress decreases leaf area index in Mungbean (Jordan and Ritichie, 2002). The pre-monsoon period has erratic rainfall leading to water stress in some years and water logging in some years which affects emergence, establishment, growth and productivity of legume crops (Belayet et al., 2010).

Mungbean is reported to be more susceptible to water deficits than many other grain legumes (Pandey et al., 1984). Water stress reduces photosynthesis; the most important physiological processes that regulate development and productivity of plants (Athar and Ashraf, 2005). Reduction in leaf area causes reduction in crop photosynthesis in plants leading to dry matter accumulation (Pandey et al., 1984). Water stress imposed at any growth stage causes reduction in dry matter accumulation depending on the growth stage exposed to stress (Sadasivan et al., 1988). According to Sadasivan et al. (1988), water stress during vegetative phase reduces grain yield through restricted plant size leaf area and root growth which subsequently reduces the dry matter accumulation, number of pods per plant and low harvest index. Water deficits at the flowering and the post-flowering stages have been found to have a greater adverse impact than that at the vegetative stage (Rafiei Shirvan and Asgharipu, 2009). The reproductive stage is the most sensitive growth phase to drought (Brown et al., 1985) resulting to less yield and poor harvest index under drought stress (Uprety and Bhatia, 1989). Identification of genotypic differences in cultivars in tolerance to drought stress is needed for development of Mungbean with reasonably high yield under water deficit. The present study was conducted to identify yield related parameters that are tightly related to grain yield under water stress condition. Mungbean cannot tolerate drought stress (Rafiei Shirvan and Asgharipur, 2009) but there are little reports on negative effects of water stress on yield and physiological characteristics of Mungbean (Asaduzzaman and Hasanuzzaman, 2008). This experiment was carried out to understand the effect of water stress during vegetative and reproductive stages on some physiological traits, yield and yield components of Mungbean.

MATERIALS AND METHODS

The study was carried out at Crop Research Centre of Sardar Vallabhbhai Patel University of Agriculture and Technology in Meerut, Uttar Pradesh, India (28°40'07"N to 29°28'11"N, 77°28'14"E to 77°44'18"E) in 2010 and in 2011. Experimental design was the split plot design while the moisture regime was the main plot and varieties were the sub-plot with three replications. Each variety was tested for soil moisture regime. The climate of the area is semi-arid. Minimum temperature in the region is 4°C in January and maximum temperature is 41–45°C in June. Relative humidity ranges in between 67–83% throughout the year. The soils are generally sandy loam to loam in texture and low to medium in organic matter con-

tent. The treatments were applied as follows: Soil moisture regime as main plot: Irrigation at 0.2 bar soil moisture tension I_1; irrigation at 0.4 bar soil moisture tension I_2; irrigation at 0.6 bar soil moisture tension I_3; irrigation at 0.8 bar soil moisture tension I_4; no post planting irrigation (control) I_5. Varieties as sub plot: Pant mung -1 (V_1); pant mung -2 (V_2); pant mung-3 (V_3); SML-668 (V_4).

Measurement of transpiration rate and leaf diffusive resistance

Transpiration rate and leaf diffusive resistance during 1300-1330 h with the help of a LI 1600 steady state porometer were measured from 30 days after sowing till maturity. Day time measurements of transpiration rate and leaf diffusive resistance were also recorded on the 30th, 45th and 60th days after sowing (DAS) from morning (0700-0800h) till evening at an interval of about 2 h. Each value was an average of 3 individual measurements. For measuring transpiration rate and leaf diffusive resistance, well developed, expended leaf was randomly selected in the plot. The resistance to diffusion of water vapour from the sub-stomatal cavities through stomata and the transpiration rate for dorsal leaf surface was measured. This is done by clamping to the leaf (intact with the plant) a small cup containing sensor responsive to change in humidity caused by transpiration.

Measurement of canopy temperature

Canopy temperature and canopy minus air temperature were measured during 1330-1400 h with the help of a Tele temp model AG-42 infrared thermometer from 30 DAS after sufficient canopy was established. Three to four separate measurements were made in each treatment and mean value for each plot was calculated. Measurements were made by directing the instrument towards canopy surface held 50 cm above the crop surface and at an angle of 30°C from the horizontal so as to view the plant parts only. On holding the grip at the above position, the instrument gave canopy temperature (Tc) and on pressing the trigger at the front of the hand grip, it gave canopy minus air temperature differential (Tc- Ta).

Rate of photosynthesis

Rate of photosynthesis was measured by the method of O' Toole et al. (1977). The measurements were made on the leaves from 3rd and 5th node from the top were cut under water at their base and cut ends were inserted in separate funnels of two liters capacity, having water in the bottom. Three trifoliates leaves during vegetative stage and two trifoliates leaves after flowering were used for measurements of photosynthetic rates. The net photosynthetic rates were measured under light intensity of 9000 lux (9 k lux) photosynthetically active radiation (PAR) on the surface of chamber. The light was supplied with three mercury vapor lamps of 400 W capacity. The assimilation chamber was fanned with cool air in order to intercept the excessive heat and regulate the air temperature in the chamber to 30 + 1°C. The relative humidity in the chamber was 90 +2% with the air flow rate of 300 ml/minute during the course of measurements. The time taken to deplete 50 ppm CO_2 in chamber was recorded by an infrared Gas Analyser (Model Type 225 MxII), till CO_2 was depleted from 350 to 100 ppm. The rate of net photosynthesis was computed by the following formula:

Where, V, volume of the chamber (ml); L, leaf area under test (cm^2). The measurements were made during vegetative stages (40 days after sowing), early pod setting and active pod filling stages of the crop growth.

Table 1. Seasonal variation in xylem water potential (XWP, - bar) during 0630-0700 h under I_1,I_2,I_3,I_4 and I_5 condition.

	Transpiration rate (XWP- bar)																			
	Days after sowing (DAS)																			
Variety	30					40					50					60				
	I_1	I_2	I_3	I_4	I_5	I_1	I_2	I_3	I_4	I_5	I_1	I_2	I_3	I_4	I_5	I_1	I_2	I_3	I_4	I_5
V_1	1.0	1.5	2.0	2.0	2.0	1.5	1.5	2.0	2.5	3.0	1.5	2.0	2.0	3.0	3.5	2.0	3.0	3.5	3.5	4.5
V_2	1.0	1.5	1.5	2.0	2.0	1.0	1.5	2.0	2.0	3.0	1.5	2.0	2.0	3.0	3.5	2.0	3.0	3.0	3.5	4.5
V_3	0.5	1.0	1.5	1.5	2.0	1.0	1.0	1.5	1.5	2.5	1.5	1.5	2.0	2.5	3.0	2.0	2.5	3.0	3.5	4.5
V_4	0.5	1.0	1.0	1.5	1.5	0.5	1.0	1.5	1.5	2.5	1.0	1.5	1.5	2.5	3.0	1.5	2.5	2.5	3.0	4.0

$$\text{Rate of net photosynthesis (mg CO}_2\text{ dm}^{-2}\text{ hr}^{-1}) = \frac{50 \times v \times x \times 2 \times 3600 \times 100}{10^6 \times L \times t}$$

Determination of free proline

Proline was estimated using the method of Bates et al. (1973). Samples were homogenized in 10 mL 3% (w/v) sulfosalicylic acid, and proline was assayed by the acid ninhydrin method. The absorbance was measured spectrophotometrically at 520 nm. Proline was calculated based on $\mu M.\ g^{-1}$. FW.

RESULTS

Biophysical processes

Seasonal variation in xylem water potential (XWP)

Seasonal changes in XWP during morning (0630-0700 h) were measured from 30 to 60 DAS in various treatments (Table 1). Treatments I_2, I_3 and I_4, follow the trend of XWP during the season but the magnitudes were different. Drought resistant varieties maintained higher XWP than drought sensitive varieties. The highest XWP of varieties (SML-668 and Pant Mung - 3) under I_5 "control" treatment was associated with extensive root system.

Seasonal variation in transpiration rate (TR)

Seasonal changes in TR during post noon (1300-1330 h) is shown in Table 2. Results showed that TR was always higher under I_1 treatment than under I_5 treatment. Under I_1 treatment, TR of V_1 Pant Mung-1, V_2 Pant Mung-2, V_3 Pant Mung-3 and V_4 SML-668 ranged from 24.3 to 26.5, 25.1 to 28.5, 25.5 to 30.2 and 25.3 to 31.5 $\mu g\ cm^{-2}\ sec^{-1}$ during 30 to 60 DAS, respectively. In I_2, I_3, I_4 and I_5, the trend of TR during the season was same but the magnitude was different. In I_1 and I_5, TR ranged from 24.3 to 26.5 and 23.0 to 17.2 in V1; 25.1 to 28.5 and 24.0 to 17.0 in V2; 25.5 to 30.2 and 23.0 to 17.5 in V3 and 25.3 to 31.5 and 23.5 to 19.0 $\mu g\ cm^{-2}\ sec^{-1}$, respectively, from 30 to 60 DAS. Variety SML-668 was more tolerant to water stress over all other varieties. These were in a sequence of SML-668>Pant Mung-3>Pant Mung-2>Pant Mung-1 in relation to water stress tolerance. Results show that TR increased gradually from 30 to 50 DAS and thereafter decreased. The drought tolerant varieties maintained highest TR and drought susceptible varieties maintained lowest TR during the 1300-1330 h.

Seasonal variation in leaf diffusive resistance (LDR)

Seasonal variations in LDR were recorded during 1300-1330 h as shown in Table 3. LDR was measured from 30 to 60 DAS in various irrigation treatments. In general, LDR of all varieties was low in I_1 than I_5. Treatment I_1, LDR of Pant mung-1, Pant Mung-2, Pant Mung-3 and SML-668 ranged from 0.23 to 0.17 sec cm^{-1}, 0.21 to 0.18 sec cm^{-1}, 0.20 to 0.15 sec cm^{-1} and 0.17 to 0.14 sec cm^{-1} during 30 to 60 DAS. The LDR of Pant mung-1, Pant Mung-2, Pant Mung-3 and SML-668 in I_2,I_3 and I_4 ranged from 0.21 to 0.26, 0.23 to 0.14, 0.21v to 0.23 and 0.19 to 0.21 sec cm^{-1}, respectively, from 30 to 60 DAS. In I_5, the trend of LDR during the season was same but with higher magnitude recorded from 45 and 60 DAS. In I_1, I_5 LDR ranged from 0.23 to 0.17, 0.25 to 0.31 in V1; 0.21 to 0.18, 0.24 to 0.27 in V2; 0.20 to 0.15, 0.23 to 0.26 in V3 and 0.17 to 0.14, 0.20 to 0.24 sec cm^{-1} in V4, respectively. Variety SML-668 was more tolerant to water stress over all other varieties. The drought tolerant varieties maintained lowest LDR and drought susceptible varieties maintained highest LDR during the 1300-1330 h.

Table 2. Seasonal variation in transpiration rate (TR µg cm⁻² sec⁻¹) during 1300-1330 h under I1,I2,I3,I4 and I5 condition.

Variety	Transpiration rate ($\mu g\ cm^{-2}\ sec^{-1}$)																			
	Days after sowing (DAS)																			
	30					40					50					60				
	I1	I2	I3	I4	I5	I1	I2	I3	I4	I5	I1	I2	I3	I4	I5	I1	I2	I3	I4	I5
V1	24.3	23.4	23.0	23.5	23.0	26.2	24.0	22.5	22.0	18.3	28.0	26.0	23.5	20.2	21.5	26.5	24.5	21.0	17.5	17.2
V2	25.1	24.3	24.0	24.2	24.0	27.0	24.5	23.6	23.2	18.2	31.0	29.0	24.0	20.5	22.3	28.5	25.5	22.5	18.5	17.0
V3	25.5	24.6	24.2	23.5	23.0	28.0	25.0	24.0	23.0	18.5	31.5	29.5	24.5	21.0	22.5	30.2	26.5	23.5	19.0	17.5
V4	25.3	24.5	24.0	24.0	23.5	28.5	26.5	23.5	21.5	19.5	33.5	31.5	25.5	21.5	21.3	31.5	28.5	24.3	20.0	19.0

Table 3. Seasonal variation in leaf diffusive resistance (LDR) (sec cm⁻¹) during 1300-1330 h under I1,I2,I3,I4 and I5 condition.

Variety	Leaf diffusive resistance ($sec\ cm^{-1}$)																			
	Days after sowing (DAS)																			
	30					40					50					60				
	I1	I2	I3	I4	I5	I1	I2	I3	I4	I5	I1	I2	I3	I4	I5	I1	I2	I3	I4	I5
V1	0.23	0.24	0.24	0.25	0.25	0.19	0.21	0.30	0.31	0.33	0.19	0.20	0.28	0.32	0.34	0.17	0.20	0.25	0.30	0.31
V2	0.21	0.23	0.22	0.23	0.24	0.18	0.19	0.26	0.27	0.29	0.19	0.19	0.24	0.27	0.29	0.18	0.17	0.22	0.26	0.27
V3	0.20	0.21	0.21	0.22	0.23	0.17	0.19	0.23	0.25	0.26	0.16	0.18	0.21	0.26	0.27	0.15	0.16	0.20	0.23	0.26
V4	0.17	0.19	0.18	0.19	0.20	0.17	0.18	0.21	0.22	0.24	0.15	0.17	0.20	0.23	0.26	0.14	0.17	0.19	0.21	0.24

the 1300-1330 h.

Seasonal variation in canopy temperature (Tc) and canopy minus air temperature differential (Tc- Ta °C)

Measurements of canopy, canopy minus air temperature differential (T_C), (T_C- T_a °C) during 1300-1400 h from 30, 40, 50 and 60 DAS in various treatments is shown in Table 4. Results showed that T_C, T_C- T_a under I1 treatment was lower than those under I5 "control" treatment. Whenever the differences in T_C- T_a among the varieties was equal to or less than the difference among the replicated measurements within the treatments, the varieties were considered similar in character and only average values were reported (Table 4). Treatment I1 T_C, T_C- T_a of Pant Mung-1,Pant Mung-2, Pant Mung-3 and SML-668 ranged from 31.2 to 36.5, 30.8 to 36.0, 30.6 to 35.5 and 30.2 to 35.0°C, -5.4 to -4.2, -5.6 to -4.7, -6.0 to -5.2 and -6.4 to -5.7°C during 30 to 60 DAS. In I2, I3, I4 and I5 treatments, the trend of T_C, T_C- T_a during the season was same but the magnitude was different. Among the varieties, SML-668 was superior to all other varieties. These all were in a sequence of Pant Mung-1>Pant Mung-2> Pant Mung-3>SML-668 in relation to drought tolerant.

Physiological processes

Rate of net photosynthesis

The photosynthetic rates varied significantly among the varieties at all three stages, viz., vegetative, early pod setting and active pod filling at all the external CO_2 concentrations (Table 5). The rate of net photosynthesis increased with the advancement of crop stage. Improved moisture supply increased the rate of net photosynthesis over I5 "control". At all the stages, I1 and I2 were found to be superior to other treatment at various levels of external CO_2 concentrations. The rate of net photosynthesis decreased in all the varieties and treatments at various stages with the decrease in external CO_2 concentrations.

The large differences are said to be caused by chlorophyll content, the conductivity of CO_2 particularly the mesophyll resistance and the differential response of varieties to light and the external conditions such as light intensity and its quality, temperature and CO_2 concentrations and finally the utilization or translocation of photosynthesis from leaves. In spite of the several factors influencing the rate of photosynthesis, occasional but

Table 4. Seasonal variation in canopy temperature (T_C) and canopy minus air temperature differential (T_C- T_a °C) during 1300-1330 h under I_1, I_2, I_3, I_4 and I_5 conditions based on infrared thermometry from 30 to 60 DAS.

Variety	Canopy temperature T_C (°C) and canopy minus air temperature differential T_C- T_a (°C)							
	Days after sowing (DAS)							
	30		40		50		60	
	T_C	T_C- T_a	T_C	T_C- T_a	T_C	T_C- T_a	T_C	T_C- T_a
1	2	3	4	5	6	7	8	9
I_1								
V_1	31.2	-5.4	36.0	-0.4	37.0	-3.1	36.5	-4.2
V_2	30.8	-5.6	35.4	-1.2	36.0	-4.1	36.0	-4.7
V_3	30.6	-6.0	35.2	-1.4	35.5	-4.6	35.5	-5.2
V_4	30.2	-6.4	33.5	-3.1	34.5	-5.6	35.0	5.7
I_2								
V_1	32.0	-4.6	36.7	+0.1	37.4	-2.7	37.8	-2.9
V_2	31.5	-5.1	35.7	-0.9	36.7	-3.4	37.2	-3.5
V_3	31.5	-5.1	35.5	-1.1	36.0	-4.1	37.0	-3.7
V_4	30.5	-6.1	33.6	-3.0	35.0	-5.1	36.0	-4.7
I_3								
V_1	32.5	-4.1	36.2	-0.4	37.5	-2.6	37.7	-3.0
V_2	32.0	-4.6	35.0	-1.6	36.8	-3.3	37.0	-3.7
V_3	31.7	-4.9	34.5	-2.1	36.5	-3.6	36.8	-3.9
V_4	31.0	-5.6	33.5	-3.1	36.0	-4.1	36.0	-4.7
I_4								
V_1	32.7	-3.9	37.5	+0.9	38.8	-1.3	39.2	-1.3
V_2	32.2	-4.4	36.5	-0.1	38.2	-1.9	38.5	-2.0
V_3	32.0	-4.6	36.0	-0.6	38.0	-2.1	38.2	-2.5
V_4	31.2	-5.4	35.0	-1.6	37.0	-3.1	37.5	-3.2
I_5								
V_1	32.5	-3.9	37.5	+0.9	39.0	-1.1	39.5	-1.2
V_2	32.0	-4.4	36.7	+0.1	38.5	-1.6	39.0	-1.7
V_3	31.7	-4.7	36.2	-0.4	38.2	-1.9	38.5	-2.2
V_4	31.2	-5.4	35.5	-1.1	37.5	-2.6	38.0	-2.7

positive association of photosynthetic rate with productivity and good heritability of carbon exchange rate was noted.

Free proline content

Increase in free proline was observed from vegetative to active pod filling stages in all varieties (Table 6). This is because of decrease in internal water status of the plant with advancement in crop age, which could be evident from decrease in leaf water potential with crop age. Irrigation significantly decreased free proline at all the stages. In the field, plants are usually subjected to various degree of water deficit even when soil water is considered as adequate. So this might be the reason for some accumulation of free proline in irrigated condition. Improved moisture supply through various treatments decreased the free proline content over I_5 "control" treatment. At all the stages I_5 and I_4 were found to be superior to all the treatments. I_3 was superior to the remaining treatments,

respectively. Among varieties, SML-668 had highest free proline content than all other varieties. Varieties having different degree of drought resistance differ in their capacity to accumulate proline under stress. Resistant varieties accumulate higher level of proline under water stress. A similar trend was obtained by Ashraf and Ibram (2005), Ashraf and Foolad (2007) and Tawfik (2008) who found that osmoprotectants such as proline and glycine betaine (GB) were increased under drought stress.

Yield and yield contributing parameters

Improved moisture supply through various irrigation treatments increased the yield attributes and grain yield significantly over I_5 "control" treatment (Table 6). The maximum yield attributes and grain yield were recorded under I_2 and minimum under I_5 treatment. Among varieties, V_4 SML-668 produced significantly higher number of pods per plant, grains per pod, 1000 grain weight and grain yield than all other varieties, respectively.

Table 5. Effect of different treatments on the rate of net photosynthesis (mg CO_2 dm^{-2} hr^{-1}) at different external CO_2 concentrations (ppm) during vegetative, early pod setting and active pod filling stages.

Treatment	Photosynthesis (mg CO_2 dm^{-2} hr^{-1}) at different external CO_2 concentrations (ppm)																			
	Varieties																			
	Pant mung -1					Pant mung -2					Pant mung -3					SML -668				
	350-300	300-250	250-200	200-150	150-100	350-300	300-250	250-200	200-150	150-100	350-300	300-250	250-200	200-150	150-100	350-300	300-250	250-200	200-150	150-100
Vegetative stage																				
I_1	5.74	3.63	3.15	1.57	0.75	4.49	3.63	3.08	1.60	1.28	6.70	5.56	3.74	2.26	1.24	6.38	4.88	3.64	2.26	2.11
I_2	7.31	4.86	4.26	2.87	1.12	9.96	5.90	3.49	2.64	1.75	9.72	7.62	6.20	4.25	2.43	7.75	6.10	4.47	3.67	2.54
I_3	3.32	2.85	2.30	0.90	0.22	4.37	2.40	2.07	1.21	1.13	5.41	3.82	3.30	1.86	0.90	7.12	3.82	3.24	2.23	1.25
I_4	3.20	2.60	1.78	0.80	0.20	4.13	2.63	1.56	1.10	0.75	3.66	3.55	2.51	1.72	0.82	4.43	2.11	1.63	1.18	0.94
I_5	2.16	2.11	1.28	0.38	0.10	2.56	1.44	0.90	0.86	0.55	2.77	2.22	1.36	0.62	0.13	3.87	2.52	2.13	1.15	0.65
Mean	4.34	3.21	2.55	1.30	0.48	5.10	3.20	2.22	1.48	1.09	5.65	4.55	3.42	2.14	1.10	5.91	3.89	3.02	2.10	1.50
Early pod setting stage																				
I_1	7.83	6.41	4.87	3.37	1.44	7.79	6.41	4.63	3.20	1.62	9.13	7.24	6.05	4.68	2.63	9.95	8.22	5.85	5.09	3.08
I_2	9.10	7.51	5.89	4.56	2.09	9.10	7.15	5.83	4.53	2.13	9.66	7.61	6.25	5.09	3.33	10.35	8.95	7.77	6.39	3.91
I_3	5.62	3.90	3.21	2.43	1.23	6.50	5.71	4.25	3.37	1.92	8.75	7.40	6.09	4.58	3.10	8.73	7.14	5.94	4.70	2.66
I_4	5.62	4.33	3.12	2.43	1.14	5.25	4.51	3.17	2.35	1.35	7.26	5.43	4.28	2.96	1.24	7.26	5.46	4.26	2.96	1.30
I_5	3.20	2.40	1.84	1.33	0.59	3.21	2.38	1.91	1.32	0.59	5.64	4.13	3.60	2.50	1.22	5.67	4.16	3.58	2.35	1.25
Mean	6.27	4.91	3.79	2.82	1.30	6.37	5.23	3.96	2.95	1.52	8.09	6.36	5.25	3.96	2.30	8.71	6.79	5.48	4.30	2.44
Active pod filling stage																				
I_1	8.91	7.73	5.87	4.36	2.36	9.32	7.26	6.11	4.46	2.44	9.15	7.20	6.11	4.27	1.94	9.85	8.36	6.41	5.08	2.09
I_2	9.85	8.36	6.41	5.08	2.46	10.23	9.85	8.26	6.46	3.10	11.15	8.16	6.71	5.11	2.64	112.26	9.85	8.23	6.46	3.09
I_3	7.77	6.86	5.44	4.11	2.15	8.94	7.73	5.87	4.39	2.39	9.07	7.40	5.50	4.21	1.87	9.33	7.96	6.10	4.46	2.45
I_4	6.51	5.52	4.61	3.23	1.45	6.65	6.09	4.09	3.09	1.15	8.36	6.66	5.10	4.09	1.66	8.97	7.76	5.87	4.34	2.37
I_5	5.67	4.72	3.38	2.59	1.24	5.93	4.68	3.48	2.41	1.28	5.43	4.45	2.45	1.61	0.90	5.93	4.71	3.48	2.37	1.27
Mean	7.74	6.64	5.14	3.87	1.93	8.21	7.12	5.56	4.16	2.07	8.63	6.78	5.17	3.86	1.80	9.27	7.73	6.02	4.54	2.25

The yield per hectare was primarily improved due to improvement in moisture supply and its beneficial effect on the per plant yield. The grain yield per plant improved with increase moisture supply mainly through improvement in number of pods per plant, number of grains per pod and 1000 grain weight. Thomas et al. (2004) reviewed that Mung bean plants in the rain shelter and rain fed treatments attained maturity earlier than the well-watered treatment.

DISCUSSION

The results indicated that the XWP and LDR decreased with the beginning of transpiration in the morning. The transpiration was maximum during 1300-1330 h (peak radiation load) under I_1 treatment (irrigation at 0.2 bar soil moisture tension) when the LDR and XWP were minimum. As LDR increased, the TR decreased and XWP increased. Decrease in XWP was a result of excess transpiration over water absorption due to rapidly rising radiation load and saturation deficit. As radiation load increased during the day, canopy temperature (T_C) increased from a minimum at sunrise to a maximum during 1300-1330 h and declined gradually thereafter. The canopy minus air temperature differential (T_C-T_a) on the other hand, reduced during 1300-1330 h because of increasing transpiration rate which resulted into cooling of the canopy relative to air temperature. Result indicated

Table 6. Effect of different treatments on free proline content µg/g fresh weight at vegetative stage, early pod setting stage and active pod filling stage.

Treatment	Free proline content (µg/g) fresh weight											
	Vegetative stage				Early pod setting stage				Active pod filling stage			
	Pant Mung-1	Pant Mung-2	Pant Mung-3	SML-668	Pant Mung-1	Pant Mung-2	Pant Mung-3	SML-668	Pant Mung-1	Pant Mung-2	Pant Mung-3	SML-668
I_1	35.2	35.0	32.5	31.6	58.7	56.6	53.5	49.5	68.5	66.3	57.6	50.6
I_2	45.0	44.6	41.5	39.5	66.2	62.5	60.3	56.3	87.2	85.8	70.2	58.5
I_3	55.0	54.5	50.5	48.0	154.8	152.6	145.7	140.8	355.4	350.3	258.6	210.2
I_4	75.5	74.3	68.5	65.6	178.0	175.6	161.2	150.3	415.3	407.0	315.3	295.3
I_5	87.0	86.2	82.5	77.5	215.5	210.3	196.5	184.5	535.5	528.5	420.5	397.7
Mean	59.5	58.9	55.1	52.4	134.6	131.5	123.4	116.3	292.4	287.6	224.4	202.5

Table 7. Yield attributes and grain yield (q/ha) as affected by different treatments.

Treatment	Number of pods plant⁻¹				Number of grains pod⁻¹				1000 grain weight (g)				Grain yield (q/ha)			
	Varieties															
	V_1	V_2	V_3	V_4	V_1	V_2	V_3	V_4	V_1	V_2	V_3	V_4	V_1	V_2	V_3	V_4
I_1	12.9	14.0	14.2	14.5	7.2	7.1	7.6	7.5	36.2	37.4	37.3	37.9	9.86	10.23	11.25	11.67
I_2	21.7	21.5	22.3	22.7	7.5	7.8	7.9	8.8	36.4	37.6	37.7	38.1	13.87	15.24	18.18	18.58
I_3	10.8	11.0	12.6	12.6	6.6	6.8	6.9	8.0	35.5	36.5	36.7	37.1	6.86	7.08	9.38	9.58
I_4	10.0	10.0	10.3	10.8	5.9	6.1	6.4	6.8	35.2	36.8	36.9	36.6	5.90	5.95	6.58	6.66
I_5	9.0	9.0	8.9	9.3	3.7	4.2	6.1	6.2	34.8	36.0	36.0	36.2	3.37	3.50	3.61	4.15
C D at 5%	0.43	0.45	0.21	0.08	0.18	0.24	0.12	0.15	1.10	1.02	1.01	0.61	3.96	2.18	3.94	5.94

that period between 1300-1400 h was best for canopy temperature measurements as irrigation treatments had little influence on canopy temperature at sunrise and sunset when the solar radiation was low. Genotypic differences were observed in case of XWP and TR in plants. Drought resistant varieties maintained higher XWP than drought sensitive varieties. The higher XWP and TR were recorded in SML-668 and Pant Mung-3 under I_5 "control" treatment. The genotype SML-668 and Pant Mung-3 had lower Tc-Ta both under I_1 and I_5 treatments during 1300-1400h as compared to the other varieties.

Physiological changes occur in response to low water condition in different varieties. The increase of free proline occurs in decrease in water supply. The synthesis of proline in plants extensively protects cell membrane and protein content in plant leaves. The results of this study are in agreement with other investigations (Behnamnia et al., 2009; Zhang et al., 2006). Proline acts as an osmolite beside enzymes and other macromolecules and therefore, protects the plant against low water potential and causes osmotic regulation in plant organs. Also, proline can act as electron receptor preventing photosystems injuries in dealing with reactive

oxygen species (ROS) function. Proline accumulation facilitates the permanent synthesis of soluble substances in closing stomata. Stress, especially in the growing stage, reduces the capacity of the source plants for the source and sink is forced to balance the number of flowers and pod production to reduce the stress that can handle grain filling period and also reduced the final yield. Asaduzzaman et al. (2008) also believe that moisture stress reduces grain yield of Mungbean and maximum negative effects of drought obtained with once irrigation during growth season. Pandey et al. (1984), Reddy et al. (2004) and Rafiei Sadasivan et al. (1988), Reddy et al. (2004) and Rafiei

Shirvan and Asgharipur (2009) also obtained the similar results.

Conclusion

Drought tolerance consists of ability of crop to grow and produce under water deficit conditions. A long term drought stress effects on plant metabolic reactions are associated with plant growth stage, water storage capacity of soil and physiological aspects of plant. Achieving a genetic increase in yield under these environments has been recognized to be a difficult challenge for plant breeders while progress in yield grain has been much higher in favorable environments. The present results showed that plants in drought stress make changes in some of their physiological and biochemical features. Also, the results of this research showed that drought stress causes low grain yield; and in drought stress conditions cultivars that have more TR, chlorophyll content, XWP, proline content and LDR and canopy minus air temperature (Tc-Ta) are more tolerant to drought stress.

REFERENCES

Ahmed ZU, Shaikh M A Q,Khan AI and Kaul A (1978). Evaluation of local, exotic and mutant germplasm of Mungbean for varietal characters and yield in Bangladesh. SABRAO J. 10:48.

Ali M, Kumar S (2004). Prospects of mungbean in rice-wheat cropping systems in Indo-Gangetic Plains of India. In: Proceedings of the Final Workshop and Planning Meeting. Improving Income and Nutrition by Incorporating Mungbean in Cereal Fallows in the Indo-Gangetic Plains of South Asia, S. Shanmugasundaram (ed.). DFID Mungbean Project for 2002–2004, 27–31 May 2004, Ludhiana, Punjab,India, pp. 246-254.

Allahmoradi P, Ghobadi M, Taherabadi S, Taherabadi S (2011). Physiological Aspects of Mungbean in Response to Drought Stress. IPCBEE vol.9 (2011) © (2011) IACSIT Press, Singapoore Int. Conf. Food Eng. Biotechnol.

Athar H, Ashraf M (2005).Photosynthesis under drought stress. : Hand Book Photosynthesis, 2nd (ed.) by M. Pessarakli. C. R. C. Press, New York, USA, pp. 795-810.

Ashraf M, Ibram A (2005). Drought stress induced changes in some organic substances in nodules and other plant parts of two potential legumes differing in salt tolerance. Flora 200:535-546.

Ashraf M, Foolad MR (2007).Roles of glycine betaine and proline in improving plant abiotic stress resistance. Environ. Exp. Botany 59:206-216.

Asaduzzaman F K, Ullah J, Hasanuzzaman M (2008).Response of mungbean (L.) to nitrogen and irrigation management. Am. Eurasian J. Sci. Res. 3(1):40-43.

Bates LS, Waldren RP, Teare ID (1973).Rapid determination of free proline for water stress studies. Plant Soil 39:205-207.

Belayet HM, Rahman W, Rahman MN, Noorul Anwar AHM, Hossen AKMM (2010). Effects of water stress on yield attributes and yield of different mungbean genotypes. 5:19-24.

Behnamnia M,Kalantari Kh M and Rezanejad F (2009).Exogenous Application Of Brassinosteroid Alleviates Drought-Induced Oxidative Stress In L. Gen. Appl. Plant Physio. 35:22-34.

Brown EA, Caviness CE, Brown DA (1985). Response of selected soybean cultivars to soil moisture deficit. Agron. J. 77:274-278.

Jordan WR, Ritichie JT (2002). Influence of soil water stress on evaporation, root absorption and internal water status of cotton. Plant Physiol. 48:783-788.

Mahmoodian L, Naseri R, Mirzaei A (2012). Variability of grain yield and some important agronomic traits in mungbean (Vigna radiate L.) cultivars as affected by drought stress. Intl. Res. J. Appl. Basic. Sci. 3(3):486-492.

Nabizade M, Tayeb SN, Mani M (2011).Effect of irrigation on the yield of mungbean cultivars. J. Am. Sci. 7(7):86-90.

O' Toole JC, Ozbun JL, Wallace DM (1977). Photosynthetic response to water stress in Phaseolus vulgaris Physiol. Plant 40:111-114.

Pandey RK, Herrera W AT, Villegas AW, Penletion JW (1984). Drought response of grain legumes under irrigation gradient. III. Plant growth, 76:557-560.

Rafiei M,Shirvan, Asgharipur MR (2009).Yield reaction and morphological characteristics of some mungbean genotypes to drought stress. J. Mod. Agric. Knowl. 5(15):67-76.

Reddy AR,Chiatanya KV and Vivekanandan M (2004). Drought-induced responses of photosynthesis and antioxidant metabolism in higher plants. J. Plant Physiol.161:1189-1202.

Sadasivan R, Natrajaratnam N, Dabu R, Muralidharan V, Rangasmay SRS (1988).Response of Mungbean cultivars to soil moisture stress at different growth phases. Mungbean Proceeding of the Second International Symposium. AVRCD. pp.260-262.

Srivalli B, Chinnusamy V, Chopra RK (2003).Antioxidant defense in response to abiotic stresses in plants. J. Plant Biol. 30:121-139.

Thomas R, Fukai MJS, Peoples MB (2004). The effect of timing and severity of water deficit on growth, development, yield accumulation and nitrogen fixation of mungbean. Field Crops Res. 86:67-80.

Tawfik KM (2008).Effect of Water Stress in Addition to Potassiomag Application on Mungbean. Australian J. Basic Appl. Sci. 2(1):42-52.

Uprety DC, Bhatia A (1989). Effect of water stress on the photosynthesis, productivity and water status of Mungbean ((L.) Wilczek). J. Agron. Crop Sci. 163:115-123.

Zhang J, Jia W, Yang J, Ismail AM (2006). Role of ABA in integrating plant response to drought and salt stresses. Field Crops Res. 97:111-119.

14

Efficacy of defatted soy flour supplement in *Gulabjamun*

Awadhesh Kumar Singh[1], Dattatreya Mahadev Kadam[2]*, Mili Saxena[3] and R. P. Singh[4]

[1]Department of Process and Food Engineering, Punjab Agricultural University, Ludhiana-141004, Punjab, India.
[2]Food Grains and Oilseeds Processing Division, Central Institute of Post-Harvest Engineering and Technology, PO: PAU, Ludhiana-141004, Punjab, India

Study was undertaken to determine the efficacy of defatted soyflour mix levels in *Gulabjamun* (sweet dessert comprised of fried milk balls dipped in sugar syrup) and its impact on the quality parameters. Soy flour was fortified in three levels (3.33, 6.66 and 9.99%) w/w to prepare different compositions of *Gulabjamuns* by replacing wheat-flour in control recipe. 10 g spherical shape balls were made using thoroughly mixed ingredients dough and these were deep-fried in oil before soaking in sugar syrup (50°Brix for 4 h) at 70°C. Standard methods were used to estimate protein and crude fat content in *Gulabjamun*. Significant effect of raw premixes, prior to sugar syrup dipping, was observed due to addition of defatted soy flour on protein and fat content. Protein content was decreased to 18.24% from 20.66% and fat content increased to 28.36 from 12.09% in deep-fried samples from raw premix. Hardness, cohesiveness, springiness/elasticity, gumminess and chewiness values were increased with the increase in the soy flour levels. Appearance, colour, texture, flavour and overall acceptability of the *Gulabjamuns* had improved with the addition of 3.33% soy flour and decreased there after.

Key words: Fat, fortification, *Gulabjamun*, protein, sensory attributes, soy flour, TPA.

INTRODUCTION

Recently, the food industry has seen expansive growth in what is known as "functional food". Recent growth of funcional foods far outpaces that of conventional foods and supplements, and has attracted the interests of both consumers and food producers (van Poppel, 1998; Locklear, 2000). van Poppel (1998) defined functional foods as food that exerts a beneficial health effect beyond the recognized traditional nutritional value of such a food. Within the grouping of functional foods are two categories; 1) potential functional foods (those with the potential for use in human nutrition), and 2) established functional foods (those with proven benefits in human nutrition) (Heller, 2008).

Soybean is one of the nature's wonderful nutritional gifts. Soybeans have served as a major source of dietary protein for many people throughout Asia for over 1,000 years (Wiseman, 1997). In other parts of the world, however, soybeans have been sought after mostly for their oil. Soybeans contain all the three-macro nutrients required for good nutrition, complete protein (40%), carbohydrate (18%), fat (18%) and moisture (9%) as well as vitamins and minerals (5%), including folic acid, calcium, potassium and iron (National Soybean Research Laboratory, 2008). Soybean protein provides all the nine essential amino acids in the amounts needed for human health. Fortification of cereals with soy will not only improve protein quantity but also improve the quality of food nutrients such as amino acid balance. Recent developments in processing technology and a need to meet demands of new soyfood consumers have brought on the development of a new class of soyfoods, known as "the second generation of soyfoods" (Liu, 1997). This "second generation of soyfoods" includes among others soy sausages, soy yogurt, and soy cheeses, and soy-based dairy analogs. These foods utilize protein ingredients derived from defatted soybean meal including soy protein concentrate, soy protein isolate, and texturized soy protein (Liu, 1997).

Consumption of soyfoods has been on the rise since the establishment of the October, 1998 U.S.FDA-approv-

*Corresponding author. E-mail: kadam1k@yahoo.com.

ed soy health claim, which links the intake of products high in soy protein with positive health benefits such as lower risk for heart disease (Henkel, 2000; Federal Register, 1998). Soy-based dairy analogs such as soy-milk, soy yogurt, and soy ice cream are available; however there are few soy-fortified dairy-based food products that might appeal to a more traditional dairy consumer (Berry, 2002). Drake et al. (2000 and 2001) evaluated soy protein fortification of dairy yogurts. Although physical and sensory properties were altered, consumer studies indicated an interest and a potential market for soy fortified dairy yogurts and other foods (Drake and Gerard, 2003).

In addition to soybeans supplying adequate protein to the diet, studies have shown that protein from soybeans may be beneficial to human health in other ways. Aside form soy protein being low in saturated fat and cholesterol free; there may be many more advantages to consumption of soy in the diet (Messina, 1995). Research shows that, blending of soy flour with wheat flour will increase the recommended amino acid availability from 40 to 80% (Khan et al., 2005). In addition to nutritional improvements, soy fortified wheat flour will improve the functional characteristics of the end products in terms of better moisture retention and less oil absorption. It has been recognized for some time now that consumption of plant proteins often results in significant lowering of low-density lipoproteins and total cholesterol levels, which are associated risk factors for cardiovascular disease (Friedman and Brandon, 2001; Krummel, 1996).

Effect of soybean flour lipoxygenase isozymes on wheat flour dough rheological and bread making properties were studied by Cumbee et al. (1997). Also, an appropriate household/small-scale technique for the production of soy-fortified fermented maize dough by comparing different treatments, processing methods and fortification levels were investigated (Plahar et al., 1997). Torres et al. (1998) studied the sensory characteristics of soymilk and tofu made from lipoxygenase-free and normal soybeans and found that there was no difference between lipoxygenase-free and normal soybeans for milky flavor, wheat flavor, thickness and chalkiness. Massey et al. (2001) observed that the consumption of soybeans and foods made from them are increasing because of their desirable nutritional value. Effects of high and low-isoflavone soy-protein foods on lipid and non-lipid risk factors for coronary artery disease were investigated. Saxena et al. (1996) studied the soy flour – Gulabjamun premixes and ready-to-serve soy flour - Gulabjamun prepared from mixture of soy flour and milk solids and found 40% soy flour substitution of whole milk powder prepared Gulabjamuns was best.

Gulabjamun is a popular and favorite sweet dish/ /desserts comprised of khoya rounds deep fried in ghee /fat and soaked in hot saffron/cardamom seeds/rosewater flavoured sugar syrup. The khoya or mava is made by reducing low fat milk and is slightly yellowish in colour and is also loose and sticky in consistency. Frying is done on sufficiently low flame that the Gulabjamuns get cooked till the inside with golden brown colour balls. They are served warm or at room temperature. Currently, there is limited literature encompassing the utilization of soy flour for preparations of soy flour fortified Gulabjamuns. Thus, the present study was undertaken to find out the effect of defatted soyflour addition levels (0, 3.33, 6.66 and 9.99%) w/w in Gulabjamun on protein, fat, sensory and textural quality parameters.

MATERIALS AND METHODS

Materials

The raw materials are such as khoya, paneer (Indian cheese), wheat flour, semolina, baking powder, refined oil and soy flour were obtained from local market of Ludhiana (Punjab, India). The experiment was conducted in the Department of Processing and Food Engineering, Punjab Agricultural University, Ludhiana, Punjab, India.

Preliminary studies

Based on the preliminary studies carried out for the preparation of Gulabjamuns using different levels (0, 3.33, 6.66, 9.99, 13.32, 16.65 and 19.98% w/w) of defatted soy flour in the recipe. Three levels of soy flour mixing, that is, 3.33, 6.66 and 9.99% w/w were considered for further studies based on sensory evaluation. Each experiment was replicated for five times.

Gulabjamun premix

Premix raw materials were procured from Ludhiana local market for making control Gulabjamun. The control premix consists of khoya (66.66%), paneer (16.6%), wheat flour (13.3%), semolina (2.66%), baking powder (0.16%) and refined oil (0.62%). Soy flour was fortified in three levels (3.33, 6.66 and 9.99% w/w) by replacing wheat flour in control recipe to prepare three different compositions of Gulabjamuns. All the ingredients like khoya, paneer, wheat flour, semolina, refined oil and baking powder were weighed and mixed thoroughly in small quantity of water to make dough and thereafter 10 g spherical/ round shape balls were prepared and these were deep fried in fat using the electrical fryer at 130°C temperature for 15 min to get a light brown coloured surface (Rangi et al., 1985).

Sugar syrup and soaking of Gulabjamun

Boiling 250 g of sugar in 300 ml of water for 5 min made sugar syrup of 50°Brix. The total solids of sugar syrup were determined by using an ERMA (Japan) make hand refractometer having range 32 - 60°Brix. TSS value was recorded on scale at a temperature of 20°C. Deep fried Gulabjamuns were soaked in hot sugar syrup having 50°Brix TSS at 70°C for 4 h (Rangi et al., 1985).

Estimation of protein content

Protein content of Gulabjamun samples were estimated using Microkjeldahl distillation apparatus as per the method of AOAC (2002a).

$$\text{Nitrogen, } N_2 \text{ (\%)} = \frac{\text{Liter value} \times 0.0014 \times \text{volume made} \times 100}{\text{Aliquot taken} \times \text{weight of the sample (g)}}$$

Table 1. Protein content (%) in *Gulabjamuns* at different levels of soy flour mix.

Soy flour mix, % and stat. parameters	Protein content (%)		
	Raw mix	Fried balls	Fried balls soakedin syrup
0.00	16.08	15.02	14.05
3.33	17.37	15.68	14.81
6.66	22.75	20.60	18.01
9.99	26.46	21.68	19.56
Mean ± sd	20.66 ± 4.17	18.24 ± 3.80	16.60 ± 4.64
SE$_M$	2.08	1.90	2.32
CD at 5%	0.75	1.32	0.88
CV (%)	1.94	3.84	2.81

Note: SE$_M$ is standard error of the mean, sd is standard deviation, CD is critical difference and CV is critical variance.

N_2 (%) = [(A) x 0.0014×250×100] / (5×1)
∴ Protein (%) = N_2 (%) × 6.25

Estimation of fat content

Crude fat content (triglycerides of fatty acid) of *Gulabjamun* samples were estimated as per the standard method of AOAC (2002b) using fat extraction tube of soxhlet apparatus.

Fat content, % = (Amount of ether extract (g) / weight of the sample) × 100

Sensory evaluation

Gulabjamuns were evaluated for overall acceptability of samples by a randomly selected panel consists of minimum 15 persons. The panel was asked to evaluate the a, b, c …. coded samples of *Gulabjamuns* for appearance, color, texture, flavour and overall acceptability as per 9 point Hedonic scale (Rangi et al., 1985). Samples were served as per standard of sensory evaluation.

Textural behaviour

Textural profile analysis (TPA) of *Gulabjamuns* was carried out using texture analyzer (Model No: TA-Hd*i*, Stable Micro Systems, UK) in the Engineering Properties Laboratory of the Department. The texture behaviour of whole *Gulabjamuns* was estimated in terms of the TPA curve. The parameters of the brittleness, hardness, cohesiveness, chewiness, springiness and gumminess were calculated from the plot of two cyclic compression tests. The following textural parameters were estimated as follows (Bourne, 1982):

Hardness: The maximum height of curve during the first compression.
Brittleness: Height of first significant break of multi peak shape of first chew.
Cohesiveness: Ratio of area under second peak to that of first peak, that is, A_2/A_1.
Elasticity: Test speed × distance on × axis from start of second bite to its peak.
Chewiness: Hardness × cohesiveness× elasticity.
Gumminess: Hardness × cohesiveness.

Statistical analysis

The statistical analysis was carried using two-way ANOVA in Gene-ral Linear Model (GLM) using Statistical Package for Social Sciences (SPSS)- version 7.5. Five replication means were computed and tested at 5% levels of significant to arrive at the best results of the treatments.

RESULTS AND DISCUSSION

Efficacy of soy flour mix level on protein content

In Table 1, it is clear that the protein content was higher in the raw mix as compared to the fried balls. The decreases in protein after frying of raw mix balls may be due to incorporation of oil and air in the fried balls. The protein content further decreased to some extent after soaking in sugar syrup. This may be due to the incorporation of sugar syrup in the balls and increase in the weight of *Gulabjamun* that has helped to lower protein content percentage. It was observed that protein content of soy-fortified mix increase might be due to use of higher protein content of defatted soy flour, which was significant in case of 6.66 and 9.99% levels. Bongirwar et al. (1979) studied the development of high protein ready to eat foods from defatted groundnut and soybean blends and observed that defatted soy flour mixed products gave satisfactory structure, colour and appearance. Babje et al. (1992) blended soymilk with buffalo milk for obtaining good quality paneer which showed higher protein content.

There was significant effect of addition of all levels of defatted soy flour on protein content of raw premixes prior to soaking in sugar syrup. There was also a significant difference between the values of protein content for raw premix and fried samples of *Gulabjamuns*. Statistical analysis shows that critical difference (CD) at 5% in raw mix, fried balls and fried balls soaked in syrup were 0.7542, 1.3196 and 0.8803 respectively. Critical variance (CV) % was 1.94, 3.84 and 2.81 in raw mix, fried balls and fried balls dipped in syrup respectively. Pair wise comparisons of protein content for different stages of *Gulabjamuns* found significantly difference between the values of protein content for raw premix and fried samples without dipping in sugar syrup (Table 2). Mean diffe-

Table 2. Pair wise comparisons of protein content in *Gulabjamuns* at different levels of soy flour mix.

Premix I	Premix J	Mean difference (I - J)	Std. error	Sig.F	95% confidence interval	
					Lower bound	Upper bound
A	B	-0.975	1.157	0.461	-8.182	6.232
	C	-6.125*	1.157	0.013	-13.332	1.082
	D	-8.520*	1.157	0.005	-15.727	-1.313
B	C	-5.150*	1.157	0.021	-12.357	2.057
	D	-7.545*	1.157	0.007	-14.752	-0.338
	A	0.975	1.157	0.461	-6.232	8.182
C	B	5.150*	1.157	0.021	-2.057	12.357
	D	-2.395	1.157	0.130	-9.602	4.812
	A	6.125*	1.157	0.013	-1.082	13.332
D	B	7.545*	1.157	0.007	0.338	14.752
	C	2.395	1.157	0.130	-4.812	9.602
	A	8.520*	1.157	0.005	1.313	15.727

Based on estimated marginal means, *The mean difference is significant at the 0.05 level. Note: A- 0% soy flour; B- 3.33% soy flour; C- 6.66% soy flour; D- 9.99% soy flour

Table 3. Fat content in *Gulabjamun* at different levels of soyflour mix.

Soy flour Mix, % & Stat. Parameters	Fat content (%)		
	Raw Mix	Fried balls	Fried balls soaked in syrup
0.00	14.37	30.74	22.12
3.33	11.4	27.38	15.25
6.66	12.1	28.9	15.01
9.99	10.5	26.44	13.86
Mean ± sd	12.09 ± 1.43	28.36 ± 1.63	16.56 ± 3.25
SE_M	0.72	0.81	1.63
CD at 5%	1.05	1.66	0.62
CV (%)	4.54	3.12	2.02

Note: SE_M is standard error of the mean, sd is standard deviation, CD is critical difference and CV is critical variance.

rence with standard error between both were found to be 2.42 ± 0.81.

Efficacy of levels of soy flour on fat content

Fat content increased to 28.36 from 12.09% in deep-fried samples as compared to the raw premix. It is clear that the fat content of *Gulabjamun* decreased with the increase in defatted soy flour level (Table 3). The fat content of raw mix having 3.33, 6.66 and 9.99% levels of defatted soy flour were significantly different from each other. The decrease could be due to a very low fat content of defatted soy flour. Fried *Gulabjamun* soaked in sugar syrup having 0% soy flour (control) had an average fat content of 22.12%. The fat content decrease may be due to diffusion of sugar syrup in fried balls. It was found that there was significant effect of addition of defatted soy flour at all levels on fat content of raw premixes and without sugar syrup soaked samples. Bookwalter et al. (1971) reported that full-fat soy flours prepared by the extrusion process have good nutritive value, flavor and stability. From Tables 4, it is clear that pair wise comparisons of fat content has significant difference between the values of fat content of raw premix and without dipping in sugar syrup.

Textural profile analysis of *Gulabjamuns*

Textural characteristics of fresh, soy-fortified *Gulabjamuns* are given in Table 5. Hardness of *Gulabjamuns* increased with the increase in the levels of soy flour. This increase might be due to the decrease in fat content, increase in protein content and reduction in moisture content (Gulhati et al., 1992). *Gulabjamun* cohesiveness, springiness/elasticity, gumminess, and chewing value were 0.45 g, 3.9 mm, 91.4 and 356.45 g respectively. Cohesiveness, springiness/elasticity, gumminess, and chewing energy values increased with the increase in soy flour level in *Gulabjamun*. Gumminess value increase with the increase of soy flour levels may be due to higher level of hardness in the *Gulabjamuns*. It was found that

Table 4. Pair wise comparison of fat content of *Gulabjamuns* at different levels of soy flour mix.

Premix I	Premix J	Mean difference (I-J)	Std. error	Sig. F	95% confidence interval	
					Lower bound	Upper bound
A	B	2.055*	0.284	0.005	0.285	3.825
	C	3.165*	0.284	0.002	1.395	4.935
	D	4.085*	0.284	0.001	2.315	5.855
B	C	1.110*	0.284	0.030	-0.660	2.880
	D	2.030*	0.284	0.006	0.260	3.800
	A	-2.055*	0.284	0.005	-3.825	-0.285
C	B	-1.110*	0.284	0.030	-2.880	0.660
	D	0.920*	0.284	0.048	-0.850	2.690
	A	-3.165*	0.284	0.002	-4.935	-1.395
D	B	-2.030*	0.284	0.006	-3.800	-0.260
	C	-0.920*	0.284	0.048	-2.690	0.850
	A	-4.085*	0.284	0.001	-5.855	-2.315

Based on estimated marginal means, *The mean difference is significant at the 0.05 levels.
Note: A- 0% soy flour; B- 3.33% soy flour; C- 6.66% soy flour; D- 9.99% soy flour.

Table 5. Effects of soy flour levels on the textural properties of fresh *Gulabjamuns*.

Soy flour mix (%)	Hardness (g)	Cohesiveness (g)	Elasticity(mm)	Chewiness (g)	Gumminess (g)
0.00	203.13	0.45	3.9	356.50	91.40
3.33	296.15	0.51	3.8	573.93	151.04
6.66	364.01	0.53	3.6	694.53	192.92
9.99	492.62	0.67	3.5	1155.19	330.05
Mean ± sd	338.97 ± 105.49	0.54 ± 0.08	3.7 ± 0.15	695.03 ± 291.98	191.35 ± 87.82
SE$_M$	52.75	0.041	0.079	145.99	43.92

Note: SE$_M$ is standard error of the mean, sd is standard deviation.

energy required during mastication also increased with the increase in soy flour levels of *Gulabjamuns*.

Sensory evaluation of *Gulabjamuns*

The average values of appearance, colour, texture, flavor and overall acceptability of freshly prepared *Gulabjamuns* at different levels of soy flour have been given in Table 6. Addition of 3.33% soy flour had improved the appearance and colour of the *Gulabjamuns* and there after it has decreased for both the levels, that is, 6.66 and 9.99% soy flour with little variation. Among the three levels of replacements of soy flour, 3.33% mix had the best appearance and colour become darker with the increase in soy flour level. Flavour was also improved with the addition of 3.33 and 6.66% levels of soy flour but score decreased at 9.99%. However best flavour was obtained in 6.66% soy flour mixed *Gulabjamuns*. Overall acceptability was the average of appearance colour, (p<0.05) between different types of premixes. Overall acceptability was the highest in 6.66% soy flour mix. *Gulabj*It decreased with increase in soy flour level (9.99%) and remained same as control in 3.33% soy flour mix Gulab-

jamun samples. Similar results were also reported by Biswas et al. (2002) and Jenkins et al. (2002). Beneficial effects for health associated consumption of soy products include menopausal symptoms, specifically hot flushes. It was concluded that soy products warrant a greater role in the Western diet on the basis of their potential health benefits accompanied by no apparent disadvantages of their consumption, and that dietitians could help consumers identify suitable soy products to act as replacements for other foods in their diet (Messina, 2003).

Conclusion

It has been concluded that there was significant effect of addition of defatted soy flour at all four levels on protein and fat content of raw premixes as well as fried samples texture and flavour. There was significant difference of *Gulabjamuns*. It was observed from TPA that the hardness, cohesiveness, springiness or elasticity, gumminess and chewiness value of *Gulabjamuns* were increased with the increase in the soy flour levels mix in *Gulabjamun*. Appearance, colour, texture, flavour and overall

Table 6. Sensory evaluations of fresh *Gulabjamuns* of different levels of soy flour.

Soy flour mix, %	Appearance	Colour	Texture	Flavour	Overall acceptability
0.00	8.8	8.7	8.7	8.8	8.75
3.33	9.0	9.0	8.7	8.5	8.80
6.66	8.9	8.8	8.8	9.0	8.90
9.99	8.8	8.7	8.7	8.8	8.75
Mean ± sd	8.875 ± 0.08	8.8 ± 0.12	8.725 ± 0.04	8.775 ± 0.18	8.8 ± 0.06
SE$_M$	0.04	0.06	0.02	0.09	0.03

Note: SE$_M$ is standard error of the mean, sd is standard deviation.

acceptability of the *Gulabjamuns* had improved with the addition of soy flour.

REFERENCES

AOAC (2002a) Method for cereal fat. Introduction to the Chemical Analysis of foods. p 186.

AOAC (2002b) Method for micro kjeldahl method. Introduction to the Chemical Analysis of foods, Association of Official Analytical Chemists, USA, pp 210-211.

Babje JS, Rathi SD, Ingle UM, Syed HM (1992) Studies on effect of Blending soymilk with Buffalo Milk on qualities of Paneer, J. Food .Sci. Technol. 29(2): 119-120.

Berry D (2002) Healthful ingredients sell dairy foods. Dairy Foods 103(3): 54-56.

Biswas PK, Charaborty R, Chouduri UR (2002) Studies on effect of blending of Soy Milk with Cow Milk on Sensory, Textural Nutrition Qualities of Chhana Analogue, J. Food Sci.Technol. 39 (6): 702-704.

Bongirwar DR, Padwal Desai SR, Sreenivasan A (1979) Development of high protein ready to eat foods from defatted groundnut and soybean by extrusion cooking, Indian Food Packer, 33: 37-53.

Book Walter GN, Mustakas GC, Kwolek WF, Ghee JE, Mc. Albrecht WJ (1971) Studies on full-tat say flour Extrusion cooked: Properties and food Uses, J. Food .Sci.Technol. 36: 5-9.

Bourne M (1982) Food texture and viscosity: Concept and measurement, Academic Press. Inc. New York, pp 114-117.

Cumbee B, Hildebrand DF, Addo K (1997) Effects on wheat flour dough rhelogical and bread making properties. J.Food .Sci. Technol. 62 (2): 281-283.

Drake MA, Gerard PD (2003). Consumer attitudes and acceptability of soy fortified yogurts. J Food .Sci. 68 (3): 1118-1122.

Drake MA, Chen XQ, Tamarau S, Leenanon B (2000) Soy protein fortification affects sensory, chemical, and microbiological properties of dairy yogurts. J. Food. Sci. 65(7): 1244-1247.

Drake MA, Gerard PD, Chen XQ (2001) Effects of sweetener, sweetener concentration, and fruit flavor on sensory properties of soy fortified yogurt. J. Sensory Studies 16: 393-406.

Federal Register (1998) Food Labeling: Health claim; soy protein and coronary heart disease. 63 (217), Nov. 10, 1998.

Friedman M, Brandon DL (2001) Nutritional and health benefits of soy proteins. J Agri Food. Chem. 49 :1069-1086.

Gulhati HB, Rathi SD, Syed HH, Bache CS (1992) Studies on qualities of *Gulabjamun*. Indian Food Packer 46 (6): 43-46.

Heller L (2008) How consumers see functional foods. http://www.foodnavigator.com/ Product-Categories/ Dairy-based-ingredients/How-consumers-see-functional-foods, 18-Jan-2008.

Henkel J (2000) Soy: Health claims for soy protein, questions about other components. United States Food and Drug Administration (online). FDA Consumer 34(3).

Jenkins DJA, Kendall CWC, Jackson CJC, Connelly PW, Parker T, Faulkner D, Vidgen E, Cunnae SC, Leiter LA, Josse RG (2002) Effects of high and low-isoflavone soyfoods on blood lipids, oxidized LDL, homocysteine and blood pressure in hyperlipidemic men and women, Am.J. Clin. Nutr. 76(2): 365-72

Khan MI, Anjum FM, Hussain S, Tariq MT (2005) Effect of soy flour supplementation on mineral and phytate contents of unleavened flat bread (chapatis). Nutrition and Food Science, 35 (3):163 – 168, DOI: 10.1108/00346650510594912.

Krummel D (1996) Nutrition in cardiovascular disease. In Krause's Food, Nutrition, and Diet Therapy, 9th Ed, Mahan, K.L.; Escott-Stump S (Ed). W.B. Saunders Co.:Philadelphia, PA. p 509-551.

Liu K (1997) Agronomic characteristics, production, and marketing. In Soybeans: Chemistry, Technology, and Utilization, Liu K. Chapman & Hall: New York, NY. p 1-24.

Locklear M (2000). Food Product Design: Functional Foods, The Road Ahead.. September 7-12, 2000. http://www.foodproductdesign.com/ articles /0900ffa_05.html

Massey LK, Palmer RG, Horner HT (2001). Oxalate content of soybean seeds, soy foods and other edible legumes. J Agri Food Chem 49: 4262-4266.

Messina M (1995). Modern Applications for an Ancient Bean: Soybeans and the Prevention and Treatment of Chronic Disease. First International Symposium on the Role of Soy in Preventing and Treating Chronic Disease. 1994. Feb. 20-23, Mesa, AZ. J Nutr 125:567-569.

Messina M (2003). Soyfoods and disease prevention : Part II - esteoporosis, breast cancer, and hot flushes, Agro Food Industry-Hitechnol. 14(6): 11-13.

National Soybean Research Laboratory (2008). National Soybean Research Laboratory, USA, http://www.nsrl.uiuc.edu/aboutsoy/soy nutrition.html

Plahar WA, Nti LA, Annan NT (1997). Effect of soy-fortification method on fermentation characteristics and nutritional quality of fermented maize meal. Plant Foods Human Nutr. 51(4) : 365-380.

Rangi AS, Minhas KH, Sidhu JS (1985). Indigenous milk products standardization of recipe for Gulabjamun. J Food Sci. Tech. 22: 191-193.

Saxena S, Ramachandran Lata, Singh S, Sharma RS (1996). Preparation and chemical quality of *Gulabjamun* premixes and sensory quality of ready to serve *Gulabjamuns* made from soy flour and milk solids. Indian Food Packer 50 (2): 41-52.

Torres AV, Penaranda, Reitmeier CA, Wilson LA, Fehr WR, Narval JM (1998. Sensory characteristics of soymilk and Tofu made from Lipoxygenase-free and normal soybeans. J. Food .Sci. Tech. 63: 1084-1087.

von Poppel G (1998). Functional foods: Ascent or Abyss? From Functional Foods, A Healthy Future? Presented at the Food Ingredients Europe conference, Frankfort, Germany, at the AACC Symposium, Nov. 4, 1998 American Association of Cereal Chemists (online). http://www.scisoc.org/aacc/FuncFood/top.htm

Wiseman H (1997). Dietary phytoestrogens: disease prevention versus potential hazards, Nutri. Food. Sci. 97 (1):32 – 38, DOI:10.1108/00346659710157303

15

Effects of harvest time and cultivar on yield and physical properties fibers of kenaf (*Hibiscus cannabinus* L.)

Jalal Shakhes[1]*, Farhad Zeinaly[1], Morteza A. B Marandi[1] and Tayebe Saghafi[2]

[1]Department of Wood and Paper Science and Technology, Faculty of Forest and Wood Technology, Gorgan University of Agricultural Sciences and Natural Resources, (Postal code: 15339-95911), Gorgan, Iran.
[2]Department of Forestry Science and Technology, Faculty of Agricultural Sciences and Natural Resources, Tehran University, Karaj, Iran.

Kenaf (*Hibiscus cannabinus* L.) is an annual non wood plant which has shown great potential as an alternative source of papermaking fiber. No information is available on kenaf cultivation in south-Iran in spite of the need to replace imported long fibers through local production of alternative sources. The purpose of this research is to investigate the effect of kenaf cultivars and harvest times on component yield and fiber quality. Six cultivar of kenaf (Cubano, Niger, Cuba 2032, 9277, 7551 and 7566) were planted on May 19, 2007 in the research farm of Agronomy Department Gorgan University. The three harvest times are 85, 105 and 135 days after planting. Result showed that bast yield, stem yield, bast: stem ratio and total dry matter, were affected by harvest time. Bast to stem ratio in second harvest time was more than any other times (40.02). Niger cultivar was the best cultivar for stem yield, bast yield and total dry material, also Cubana 2032 and Cubano were the best cultivar for bast to stem ratio with 40.41 and 40.00% respectively. Fiber morphology results showed that interaction between cultivar and harvest time was significant. The bast fiber length increased with plant age in all cultivars except Cuba 2032. Although core fiber length increased with age in 7551, 7566, 9277 and Cuba 2032 cultivars but it decreased, in Niger and Cubano cultivars. The result indicated that kenaf bast fiber were long and slender, while the core fiber were much shorter and wider. Morphology analysis indicated that bast and core fibers were significantly different. The bast fiber dimension was better than the core in the production of quality paper.

Key words: *Hibiscus cannabinus*, kenaf, harvest time, cultivar, yield, fiber morphology.

INTRODUCTION

The demand for various forest products such as pulp and paper will certainly increase because of growth population and economic development. The principal raw materials for paper manufacturing are wood, non-wood and recycled fibers. Non-wood plants could reduce the shortage of fibrous raw materials for pulp and paper industry (Atchison, 1996; Villar et al., 2001). There is a wide variety of non-wood plant fibers that are being used in the manufacture of pulp and paper all over the world. In 1950s and early 1960s, the United States Department of Agriculture (USDA) researchers selected kenaf as the most promising annual crop source of fiber for the pulp industry; various investigations have focused on the application of farming and pulping techniques. This examinations include new cultivars have been developed with greater yield potential in a wide range of growing seasons; yield and growth performance in different locations and under different cropping situations have been described. There are also other investigations of

*Corresponding author. E-mail: jalalshakhes@yahoo.com.

changes in fiber properties of kenaf at different stages of plant growth. Their results show that it produces a very high fiber yield per hectare and has proved to be a suitable material for pulp and paper industry. (Clark et al., 1967; Muchow, 1979; Han et al., 1995; Mambelli and Grandi, 1995; Ayerza and Coates, 1996; Mcmillin et al., 1998; Bledsoe and Webber, 2001; Nkaa et al., 2007).

Kenaf (*Hibiscus cannabinus* L.) is an annual fast growing non-wood plant which has been regarded as one the most promising and suitable raw material to make a variety of paper grades (Kaldor et al., 1990). As an annual fiber crop, kenaf has high biomass production capacity (14 to 22 t /ha) and a higher pulp yield than those of other non-wood and wood species (Atchison and McGovern, 1983). The kenaf plant consists of two distinct fibers. The external part or bast which is the portion used for cordage fiber is about 35 to 40% of stem by the weight and internal part, woody core which is about 60 to 65%. The fibers in the bast section are long and similar to softwood fibers but the fibers in the core section are shorter than those in the bast and somewhat shorter than those in temperate zone hardwoods (Nkaa et al., 2007). The longer bast fibers are used to manufacture products such as high grade pulps for the pulp and paper industry, composite boards and textiles. The shorter core fibers are used for products such as animal bedding, sorbents and horticultural mixes (Fisher, 1994).

The aims of this study were to determine the suitability of harvest time and kenaf cultivar in north of Iran and to compare components yield and fiber dimension properties in six kenaf cultivars. This information is necessary to admit that kenaf is a feasible alternative for farmers in north Iran where forest resources are inadequate to supply a pulp mill of economic size.

MATERIALS AND METHODS

Cultural details and measurements

This experiment was conducted at Gorgan University (latitude 35.5 north and longitude 54.4 east) on silty clay loam. The characteristics of the soil were analyzed as EC (0.6), pH (7.9), clay (36%), sand (10%) and silt (54%).

Six kenaf varieties have been used in the experiment. Six cultivars namely Cubano, Niger, Cuba 2032, 9277, 7551, and 7566, were used for the study. The field was cleared of bush, plowed and made into beds of 9 × 9 m. Kenaf seeds were planted at a soil depth of 5 cm with a spacing distance of 5 cm between plants and 35 cm between rows to give a total of 66.66 plants/m². In order to ensure optimum crop yield, all plots received 105 kg of urea fertilizer, 85 kg of phosphate diammonium fertilizer, and 125 kg potassium sulfate fertilizer per hectare added and mix with soil.

The crop was shown on 19 May, and five irrigations were applied during the growing season using the traditional furrow system found in the region. The harvest times were 85, 105 and 135 days after planting (DAP). Sample area (1 m²) was harvested from the center rows of each subplot in each harvest time. After each harvest, the plants were separated into a core, bast and leaf fractions. These components were oven-dried at 70°C f or 48 h and

then oven- dried weight of stem, bast, core, total dry material (yield stem add to pluses yield leaf) and ratio of bast fiber to stem material were determined. Also plant height (cm), base stem diameter (mm) and stem diameter at 30 cm (mm) were determined.

Fiber dimensions

Stem samples for the fiber studies were obtained from the approximately fifth internodes counting from the base. For fiber length, fiber width, lumen width and cell wall thickness measurements, the material was macerated by Franklin's method in acetic acid and hydrogen peroxide (1:1) at 60°C for 48 h. The macerated fiber suspension was finally placed on a slid (standard, 7.5 × 2.5 cm) by means of a medicine dropper (Han et al., 1999). All fiber samples were viewed under a projection microscope. For measuring fiber length and diameter, 200 fibers were measured from 10 slides and average reading was taken (Shatalov and Pereira, 2001).

Experimental designed statistical analysis

The experiment was based on a split plot design. Three harvest time the main plot and the six Kenaf varieties as the sub-plots. Data obtained were subjected to analysis of variance and means separated using the Duncan multiple range tests.

RESULTS AND DISCUSSION

Growth parameters

Harvest time and cultivars significantly affected all agronomic traits measured except steam diameter at 30 cm (Tables 1 and 2). At 85 to 135 days after sowing, the stem yield increased from 3.5 to 8 ton ha^{-1}, and the bast yield production from 1.3 to 3 ton ha^{-1} and the plant height from 92 to 140 cm. In addition, stem base diameter increased significantly with the plant age while stem ratio decrease from 39.41 to 37.1% between the first and the third harvest time (Table 2).

Niger cultivar with 7.170 ton/ha had the highest fiber yield while 9277 cultivar with 4.851 ton/ha had the lowest yield. The ratio of bast to stem fibers was varied by cultivar. Cultivars Cuba 2032 and Cubano had the highest percentage of bast fiber and 7551 had the high percentage of core fiber (Table 1). Statistical analysis showed that there was no significant interaction between harvest time and cultivars.

Fiber dimensions

The effect of harvest time on dimensional properties in six cultivars is shown in Tables 3 and 4. In Table 3, the bast fiber length increased with age except Cuba 2032 cultivar which decreased with age. The longest fiber was from Cubano cultivar with an average of 3.14 mm in the third harvest. The shortest fiber was from 9277 cultivar,

Table 1. Effect of cultivar on yield parameters of six Kenaf cultivars.

Cultivar	Stem yield (kg/ha)	Bast yield (kg/ha)	Total dry material (kg/ha)	Bast: stem ratio	Plant height (cm)	Base stem diameter (mm)	Stem diameter at 30 cm (mm)
7551	6411.7[ab]	2370.6[ab]	10134.7[abc]	37.7[c]	121.7[ab]	8.6[ab]	6.9
7566	6248.3[abc]	2371.3[ab]	10488.8[ab]	38.2[c]	125.5[a]	8.3[b]	6.9
Cubano	5806.0[bc]	2293.8[b]	9516.4[bc]	40.0[ab]	117.2[b]	8.3[b]	6.5
Niger	7170.7[a]	2759.4[a]	11833.3[a]	38.7[bc]	124.0[a]	8.8[a]	6.9
Cuba 2032	5300.8[cd]	2138.0[bc]	8834.5[c]	40.4[a]	112.9[c]	8.6[ab]	6.6
9277	4851.2[d]	1824.7[c]	8207.6[c]	38.3[c]	113.9[c]	7.6[c]	6.6

Table 2. Effect of harvest time on yield parameters of six Kenaf cultivars.

Fiber/ age (days)	Stem yield (kg/ha)	Bast yield (kg/ha)	Total dry material (kg/ha)	Bast: stem ratio	Plant height (cm)	Base stem diameter (mm)	Stem diameter at 30 cm (mm)
At 85 day after planting	3506.7[c]	1378.8[c]	6276.2[c]	39.4[ab]	92.9[c]	7.3[c]	5.5[b]
At 105 day after planting	6196.6[b]	2460.8[b]	10346.2[b]	40.0[a]	124.6[b]	8.4[b]	7.1[ab]
At 135 day after planting	8190.9[a]	3039.3[a]	12875.3[a]	37.2[b]	140.1[a]	9.4[a]	7.4[a]

with average of 2.40 mm in the first harvest (Table 3). Cubano and Niger core bast fiber lengths were the least and both decreased with age while other cultivars increased with age. The longest core fiber was from 9277 cultivar with 0.88 mm length and the shortest core fibers from Cubano cultivar with 0.68 mm length at the third harvest.

The fiber diameter in core was much more than the bast fibers diameter in all cultivars. In most cultivars, bast and core fiber diameter increased with age with the exception of Cubano and Cuba 2032 cultivars (Table 3).

Both lumen width and cell wall thickness increased age in core tissue, additionally lumen width increased with age in bast fiber excepted Niger cultivar besides cell wall thickness increased with age of bast tissue except Cubano and Cuba 2032 cultivars. Finally, the lumen width is greater in core fibers as compared to that of bast fibers (Table 4).

DISCUSSION

This experiment has shown that kenaf productivity was highly affected by harvest time and cultivar. Fiber yield increased with growing season. These results were consistent earlier reports that fiber yield increased with plant age (Ayerza and Coates, 1996; Bledsoe and Webber, 2001).

Stem yield was positively affected with stem diameter and stem height. These results were in agreement with previously reports that stem diameter increased with increasing plant height and plant height is an important factor as a contributor to total or stem dry matter (Dempsey, 1975; Webber, 1993; Mambeli and Grandi,

1995; Mcmillin et al., 1998; Rouxlene, 2004). We found that fiber yield (bast, core and total dray material) were more closely relationship with stem diameter and plant height. These factors may have supported the yield difference found among the various cultivars.

The data presented in Table 2 indicates that during the growing cycle, the bast: stem ratio first increased, then decreased, these results were consistent with earlier works that the ratio of core fiber to bast fiber increased with age (Han et al., 1999). We found that bast: stem ratio were significantly affected by harvest time and cultivar. However, in contrast to conclusions of Muchow (1979), that revealed bast content is improved by irrigation and not affected by cultivar. The bast: stem ratio of cultivars Cubano and Cuba 2032 was similar to that reported by Ayerza and Coates (1996).

The fiber dimensional properties of both kenaf core and bast were affected by harvest time and cultivar interaction (Tables 3 and 4). There was a big difference in lengths of fibers coming from the bast and core tissues. The bast fibers average length (2.4 to 3.1 mm) is in the range of soft woods, whilst the core fiber length (from 0.68 to 0.88 mm) is like that of hardwoods (Miller, 1965; Nkaa et al., 2007). Therefore, papers made from kenaf bast fibers are expected to have increased mechanical strength and thus be suitable for writing, printing, wrapping, and packaging purposes (Ververis et al., 2004; Shakhes et al 2011). The length of bast fibers was found to be double the length of the core fibers. Fiber length and strength have been shown to be particularly important for tearing resistance (Wangaard and Williams, 1970). The results of this experiment show that bast fiber length increases with plant age, except in cuba 2032 cultivar where bast fiber length decreases with plant age.

Table 3. Effect of harvest time on fiber length and fiber diameter of six Kenaf cultivars.

Fiber/age (Days)	Variety/ fiber type	Fiber/age (Days)			
		At 85 day after planting	At 105 day after planting	At 135 day after planting	Mean
		7551			
	Core	0.763	0.810	0.852	0.808
	Bast	2.441	2.583	2.792	2.605
		7566			
	Core	0.828	0.834	0.837	0.833
	Bast	2.504	2.651	2.649	2.601
		Cubano			
	Core	0.755	0.699	0.682	0.712
	Bast	2.437	3.054	3.141	2.877
Fiber length		Niger			
	Core	0.778	0.752	0.721	0.750
	Bast	2.807	2.705	2.770	2.760
		2032			
	Core	0.785	0.759	0.815	0.763
	Bast	2.712	2.545	2.521	2.592
		9277			
	Core	0.759	0.809	0.880	0.816
	Bast	2.403	2.411	2.480	2.431
		7551			
	Core	23.04	25.64	28.81	25.83
	Bast	16.19	16.63	17.12	16.64
		7566			
	Core	20.37	22.27	24.81	22.48
	Bast	15.82	16.04	16.98	16.28
		Cubano			
	Core	24.55 [c]	26.42	28.64	26.53
	Bast	15.83	16.91	15.95	16.23
Fiber diameter		Niger			
	Core	24.30	26.85	29.89	27.01
	Bast	17.12	17.38	17.99	17.49
		2032			
	Core	24.53	25.71	27.29	25.84
	Bast	17.44	17.17	16.44	17.01
		9277			
	Core	21.91	22.61	24.36	22.96
	Bast	16.15	16.33	16.49	16.32

Clark et al. (1967), demonstrated that the bast fibers are longer than the core fibers and that both decreases in length with age. However, in contrast to findings of Han et al. (1995), that average length of bast and core fiber increased with age. Another study described that irrigation and harvest time had no significant effect on bast fiber morphological properties while cultivar had a significant influence on fiber length and wall thickness (Villar et al., 2001). In general, these findings indicate that

of both harvest time and cultivar was significant on the fiber dimensional properties.

The thickness of the fiber wall has an important bearing on most paper properties, with thick-walled fibers forming bulky sheets of low tensile, burst, and folding endurance but with a high tearing strength (Haygreen and Bowyer, 1996). The strength properties of papers were found to positively correlate with the slenderness ratio (fiber length/ fiber diameter). Lumen width and fiber diameter of

Table 4. Effect of harvest time on fiber lumen and cell wall thickness of six Kenaf cultivar.

Fiber/age (Days)	Variety/fiber type	Fiber/age (Days)			
		At 85 day after planting	At 105 day after planting	At 135 day after planting	Mean
		7551			
	Core	15.08	17.32	20.15	17.51
	Bast	4.51	5.64	5.95	5.36
		7566			
	Core	14.48	15.84	17.41	15.91
	Bast	5.30	5.18	6.53	5.67
		Cubano			
	Core	17.62	19.10	21.05	19.25
Fiber lumen diameter	Bast	4.01	6.36	6.02	5.46
		Niger			
	Core	18.65	20.54	23.03	20.74
	Bast	7.63	6.28	5.37	6.42
		2032			
	Core	16.85	17.48	17.38	17.23
	Bast	6.18[a]	5.74	4.56	5.49
		9277			
	Core	15.26	14.95	15.24	15.15
	Bast	4.66	5.02	5.24	4.97
		7551			
	Core	3.98	4.15	4.33	4.15
	Bast	5.84	5.49	5.58	5.63
		7566			
	Core	2.94	3.21	3.70	3.28
	Bast	5.26	5.42	5.22	5.30
		Cubano			
	Core	3.65	3.65	3.79	3.69
Fiber wall thickness	Bast	5.91	5.27	4.96	5.38
		Niger			
	Core	2.82	3.15	3.43	3.13
	Bast	4.74	5.55	6.31	5.33
		2032			
	Core	3.84	4.11	4.95	4.30
	Bast	5.63	5.71	5.94	5.76
		9277			
	Core	3.32	3.83	4.56	3.90
	Bast	5.74	5.67	5.26	5.55

cultivar was similar to that reported by Clark et al. (1967). Results showed that core fiber wall thickness is lower than bast fiber wall thickness and this finding is in contrast to conclusion of Clark et al. (1967).

In general these finding exhibited that the variability of bast fiber dimension in some cultivars may be due to their different reactions to environmental conditions. Unlike bast fibers, core fibers are less affected by environmental conditions and show almost constant behavior.

Conclusions

This research confirms the possibility of producing kenaf for fiber and other products in north Iran. Also a conception of the changes in yield and quality of kenaf fiber during the growing season is essential for determining suitable variety for pulp and paper industry. Our result indicates that kenaf fiber yield was significantly affected by harvest time and cultivar. Fiber yield increases with plant age. Among the cultivars, Niger was better than other cultivars from the crop yield point of view.

Result based on morphology display that bast and core fiber fractions in kenaf stem are significantly different. And both core and bast fiber were affected by harvest time and cultivar interaction. In general, the bast material has a higher fiber length, lower fiber diameter, higher cell wall thickness and lower lumen width and would be expected to give better paper properties than core material.

Harvest time seems to play an important role in determining not only the stem yield but also Fiber properties. The data strongly suggest that planting of kenaf in Gorgan (Iran) earlier (early spring) could increase growth and result in yield increase in Grogan zone.

ACKNOWLEDGMENTS

The authors thank M. Tassoji, and M. Mirmehdi at the University of Gorgan. We also acknowledge Department of Agronomy Gorgan University for use of the research farm.

REFERENCES

Atchison JE, McGovern JN (1983). History of paper and the importance of non-wood plant fibers. In: Secondary fibers and non-wood pulping. Third ed. Hamilton F, Leopold B, Tech. Eds., Pulp and paper manufacture, Vol. 3. Joint Textbook Committee of the Paper Industry, Atlanta, GA, Montreal: 3.

Atchison JE (1996). Twenty-five years of global progress in non-wood plant fiber pulping. Tappi J., 79(10): 87-95.

Ayerza R, Coates W (1996). Kenaf performance in northwestern Argentina. Ind. Crops Prod. J., 5: 223-228.

Bledsoe VK, Webber CL (2001). Crop maturity and yield component. 5th natl. symp., New Crops and New Uses: Strength in Diversity, Atlanta, GA: 64.

Clark TF, Uhr SC, Wolff IA (1967). Search for new fiber crops, kenaf storage. Tappi J., 50(11): 52.

Dempsey JM (1975). Fiber crops. The University Presses of Florida, Gainesville, FL.

Fisher G (1994). Availability of kenaf fibers for the U.S. paper industry. TAPPI pulping conference proceeding, book 2, November 6-10, Sheroton harbor island, San Diego, California, Tappi press, Atlanta, GA, USA: 91-94.

Han JS, KIM W, Rowell RM (1995). Chemical and physical properties of kenaf as a function of growth. Proceeding of the seventh annual international kenaf association conference. Irving TX: 63-83.

Han JS, Minanowaski T, Lin Y (1999). Validity of plant fiber length measurement -A review of fiber length measurement based on kenaf as a model. In: Kenaf Properties, Processing and Products, Chapter 14, Mississippi State University. Mississippi State, MS: 149-167.

Haygreen JG, Bowyer J (1996). Forest Products and Wood Science: An introduction, Third Edition, Iowa University Press.

Kaldor AF, Karlgren C, Verwest H (1990). Kenaf- A fast growing fiber source for papermaking. Tappi J., 73(11): 205-208.

Mambelli S, Grandi S (1995). Yield and quality of Kenaf stem as affected harvest date and irrigation. Ind. Crops Prod., 4: 97-104.

Mcmillin JD, Wanger MR, Webber III CL, Mann SS, Nichols JD, Jech L (1998). Potential for kenaf cultivation in south-central Arizona. Ind. Crops Prod., 9: 73-77.

Miller DL (1965). Kenaf- A potential paper making raw material. Tappi J., 48(8): 455-459.

Muchow RC (1979). Effect of plant population and season on Kenaf grown under irrigation in tropical Australia I. Influence on the component of yield. Field Crops Res., 2: 55-66.

Nkaa FA, Ogbonnaya CI, Onyike NB (2007). Effect of differential irrigation on physical and histochemical properties of kenaf (Hibiscus cannabinus L.) grown in the field in Eastern Nigeria. Afr. J. Agric. Res., 2(6): 252-260.

Rouxlene C (2004).Characterization of Kenaf cultivars in South Africa. Faculty of Natural and Agricultural Sciences Department of plant Sciences. M.Sc. Thesis. Plant Breeding University of the State Bloemfontein, 128p.

Shatalov AA, Pereira H (2001). Arundo donax l. reed: new perspectives for pulping and bleaching - 1. Raw material characterization. Tappi J., 84(1): 1-12.

Shakhes J, Zeinaly FAB, Marandi M, Saghafi T (2011). The effects of processing variables on the Soda and Soda-AQ pulping of Kenaf bast fiber. BioResources, 6(4): 4481-4493.

Ververis C, Georghiou K, Christodoulakis N, Santas P, Santas R (2004). Fiber dimensions, lignin and cellulose content of various plant materials and their suitability for paper production. Ind. Crops Prod., (19): 245-254.

Villar JC, Poveda P, Tagle JL (2001). Comparative study of kenaf varieties and growing condition and their effect on Kraft pulp quality. Wood Sci. Technol., 34(6): 543-552.

Webber III CL (1993). Yield components of five kenaf cultivars. Agron. J., 85(3): 533-535.

Wangaard FF, Williams DL (1970). Fiber length and fiber strength in relation to tearing resistance of hardwood pulp. Tappi J., 53(11): 2153.

16

Application of acid-hydrolyzed cassava (*Manihot esculenta*) and cowpea (*Vigna unguiculata*) for the production of yeast (*Saccharomyces cerevisiae*)

Nyerhovwo J. Tonukari* and Linda I. Osumah

Department of Biochemistry, Faculty of Science, Delta State University, P. M. B. 1, Abraka, Delta State, Nigeria.

Much progress has been made in the cultivation and production of cassava (*Manihot esculenta*) and cowpea (*Vigna unguiculata*) in Nigeria. In the present study, investigation was carried out on the possibility of using cassava flour as a source of glucose as well as cowpea as source of nitrogen in the production of yeast. Acid hydrolysis (using dilute H_2SO_4) of cassava and cowpea was undertaken to release the sugars and amino acids. The pH of the growth medium using hydrolyzed cassava as carbon source and cowpea as nitrogen source was varied from pH 2.5 - 6.5. The results obtained show that pH 6.5 gave optimum yeast biomass. The hydrolyzed cassava was also varied in the growth medium. The result obtained shows that increased concentrations of acid-hydrolyzed cassava increased yeast biomass, indicating that hydrolyzed cassava is a good carbon source of glucose for yeast production. It was also observed that yeast biomass using acid hydrolyzed cowpea extract as nitrogen source was high. This is due to the fact that cowpea contains 66.35% of carbohydrate in addition to about 25% protein and hence a good source of carbon and nitrogen in the culture medium. The residual glucose concentration of the yeast culture for each medium was also determined. The result obtained indicates that with increased yeast biomass, there was significant decrease in the residual glucose. Also, there was a significant decrease in pH of the culture media following yeast culture; the culture media tends to be acidic after yeast culture. Therefore, yeast can be produced using acid hydrolyzed cassava flour as carbon source with cowpea as nitrogen source.

Key words: Cassava, cowpea, yeast, acid hydrolysis, glucose, nitrogen source.

INTRODUCTION

Yeasts (*Saccharomyces cerevisae*) can grow in the presence or absence of air. Anaerobic growth which is growth in the absence of oxygen is quite slow and inefficient. For instance, in bread dough, yeast grows very slowly. In this case, the sugar that can sustain either fermentation or growth is used mainly to produce alcohol and carbon dioxide. Only a small portion of the sugar is used for cell maintenance and growth. In contrast, under aerobic conditions, in the presence of a sufficient quantity of dissolved oxygen, yeast grow by using most of the available sugar for growth and producing only negligible quantities of alcohol (Bekatorov et al., 2006). Studies show that organic nitrogen sources such as yeast extract support rapid growth and high cell yields of microorganisms because it contains amino acids and peptides, water soluble vitamins and carbohydrates (Peppler, 1982; Watson, 1976).

The basic carbon and energy source for yeast culture are sugars (Dubai and Muhammad, 2005). Starch cannot be used because yeast does not contain the appropriate enzymes to hydrolyze this substrate to fermentable sugars. Beet and cane molasses are commonly used as raw materials because the sugars present in molasses, a mixture of sucrose, fructose and glucose, are readily fermentable (Ohara et al., 1992). In addition to sugar, yeast also requires certain minerals, vitamins and salts for growth. Some of these can be added to the blend of beet and cane molasses prior to flash sterilization while others are fed separately to the fermentation (Stanbury et al., 1995). Required nitrogen is supplied in the form of ammonia and phosphate is supplied in the form of phosphoric acid (Zheng, 2005).

*Corresponding author. E-mail: tonukari@gmail.com.

Cassava (*Manihot esculenta*) ranks very high among crops that convert the greatest amount of solar energy in soluble carbohydrates per unit area of land. It is grown for its large starch filled roots, which contain nearly the maximum theoretical concentration of starch on a dry weight basis among food crops. Among the starchy stables, cassava gives a carbohydrate production which is about 40% higher than rice and 25% more than maize (Tonukari, 2004; Akinfala et al., 2002; Nwokoro et al., 2002). Fresh cassava contains very little protein and fat (Okezie and Kosikowski, 1982). The approximate compositions of the cassava tuber are: moisture (75 – 80%), starch (20 – 30%), protein (2 – 3%), ash (1.15%), fibre (1.0%) and fat (0.1%) (Ihekonronye and Ngoddy, 1985; Alais and Linden, 1999). The future demand for fresh cassava mainly depends on improved storage methods even as the market for cassava as a substitute for cereal flours in bakery products and as energy source in animal feed ratios are likely to expand (Tonukari, 2004).

Cowpea (*Vigna unguiculata*) belongs to the legume family (fabaceae – legumunosae). It is an annual legume which is commonly referred to as blackeye bean, blackeye pea (Quinn and Myers, 2002). Cowpea is one of the world's important legume food crops (Bressani, 1985; Awonaike et al., 2001). At least 12.5 million hectares of cowpea are cultivated with an annual production of over 3 million metric ton worldwide. Development of new cultivators with early maturity, acceptable grain quality and resistance to important diseases and pest has significantly increased (Miller, 1998). The international institute of tropical Agriculture (IITA) has been working on the improvement of cowpea for more than 30 yrs and over 60 countries receive cowpea cultivars improved by IITA for testing and adoption where needed (Davis et.al., 1991). Cultivars which are developed by the IITA in Nigeria are now widely grown in over 60 countries and the production has increased up to 341 from 1961 to 1995 in Nigeria.

The industrial and commercial use of yeast started at the end of the 19th century after it was identified and isolated by Pasteur (Broach and John, 1991). Yeasts are included in starter culture for the production of specific types of fermented foods like cheese, bread, fermented meat, vinegar and vegetable products (Gilland, 2002). *Saccharomyces cerevisiae* and other yeasts have industrial and medical applications beneficial to human life. *S. cerevisiae* also known as baker's yeast is the most common yeast, which is used worldwide for the production of bread and baked products (Corriher, 2001). The need to design feasible and financially viable processes and the ultilization of low-cost raw materials for edible yeast biomass production is extremely important (Albers et al., 1996). The purpose of this project work is to develop a rapid method for producing glucose and simple sugars from cassava as well as amino acids from cowpea through acid hydrolysis and combining this for yeast production. This will enhance the use of locally abundant agricultural products (cassava and cowpea) for the industrial production of yeast.

MATERIALS AND METHODS

Cassava and cowpea

Cassava tubers were purchased from Abraka market in Delta State, Nigeria. The tubers were peeled and thoroughly washed (to reduce the cyanide content), sliced into small pieces, dried under the sun, ground to flour and passed through a sieve of 0.25 mm before being stored in a dry plastic container, ready for use. Cowpea seeds also known as beans were also purchased from Abraka market in Delta State. They were ground to flour and passed through a sieve of 0.25 mm before it was stored in a dry plastic container, ready for use.

Acid hydrolysis of cassava

To get the best hydrolysis of cassava flour, the weights of cassava flour was varied at constant acid volume. 5 to 25 g of cassava flour was weighed into six (6) conical flasks. To each of the conical flask, was added 0.5 ml of 0.5% (v/v) H_2SO_4 acid. The conical flask was then made up to 100 ml with distilled water and shaken thoroughly to have an even mix. It was autoclaved at about 115°C for 20 min, cooled and filtered into different sterilized test tubes as hydrolyzed glucose (sample glucose) and stored in a cool dry place.

Acid hydrolysis of cowpea

This was done to get the best hydrolysis of cowpea flour by varying the weights of cowpea flour used at constant acid volume. 5 - 30 g of cowpea flour was weighed into six (6) conical flasks. To each of the conical flask was added 0.5 ml of conc. H_2SO_4 acid. The conical flask was then made up to 100 ml with distilled water and shaken thoroughly to have even mix. The conical flask was then autoclaved for 30 min, cooled and filtered into different sterilized test tubes using filter paper. It was then stored in a cool dry place.

Glucose estimation

The sample glucose estimation was carried out to determine the optimum hydrolysis of cassava flour. This was done using glucose estimation kit (Randox Laboratories Ltd, Diamond Road, Crumlin, Co. Antrim United Kingdom, BT29 4QY). Five test tubes were labeled test tubes 1 - 5 with tube 1 serving as the control. To each of the test tubes, 100 μl of hydrolyzed glucose filtrate was added, 2.0 ml of distilled water and 2.0 ml glucose buffer was also added to the test tubes. This was incubated for 5 min at room temperature and optical density (O.D) taken using the spectrophotometer at 500 nm to estimate the amount of glucose as well as the optimum hydrolysis of cassava flour (Barham and Trinder, 1972).

Yeast tablets

Yeast (*S. cerevisiae*) tablets were purchased from a chemist in Abraka, Delta State, Nigeria.

Preparation of yeast culture (YPD media) for inoculation

To prepare an uncontaminated yeast culture which was used for

Table 1. Proximate analysis of cowpea and cassava flours.

Parameter	Cowpea	Cassava
Lipid	2.23 ± 0.153	0.92 ± 0
Crude protein	25.38 ± 0.290	1.93 ± 0.03
Ash	3.66 ± 0.025	4.83 ± 0.106
Fibre	2.35 ± 0.132	1.99 ± 0.014
NFE (carbohydrate)	66.35 ± 0.300	90.34 ± 0.163
Moisture	8.4 ± 0.158	6.40 ± 0.141
Energy (KCal)	387.01 ± 1.140	377.34 ± 0.481

Values are mean ± standard deviation of triplicate experiments; values are in % (except for energy); NFE, nitrogen free extract.

inoculation of different reaction media for yeast production, 2 g peptone, 2 g D-glucose and 1 g yeast extract were weighed into a 250 ml autoclaved conical flask. The content was mixed with a little amount of distilled water and the solution was made up to 100 ml with distilled water. The content was autoclaved for 30 min; it was brought out, cocked with cotton wool to make it air tight and then autoclaved again for another 30 min. After which, it was brought out and allowed to cool at room temperature. After cooling, 100 µl of 20% (w/v) antibiotics (ampiclox) was added to the solution to prevent bacteria growth before adding two tablets of yeast. Its content was shaken very well to allow the tablet dissolve and the culture was incubated for 6 - 12 h before using the culture for inoculation of other substrate for yeast culture.

Yeast culture with varying acid-hydrolyzed cowpea extract

To determine the yeast biomass using acid hydrolyzed cassava as carbon source and acid-hydrolyzed cowpea extract as nitrogen source, 20 ml of the acid-hydrolyzed cassava was measured into six (6) conical flasks having the glucose control flask; the acid-hydrolyzed cowpea extract control flask containing 10 ml of acid-hydrolyzed cowpea extract and four other flasks containing 2.5, 5.0, 7.5 and 10.0 ml of acid-hydrolyzed cowpea extract, respectively. To the acid-hydrolyzed cowpea extract control flask, there was no addition of 20 ml acid hydrolyzed cassava and no cowpea extract was added to the glucose control flask. The pH in the conical flasks was adjusted to 6.5. This was done using the pH meter and NaOH solution (1.0 M) as alkaline medium. The total volume in each conical flask was made up to 100 ml with distilled H$_2$O. The flasks were then autoclaved for 30 min, allowed to cool, corked with cotton wool and re-autoclaved for another 30 min. After autoclaving the second time, it was allowed to cool and 100 µl of 20% (w/v) antibiotics (ampiclox) was added to each flask to prevent bacteria growth. 2 ml of the incubated yeast culture was then used to inoculate each of the conical flasks. The flask was shaken very well and allowed to grow for 36 h. After the growth period, the optical density (OD) was taken at 600 nm to determine the yeast biomass.

Yeast culture with varying acid-hydrolyzed cassava

To determine the yeast biomass using varying amount of hydrolyzed cassava as carbon source with acid-hydrolyzed cowpea extract as nitrogen source, a varied amount of the acid-hydrolyzed cassava was added from 2.5 to 20 ml into different conical flasks (flasks 2 – 9). To flask one, there was no addition, serving as control. 5% of acid-hydrolyzed cowpea extract was then added to the different conical flasks. After these additions, the subsequent procedures following the yeast determination is as previously

described. In estimating the glucose content of the yeast culture using varied acid-hydrolyzed cassava, nine test tubes were used and labeled test tubes 1 – 9. All other procedures were followed as previously described. After the yeast biomass has been determined, the pH readings were also determined as previously described.

Yeast culture with varying pH

To determine the yeast biomass with varying pH using acid hydrolyzed cassava as carbon source and acid hydrolyzed cowpea as nitrogen source, 20 ml of the acid-hydrolyzed cassava was measured into six (6) conical flasks followed by the addition of 5 ml of acid-hydrolyzed cowpea extract to the conical flasks. The pH of contents of the conical flasks labeled 2-6 was adjusted to 2.5, 3.5, 4.5, 5.5, and 6.5, respectively, with the exception of conical flask 1, which served as the control. This was done using the pH meter and NaOH (1.0 M) solution as alkaline medium. The total volume in each conical flask was made up to 100 ml with distilled H$_2$O after adjusting the pH. It was then autoclaved for 30 min, allowed to cool, corked with cotton wool and re-autoclaved for another 30 min. After autoclaving the second time, it was allowed to cool and 100 µl of 20% (w/v) antibiotics (ampiclox) was added to each flask to prevent bacteria growth. 2 ml of the incubated yeast culture was used to inoculate each of the conical flasks. The flask was shaken very well and allowed to grow for 36 h. After the growth period, the optical density (OD) was taken at 600 nm to determine the yeast biomass.

RESULTS

Proximate analysis

Prior to the commencement of yeast culture using locally available raw materials, proximate analysis of experimental samples was carried out to determine the biochemical components of cassava and cowpea. The result of the analysis carried out on the lipid, crude protein, ash, fibre, nitrogen free extract (NFE), moisture and energy level of cassava and cowpea is as shown in Table 1.

Acid hydrolysis of cowpea flour

Acid hydrolysis of the cowpea flour was also carried out

Table 2. Glucose estimation after acid-hydrolysis of cassava.

Acid-hydrolyzed cassava	5%	10%	15%	20%	25%
O.D (500 nm)	1.757	1.723	1.707	1.785	1.268
Glucose concentration (mmol/L)	9.800	9.611	9.521	9.956	7.072
Amount of glucose (g)/100 ml	1.764	1.730	1.714	1.792	1.273

Table 3. Yeast biomass estimation after 36 h culture* with varying amount of acid-hydrolyzed cassava.

Acid-hydrolyzed cassava (%)	Acid-hydrolyzed cowpea (%)	Yeast biomass (O.D 600 nm)**	Residual glucose (mmoles/L)**	pH after 36 h of yeast culture
0	10	1.023 ± 0.006	1.473 ± 0.002	5.26
2.5	10	1.132 ± 0.006	1.445 ± 0.002	5.3
5.0	10	1.149 ± 0.002	1.210 ± 0.002	5.28
7.5	10	1.167 ± 0.004	1.244 ± 0.003	5.25
10.0	10	1.185 ± 0.005	1.149 ± 0.001	5.22
12.5	10	1.206 ± 0.003	1.132 ± 0.002	4.92
15.0	10	1.213 ± 0.004	1.132 ± 0.002	4.79
17.5	10	1.225 ± 0.001	1.032 ± 0.001	4.67
20.0	10	1.275 ± 0.006	1.071 ± 0.001	4.62

*The medium contained varying amount (%) of 20% acid-hydrolyzed cassava (20 ml of acid-hydrolyzed cassava contains 0.36 g glucose) and 5% poultry manure extract. The pH in the conical flasks was adjusted to 6.5. The total volume in each conical flask was made up to 100 ml with distilled H_2O and autoclaved for 30 min. 100 µl of 20% antibiotics (ampiclox) was added and inoculated with 2 ml of yeast culture ($OD_{600\,nm} = 1$).
**Values are mean ± standard deviation of triplicate experiments.

to determine the best hydrolysis of cowpea flour and from the results obtained, 10% cowpea flour gave the optimum hydrolysis using 0.5% concentrated H_2SO_4 with 30 min autoclaving. It was easier to filter compared to higher cowpea flour amounts. Thus, subsequent analysis for yeast production was carried out using 10% cowpea flour for acid hydrolysis, which was cooled and stored as cowpea extract serving as nitrogen source for yeast culture.

Acid hydrolysis of cassava flour

Acid hydrolysis of the cassava flour was carried out to determine the best hydrolysis of cassava flour varying the weights of cassava flour from 5 to 25 g at constant acid volume. At the end of the hydrolysis, the hydrolyzed glucose (sample glucose) was filtered into different sterilized test tubes and stored in a cool dry place. Glucose estimation of the varied hydrolyzed cassava was carried out as well as standard glucose estimation in comparison. This was to test for the optimal hydrolysis of cassava. The glucose estimation was carried out using glucose kit and the result shows that 20% of the cassava flour gave the optimum hydrolysis as shown in Table 2. This gave the highest amount of glucose on hydrolysis. Thus, subsequent analysis for yeast production was carried out using 20% cassava flour for acid hydrolysis and the glucose produced was used for yeast production.

Yeast biomass estimation with varying acid-hydrolyzed cassava

The yeast biomass estimation with varied acid-hydrolyzed cassava (glucose) was aimed at determining the yeast biomass level using various percentage of acid-hydrolyzed cassava (glucose). The amount of residual glucose and the pH level (whether acidic or alkaline) of the yeast culture was also determined. The result is shown in Table 3.

From Table 3, it was observed that, as the percentage of acid-hydrolyzed cassava increased from 0 – 20%, there was also a concurrent increase in the yeast biomass. The yeast biomass observed in the control is due to the extra carbon content contained in the cowpea. After the yeast culture had been incubated at the specified period, the residual glucose concentration as well as the pH of the media after growth was measured. From Table 3, it was observed that yeast biomass is inversely related to the residual glucose after yeast culture. Thus, indicating that, the more the yeast biomass, the lesser the residual glucose after yeast culture while the lesser the yeast biomass, the more the residual glucose after yeast culture.

Also measured was the pH of the yeast culture, which was seen to be acidic (reduced pH). The reduced pH observed indicates that, after the optimal growth of yeast, the media in which the growth occurred becomes more acidic because of the production of organic acids like

Table 4. Yeast biomass estimation after 36 h culture* with varying acid-hydrolyzed cowpea.

Acid-hydrolyzed cassava (%)	Acid-hydrolyzed cowpea (%)	Yeast biomass (O.D 600 nm)*	Residual glucose (mmoles/L)*	pH after 36 h of yeast culture
20	0	0.507 ± 0.002	5.438 ± 0.002	4.39
20	5	0.535 ± 0.003	7.212 ± 0.008	4.44
20	10	0.693 ± 0.002	3.984 ± 0.002	4.46
20	15	0.726 ± 0.006	3.720 ± 0.002	4.48
20	20	0.746 ± 0.009	3.564 ± 0.004	4.49
0	20	0.684 ± 0.014	0.039 ± 0.000	4.51
YPD		0.992 ± 0.002	7.351 ± 0.045	4.79

*The medium contained 20 ml of the 20% acid-hydrolyzed cassava (20 ml of acid-hydrolyzed cassava contains 0.36 g glucose) and 5% poultry manure extract with varying pH. The total volume in each conical flask was made up to 100 ml with distilled H_2O and autoclaved for 30 min. 100 µl of 20% antibiotics (ampiclox) was added and inoculated with 2 ml of yeast culture ($OD_{600\,nm}$ = 1).
** Values are mean ± standard deviation of triplicate experiments; **YPD,** Yeast peptone dextrose culture medium containing 2% glucose.

lactic and malic acids (Roble et al., 2003).

Yeast biomass estimation with varying acid-hydrolyzed cowpea

The yeast biomass estimation analysis was aimed at determining the amount of yeast as well as determining the amount of residual glucose and the pH level (whether acidic or alkaline) of the yeast culture. This was done using the acid-hydrolyzed cassava (glucose) as carbon source and cowpea extract as nitrogen source. The result obtained is shown in Table 4.

Yeast grows very well in the presence of carbon and nitrogen. From Table 4, it is observed that, under the normal yeast culture (YPD), the yeast grew optimally compared to when other carbon (acid-hydrolyzed cassava filtrate – glucose) and nitrogen sources (acid-hydrolyzed cowpea) were used. Using constant percentage of acid-hydrolyzed cassava filtrate as carbon source together with varied percentage of acid-hydrolyzed cowpea as nitrogen source, increased yeast biomass was observed as the percentage of the cowpea extract increases (Table 4). This indicates that, Yeast grows very well in the presence of high amount of nitrogen and carbon, with cowpea containing extra amount of carbon from the carbohydrate it contains. The plot of the yeast biomass using cowpea extract as nitrogen source.

Also, in Table 4, there is glucose filtrate (acid-hydrolyzed cassava) control and cowpea extract control. It was observed from the table that, in the presence of high amount of glucose filtrate (acid-hydrolyzed cassava) and no amount of cowpea extract (cowpea extract control), there was high amount of yeast biomass. The high yeast biomass observed with cowpea accounts for the extra carbon that is contained in cowpea as carbohydrate together with the nitrogen content.

The residual glucose as well as the pH of the media after yeast culture was also measured. From Table 4, it

was observed that, with increase in yeast biomass, there was significant decrease in the residual glucose concentration, which indicates that, much of the glucose have been used during the yeast growth process.

Yeast biomass estimation with varying pH

The yeast biomass estimation with varying pH was aimed at determining the pH at which optimal growth is observed. This experiment is necessary because, after acid hydrolysis, the medium was very acidic (about pH 1.3); thus, varying the pH helps to ascertain the minimum amount of NaOH needed to adjust the pH. Hence, neutralizing with least amount (few drops) of base will save cost using this method. The amount of residual glucose and the pH level (whether acidic or alkaline) after the yeast biomass was also determined. The results obtained are shown in Table 5.

Yeast grows best (optimally) at pH 6.5. From Table 5, the pH of the yeast media was adjusted to varying pH range, from pH 2.5 to 6.5 using 1 M sodium hydroxide (NaOH) solution as the alkaline medium while the acid-hydrolyzed cassava (glucose filtrate) was used as the acidic medium as well as carbon source and the cowpea extract was used as the nitrogen source. The pH of the medium before adjusting with 1 M NaOH (serving as the control for the analysis) was 1.40.

Using constant percentage of acid-hydrolyzed cassava and cowpea, increased yeast biomass was observed as the pH variation increases from pH 2.5 to 6.5 and pH 6.5 gave the best (optimum) yeast biomass (Table 5).

After the growth of the yeast at the specified incubation period, the residual glucose concentration as well as the pH of the media after growth was measured. From Table 5, it was observed that, as the pH increases, the residual glucose concentration of the yeast culture decreases.

This indicates that there was increased yeast biomass which resulted in decreased residual glucose concentration due to the high utilization of the carbon

Table 5. Yeast biomass estimation after 36 h culture* with varying pH using acid-hydrolyzed cassava and cowpea.

Acid-Hydrolyzed cassava (%)	Acid-hydrolyzed cowpea (%)	pH	Yeast biomass (O.D 600 nm)**	Residual glucose (mmoles/L)**	pH after 36 h of yeast culture
20	10	1.40***	0.220 ± 0.003	11.557 ± 0.011	2.35
20	10	2.5	0.306 ± 0.002	12.433 ± 0.005	3.24
20	10	3.5	0.478 ± 0.005	12.322 ± 0.021	4.13
20	10	4.5	0.552 ± 0.002	11.368 ± 0.024	4.35
20	10	5.5	0.652 ± 0.003	10.882 ± 0.001	5.39
20	10	6.5	1.046 ± 0.003	7.435 ± 0.004	5.86

*The medium contained 20 ml of the 20% acid-hydrolyzed cassava (20 ml of acid-hydrolyzed cassava contains 0.36 g glucose) and 5% poultry manure extract with varying pH. The total volume in each conical flask was made up to 100 ml with distilled H_2O and autoclaved for 30 min. 100 µl of 20% antibiotics (ampiclox) was added and inoculated with 2 ml of yeast culture ($OD_{600\ nm} = 1$).
** Values are mean ± standard deviation of triplicate experiments;
*** = pH after acid hydrolysis and before adjusting with NaOH (1 M).

content in the acid-hydrolyzed cassava (Stewart and Russel, 2002). Also observed was the pH of the yeast culture, which was seen to be acidic (reduced pH). However, it was observed that, at other adjusted values of pH ranging from 2.5 to 4.5 with the controls inclusive, the pH values of the yeast culture was higher than the adjusted values, which indicates that, yeast do not grow very well in very acidic medium (Gaudreau et al., 1997). Fair yeast growth was observed for pH 5.5, as the pH value after the yeast culture was seen to be reduced (acidic) but not as good as pH 6.5 (Table 5).

DISCUSSION

Cassava is basically made up of starch. Cassava is rich in carbohydrate and can be used both industrially and as an important food source (El-Sharkawy, 2004). Various industries have exploited the use of cassava in the production of many items such as textiles, cosmetics, glue and adhesive, pharmaceutical and cement (Tonukari, 2004; Altschul and Von, 1973; Nduele et al., 1993). The acid catalyzed hydrolysis of starch is a complex heterogenous reaction. It involves physical factors as well as the hydrolytic chemical reaction. The hydrolysis is therefore controlled by both the reaction conditions (which are acid concentration and temperature) and the physical state of starch (Oboh and Akindahunsi, 2003). ∝-Amylose and amylopectin glycosidic bonds are broken to produce glucose and oligosaccharide residues (Burelli, 2003). The major hydrolytic product of cassava starch (glucose) can serves as a suitable carbon source for the production of yeast. The production of yeast is an important step to the commercial use of the product in baking, brewing and other applications. Yeast has been known to humans for thousands of years as they have been used in fermentation processes including the production of alcoholic beverages and bread leavening (Chao et al., 2001; Hough, 1998).

From the results of the present study, it was shown that, acid hydrolyzed cassava flour (glucose) can serve potentially as a good and locally available carbon source for the production of yeast. Moreover, there is relative high availability of cassava (Ihekonronye and Ngoddy, 1985; Alais and Linden, 1999). The processes involved in acid hydrolysis of cassava flour is fast and economically feasible, as little amount of the acid can hydrolyzed large amount of the cassava flour within 2 h. The results obtained showed that the hydrolysis of cassava flour gave a high yield of glucose which serves as good and locally available carbon source for yeast production. This is in agreement with the findings of Oboh and Akindahunsi (2003), who observed that, the hydrolytic product of cassava starch – glucose, serves as a suitable substrate in providing carbon source for the growth of yeast.

Cowpea is rich in protein which can be used industrially as well as an important food source (Quinn and Myers, 2002). The results obtained from proximate analysis of cowpea and cassava showed that cowpea is a good source of nitrogen since the protein content is 25.38%. Also, cassava is a good source of carbohydrate (90.36%). This indicates that cowpea and cassava can be good sources of nitrogen and carbon, respectively, for yeast culture. This present research shows that yeast culture can be enhanced with the use of cowpea which serves as an alternative to peptone. Peptone is commonly used as nitrogen source for the growth of yeast. Cowpea also contains carbohydrate in addition to protein.

The results of the present study showed that yeast grows best closer to neutral pH with a sharp increase at pH 5.5 - 6.5. This result is in agreement with the findings of Glen and Dilworth (2002). Yeast grows poorly when the pH of the medium is lower than 4. Yeast biomass was also determined at varied values of the acid-hydrolyzed cassava. The results showed that yeast biomass increases with increasing amount of acid-hydrolyzed cassava as carbon source, thus, confirming

Dubai and Muhammad's (2005) observation that the basic carbon and energy source for yeast culture are sugars. Increasing acid-hydrolyzed cowpea also increased yeast biomass after 36 h of culture.

Conclusion

The methods described in this work can be used in the development of a rapid method for producing glucose and simple sugars from cassava through acid hydrolysis and combining this with cowpea for yeast production. It is of significance that locally abundant agricultural products such as cassava and cowpea can be used for the production of an industrial raw material, yeast. This research has taken the advantage of cassava and cowpea flour as alternative carbon and nitrogen sources because of its availability. Yeast is currently imported into Nigeria for various uses. Local industries should take advantage of the results presented here to start yeast processing plants using locally and readily available cassava and cowpea.

REFERENCES

Akinfala EO, Aderibigbe AO, Matanmi O (2002). Evaluation of the nutritive value of whole cassava plant as replacement for maize in the starter diets for broiler chicken; Livest. Res. Rural Dev. pp. 14-16. Retrieved March 17, 2009, from http://www.cipav.org.co/lrrd/lrrd14/6/akin146.htm.

Alais C, Linden G (1999). Food Biochemistry. Aspen publishers Inc. Maryland.

Albers E, Larsson C, Lidén G, Niklasson C, Gustafsson L (1996). Influence of the nitrogen source on Saccharomyces cerevisiae anaerobic growth and product formation. Appl. Environ. Microbiol. 62(9): 3187–3195.

Altschul S, Von R (1973). Drugs and Food from Little known plants Harvard Univ. Press, Cambridge.

AOAC (1995). Association of Official Analytical Chemists. Official Methods of Analysis. Washington DC.

Awonaike MJ, Ferrer J, Ejiofor AO (2001). Nitrogen contribution in cowpea. Biores. Technol. 57: 275-288.

Barham D, Trinder P (1972). An improved colour reagent for the determination of blood glucose by the oxidase system. Analyst 97(151): 142-145.

Bekatorov A, Psarianos C, Athanasios A (2006). Production of food grade yeasts. Food Technol. 44(3): 407-415.

Bressani R (1985). Nutritive Value of Cowpea. John Wiley and Sons, Chinchester.

Broach JR, John R (1991). Pringle and biology of the yeast Saccharomyces cerevisiae: Genome dynamics protein synthesis and energetics, Cold Spring Harbour Laboratory Press, New York.

Burelli MM (2003). Starch: the need for improved quality or quantity – an overview. J. Exp. Bot. 54(382): 451-456.

Chao HJ, Joo Ms (2001). Utilization of Brewer's Yeast Extract Part 1: Effects of Different Enzymatic Treatment on Solid and Protein Recovery and Flavour Characteristics. Biores. Technol. 76: 253-258.

Corriher S (2001). Yeast's crucial roles in bread making. Fine cooking. 43: 80-81.

Davis DW, Oelke EA, Oplinger ES, Doll JD, Hanson CV Putnam DH (1991). Cowpea alternative field crops manual. University of Wisconsin- Madison, United States pp. 1-10.

Dubai YU, Muhammad S (2005). Cassava starch as an alternative to agar-agar in microbiological media. Afr. J. Biotechnol. 4(6): 573-574.

El-Sharkawy MA (2004). Cassava biology and physiology. Plant. Mol. Biol. 56: 481-501.

Gaudreau H, Champagne CP, Goulet J, Conway J (1997). Lactic acid fermentation of media containing high concentration of yeast extracts. J. Food Sci. 62: 1072-1075.

Gilland B (2002). World population and food supply In:'can food production keep pace with population growth in the next half century'. Food policy.

Glen ST, Dilworth EA (2002) Growth and Survival of Yeast in dairy product. Food Res. Int. 34: 791-796.

Hough JS (1998). The biotechnology of malting and brewing pp. 93-96.

Ihekonronye AI, Ngoddy PO (1985). Integrated food science and technology for the tropic. Macmillian Education Ltd. Oxford 2nd ed. pp. 15-35.

Miller P (1998). Growth rate of cowpea. J. Appl. Food Sci. 29: 253-255.

Nduele M, Ludwig A, Van Ooteghem M (1993). The use of cassava starch in the formulation of gelatin capsules. J. de Pharm. Belg. 48(5): 325-334.

Nwokoro SO, Orheruata AM, Ordiah PI (2002). Replacement of maize with cassava sieviates in cockerel starter diets: effects on performance and carcass characteristics. Trop. Anim. Health Prod. 34(2): 103-107.

Oboh G, Akindahunsi AA (2003). Biochemical changes in cassava products (flour and garri) subjected to Saccharomyces cerevisiae solid media fermentation. Food Chem. 52(4): 599-602.

Ohara H, Hiyama K, Yoshida T (1992). Non-competitive product inhibition in lactic acid fermentation from glucose. Appl. Microbiol. Biotech. 36: 773-776.

Okezie BO, Kosikowski FV (1982). Cassava as a food. Crit. Rev. Food Sci. Nutr. 17(3): 259-275.

Peppler HJ (1982). Yeast extract. In fermented foods ed. Rose, A.H. London: Academic press pp. 93-312.

Quinn J, Meyers R (2002). Cowpea: A versatile legume for hot, dry conditions In: "Alternative crop guide". Jefferson Institute pp 4. Singh, B. B. 1987.

Roble ND, Ogbonna JC Tanaka H (2003). L-lactic acid production from raw cassava starch in a circulating loop bioreactor with cells immobilized in loofa (Luffa cylindrical). Biotechnol. Lett. 25(13): 1093 -1098.

Stanbury PF, Whitaker A, Hall SJ (1995). Media for industrial fermentations. In principles of fermentation Technol. Oxford; Pergamon Press pp. 93-121.

Stewart G, Russel I (2002). Biochemistry and genetics of carbohydrates utilization by industrial yeast strains. Pure Appl. Chem. 59: 1493-1500.

Tonukari NJ (2004). Cassava and the future of starch. Electron. J. Biotechnol. 7(1). www.ejbiotechnology.info//content/vol7/issue1/issues/2/.

Watson TG (1976). Amino acid pool composition of Saccharomyces cerevisiae as a function of growth rate and amino acid nitrogen source. J. Gen. Microbiol. 96: 263-268.

Zheng S, Yang M, Yang Z. (2005). Biomas production of yeast isolate from salsd oil manufacturing waste water. Bio. Res. Technol. 96: 1183-1187.

17

The response to iron deficiency of two sensitive grapevine cultivars grafted on a tolerant rootstock

M. A. Russo*, F. Sambuco and A. Belligno

Department of Agrochemistry, University of Catania, Italy.

Two sensitive cultivars were examined, *Vittoria* and *Italia,* grafted on a tolerant rootstock (140 Ruggeri). Two levels of iron chlorosis in scions were selected, initial and evident, and compared to the healthy rootstock (control). The fractions of extracellular and cytoplasmatic cations, chloroplastic mobile, loosely linked, strongly linked and residual cations as well as the active fraction were extracted from the fresh matter. In the chlorotic the plant inability to use Fe^{2+} uptaken by the rootstock was highlighted, with a different response from *Vittoria* and *Italia*, as evidenced by the decrease in the available fraction of Fe^{2+}, particularly in the case of evident chlorosis. The increase in leaf content of the active forms of K^+ and Ca^{2+} resulted directly correlated to the intensity of iron deficiency, since they bring about a higher pH and a destabilization of membranes, respectively, both hindering iron utilization. The modified response in terms of reduced photosynthetic activity in chlorotic scions was evidenced through the decrease in the active form of Mg^{2+} and consequently in chlorophyll content.

Key words: Fe and nutrition unbalance, active Fe fraction, iron chlorosis.

INTRODUCTION

The vast variability of soil properties requires that appropriate grapevine rootstocks be selected able to adapt to specific soil conditions; developing rootstocks able to uptake iron ions under conditions of deficient availability is one of the present challenges to genetic improvement (Gupton and Spiers, 1992).

However, since iron uptake must be followed by its translocation to leaves, the tolerance of rootstock to iron deficiency is not sufficient to prevent leaf chlorosis, which can be brought about by iron inactivation in leaves (Mengel and Malissovas, 1981).

Until now, research has focused mainly on the soil-root interface but not enough is known on the relationships between rootstock and scion: consequently the present research addressed the response to iron deficiency of two different cultivars of *Vitis vinifera* grafted on a tolerant rootstock (*Vitis labrusca*).

Chlorosis in fact can be evidenced even in some scions grafted on tolerant rootstocks: even if the rootstock is able to mobilize iron at root level, the reduction in mobility which can be formed at the grafting level can impair plant

ability to satisfy its metabolic requirements.

Many pedological factors as well as anthropic interventions can impair plant iron uptake: example it has been highlighted that high levels of potassium or cultivation practices can reduce iron availability by raising soil pH reaction thus bringing about conditions unfavourable to maintain iron in its reduced form (Lucena et al., 1990; Pal et al., 1990; Szlek et al., 1990). Furthermore, even concentrations are at an optimal level and it is present in an available form, an unbalance due to excess in Mn^{2+} and Cu^{2+} can cause iron deficiency (Lucena et al., 1990; Mench and Fargues, 1994; Pich et al., 1994; Welch et al., 1993).

Some rootstock cultivars are able to reduce Fe^{3+} to $Fe^{2+,}$ making the ions mobile in the soil and enhancing their uptake (Brown and Draper 1980; Brancadoro et al., 1995; Tagliavini et al., 1995). Also roots of some cultivars can reduce Fe^{3+} to Fe^{2+} encouraging its migration from roots to leaves (Cinelli, 1995).

It can be presumed that such plants have an enzymatic redox equipment depending on the Fe^{2+}/Fe^{3+} ratio (Nenova and Stoyanov, 1995), able to make the microelement in its active form available to the plant. Some varieties of grapevine, particularly rootstocks of *V. labrusca* and scions grafted on them, achieve a higher

---

*Corresponding author. E-mail: marcoanton.russo@tiscali.it.

Table 1. Selected properties of the soils.

Farm n.	1	2	3	4
Sand (%)	68.10	67.30	67.90	66.90
Silt (%)	13.00	13.50	13.10	13.60
Clay (%)	18.90	19.20	19.00	19.50
pH	7.89	7.98	7.79	7.86
Total ca (%)	47.30	48.20	49.00	46.90
Active ca (%)	11.80	12.60	13.20	11.40
Organic C (%)	1.34	1.36	1.39	1.40
C/N	7.44	8.00	7.85	7.61
Total N (%)	1.80	1.70	1.77	1.84
Available P (kg ha^{-1})	1361.00	1257.00	1385.00	1360.00
Available K (kg ha^{-1})	997.00	986.00	1005.00	1097.00

ability in uptaking iron, even in markedly alkaline soils. Such tolerant varieties can mobilize iron by reducing soil pH at root level, thanks to their ability to emit H^+ and/or organic acids; in the latter case, iron is absorbed and transferred as a complex (Brancadoro et al., 1995).

In this research the trade-offs between mobile and non-mobile forms of iron during plant development were evaluated, considering that the fraction of chloroplastic mobile iron is largely represented, at least in those leaves where no chlorosis is evidenced, by the iron-protein-chlorophyll complex at basically constant levels, whereas it seems that iron in chlorotic leaves is stored as a non-active state, possibly such as ferritin, a protein which captures iron as Fe-phosphate in a non-readily usable form (Grossman et al., 1992).

Since all nutrients concur to the development of plants throughout all their life cycle, this research was aimed at assessing the evolution during a full season of the impact of different Fe levels on the mineral nutrition of two iron deficiency-sensitive grapevine cultivars grafted on a tolerant rootstock. The results were compared to the response of the tolerant rootstock.

While this paper deals with the evolution of mineral components, total chlorophyll and proteins throughout a vegetative cycle, in a companion paper the impact of iron deficiency on organic components at harvesting time will be described.

MATERIALS AND METHODS

The field investigation was conducted in four farms, representative of the typical conditions for table grape production, with uniform pedologic conditions; soils are loamy sands and their averaged main characteristics are reported in Table 1. Two iron deficiency-sensitive cultivars of table grape *V. vinifera*, "Italia" and "Vittoria" grafted on a tolerant rootstock, *V. labrusca* ("140 Ruggeri"), were examined.

Six plants per cultivar in each of the four farms were labelled on healthy rootstocks (2.27 - 3.84 mg total chlorophyll g^{-1} fresh matter (f.m.): chlorosis absent) and two levels of chlorosis were identified in the scions, namely incipient (1.47 to 1.82 total chlorophyll g^{-1} f.m.) and evident (0.77 - 1.30 total chlorophyll g^{-1} f.m.).

Plant development was followed from May to August, prior to fruit ripening time, with four monthly leaf samplings from apical shoots of rootstocks and scions, in three replications.

The sampled leaves were dried and the dry matter (d.m.) was mineralized at 500 - 550°C; after that, the content in Fe, K, Ca, Mg, Mn, Cu was determined by a atomic absorption spectrophotometry (AA Perkin Elmer mod, 4000).

The nitrogen (Kjeldahl), after wet-ashing in conc. HNO_3, was determined also by means of atomic absorption spectroscopy (Perkin Elmer, 4000).

In fresh plant matter, total protein content was determined (Bradford, 1976) as well as cations in their fractions a) available, on the extract obtained with N HCl (Köseoglu and Açikgöz, 1995); and as components: 1- extra-cellular and cytoplasmic; 2- mobile and loosely linked chloroplastic; b) unavailable (strongly linked and residual).

The three fractions were obtained by applying an exhaustive extraction, using in succession solvents with growing extracting strength, namely NaCl 0.35 M; NaEDTA 3×10^{-3} M for the available fraction and Triton-X 1.5% in water for the unavailable fraction (Machold, 1968). Additionally total chlorophyll content (Arnon, 1964) was determined.

The extracts from fractions a) and b) as well as the residues were subdivided in two parts: one part was dried and mineralized and the cations were determined as described above; the second part was used to determine the protein content. Finally, also the cation and protein content in the residues was determined.

RESULTS AND DISCUSSION

The values of dry matter, total proteins and chlorophyll were always higher in rootstock leaves, that showed no chlorosis, and opposite to this, nitrogen content was higher in leaves showing iron chlorosis (Figures 1 - 2).

Total and available iron content (Table 2) was in chlorotic leaves compared to the unstressed leaves, with the iron content of the active form proportional to chlorosis intensity, in accordance to results obtained by Brancadoro, 1995 with the roots of *V. vinifera*.

Throughout the period of plant development, the iron content in its two forms, total and active, increased only in non-chlorotic plants (Table 2). Parallel to this, a considerable decrease in the forms "strongly linked" and "residual" was found in non-chlorotic plant matter.

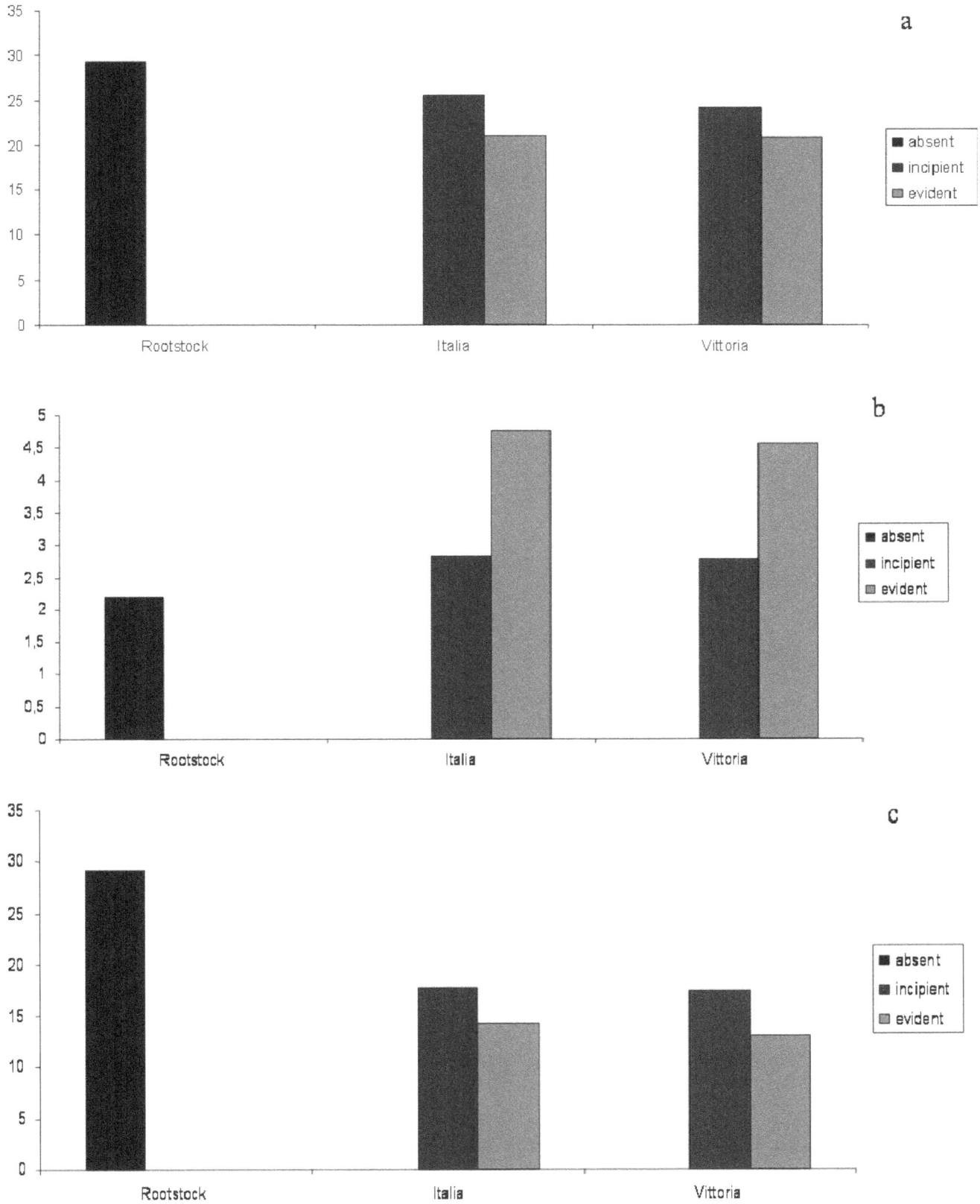

Figure 1. Dry matter (%, a), total nitrogen (% dm, b) and total proteins content (mg g^{-1} dm, c) in unstressed (rootstock) and variously stressed scions.

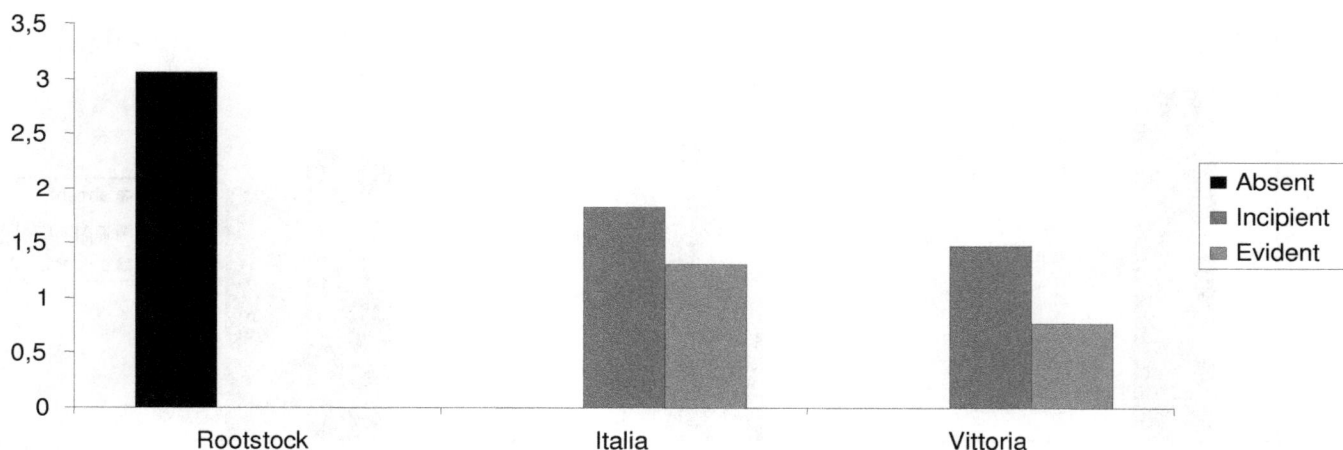

Figure 2. Total chlorophyll content (mg g^{-1} fm) in unstressed (rootstock) and variously stressed scions.

Table 2. Content of the various forms of iron in the rootstock and the variously stressed scions.

Parameters		Iron (mg g^{-1} d.m.)			Iron (% on total)			
					Available		Unavailable	
Cultivars and chlorosis intensity	Sampling date	Total	Available	Unavailable	Extra cell and cytopl.	Chloroplastic mobile	Strongly linked	Residual
Rootstock (absent)	06-May	1.12 d	0.98 e	0.16 f	34.52	39.83	10.93	14.72
	10-Jun	1.27 c	1.13 d	0.15 f	38.20	43.02	8.11	10.69
	13-Jul	1.45 b	1.32 c	0.13 g	41.60	45.14	5.31	7.95
	10-Aug	1.62a	1.53 a	0.12 g	42.40	47.29	3.39	6.92
	Means	1.37 a	1..24 b	0.14 c				
Italia (incipient)	06-May	1.16 a	0.42 e	0.25e	33.74	35.24	12.27	18.75
	10-Jun	1.14 a	0.80 cd	0.21 f	29.81	32.72	16.63	20.84
	13-Jul	1.14 a	0.85 c	0.23 ef	26.34	28.59	19.21	25.86
	10-Aug	1.05 b	0.76 d	0.42 d	25.80	24.97	23.21	26.02
	Means	1.12 a	0.71 b	0.28 c				
Vittoria (incipient)	06-May	1.08 a	0.54 c	0.34 e	29.56	26.10	15.43	23.08
	10-Jun	1.12 a	0.55 c	0.34 e	29.42	25.33	18.72	26.53
	13-Jul	1.12 a	0.63 b	0.34 e	25.71	22.05	20.57	31.67
	10-Aug	1.02 a	0.60 b	0.32 e	22.70	22.76	22.28	32.26
	Means	1.09 a	0.58 b	0.34 c				
Italia (evident)	06-May	1.06 a	0.56 d	0.20g	21.20	17.57	20.81	40.42
	10-Jun	1.09 a	0.69 c	0.20g	17.38	16.27	24.53	41.82
	13-Jul	1.11 a	0.68 c	0.20g	14.10	11.40	30.82	43.68
	10-Aug	1.09 a	0.60 d	0.24f	11.90	9.70	32.19	46.21
	Means	1.09 a	0.63 b	0.21 c				
Vittoria (evident)	06-May	0.97 b	0.53 b	0.23 f	9.21	14.01	26.44	50.34
	10-Jun	1.01 a	0.53 b	0.28 e	8.52	11.14	29.11	51.23
	13-Jul	1.02 a	0.49 b	0.33 d	6.71	8.42	31.74	53.13
	10-Aug	1.00 a	0.38 c	0.30 e	3.43	4.48	35.37	56.72
	Means	1.00 a	0.48 b	0.29 c				

The values of the "total" form may differ from the summation of "available" and "unavailable" forms due to the different analytical methodology.

Table 3. Content of the various forms of potassium in the rootstock and the variously stressed scions.

Parameters		Potassium (mg g^{-1} d.m.)			Potassium (% on total)			
					Available		Unavailable	
Cultivars and chlorosis intensity	Sampling date	Total	Available	Unavailable	Extra cell and cytopl	Chloroplastic mobile	Strongly linked	Residual
Rootstock (absent)	06-May	13.37 a	12.83 ab	0.54 g	93.91	1.80	1.55	2.80
	10-Jun	12.23 b	11.69 bc	0.54 g	92.65	2.55	1.90	2.75
	13-Jul	10.61 c	9.79 c	0.82 f	88.71	3.80	4.45	3.05
	10-Aug	5.46 d	4.56 e	0.91 f	77.49	8.60	7.90	6.00
	Means	10.42 a	9.72 a	0.70 b				
Italia (incipient)	06-May	15.05 a	14.63 ab	0.41 g	95.40	1.50	1.00	2.00
	10-Jun	13.27 bc	12.83 c	0.44 fg	94.30	1.90	1.80	2.00
	13-Jul	11.99 cd	11.57 d	0.42 g	93.90	2.40	2.00	1.80
	10-Aug	9.51 e	9.00 e	0.51 f	87.10	5.80	4.80	2.30
	Means	12.46 a	12.01 a	0.45 b				
Vittoria (incipient)	06-May	13.88 a	13.58 a	0.30 h	96.30	1.20	1.10	1.40
	10-Jun	11.39 b	11.00 bc	0.39 g	93.00	2.70	2.20	2.10
	13-Jul	10.60 c	10.06 c	0.54 f	91.50	3.10	4.10	1.30
	10-Aug	8.57 d	7.97 d	0.60 e	82.90	7.60	7.80	1.70
	Means	11.11 a	10.65 a	0.46 b				
Italia (evident)	06-May	16.46 a	16.04 a	0.42 f	96.10	1.20	1.20	1.50
	10-Jun	14.94 b	14.54 b	0.40 fg	95.30	1.50	1.60	1.60
	13-Jul	13.45 c	13.12 c	0.33 g	94.80	2.40	1.50	1.30
	10-Aug	11.90 d	11.06 d	0.84 e	91.20	4.80	2.90	1.10
	Means	14.19 a	13.69 a	0.50 b				
Vittoria (evident)	06-May	15.08 a	14.72 a	0.36 fg	96.30	1.00	1.20	1.50
	10-Jun	13.06 ab	12.67 b	0.39 f	95.40	1.30	1.78	1.50
	13-Jul	11.78 bc	11.47 c	0.31 g	93.40	3.40	2.10	1.10
	10-Aug	9.44 d	8.62 de	0.82 e	89.30	6.30	3.50	0.90
	Means	12.34 a	11.87 a	0.47 b				

The values of the "total" form may differ from the summation of "available" and "unavailable" forms due to the different analytical methodology.

Total and active potassium percentage decreased during plant development in all the sampled leaves; also the percentage of active form referred to the total was decreasing (Table 3).

The percentages of K increased parallel to chlorosis intensity, and in the cv. Italia were higher than in Vittoria. The higher K content in chlorotic plants could be related to a higher pH in leaf apoplast (Nikolic and Römheld, 1999) which impairs Fe mobilization (Monge et al., 1993; Singh et al., 1995; Szlek et al., 1990). The different levels of K could depend on the unbalance in respiration and photosynthesis typical of sensitive cultivars, where K ions are accumulated to activate stomatal openings(Ward and Schroeder,1994; (solo Blatt), as demonstrated by Lucena et al. (1990) and Pal et al. (1990) for other plant species.

Such assumption is confirmed by the variations in the active form of Ca^{2+} (Table 4) needed to balance ions as required to regulate stomatal openings in response tostress (Lucena et al., 1990; Ward and Schroeder, 1994; McAinsh et al., 1995). In fact under iron stress conditions the ratio K^+/Ca^{2+} in their active form decreased about 15% in both sensitive cultivars compared to the tolerant rootstock (Figure 3).

The different percentages of the active Ca^{2+} fraction in chlorotic and non-chlorotic plants (Table 4) indicate a different plant ability in the cation mobilization: as a consequence the active fractions of Fe^{2+} and Ca^{2+} result inversely correlated as found by Pal et al. (1990) in sugarcane.

Variations in Mg^{2+} content as a response to chlorosis (Table 5) are mainly reflected in the photosynthetic process: the lower amounts of its active form in the chlorotic plants demonstrate their lower photosynthetic ability. The active forms of this ion in fact are very significantly correlated to the chlorophyll (Chl) content, and also significantly correlated to Fe^{2+}. Iron in turn, although not present in the chlorophyll molecule, is highly

Table 4. Content of the various forms of calcium in the rootstock and the variously stressed scions.

| Parameters | | Calcium (mg g^{-1} d.m.) | | | Calcium (% on total) | | | |
| Cultivars and chlorosis intensity | Sampling date | Total | Available | Unavailable | Available | | Unavailable | |
					Extra cell and cytopl	Chloroplastic mobile	Strongly linked	Residual
Rootstock (absent)	06-May	33.83 c	15.80 g	18.03 f	43.70	2.11	3.20	51.01
	10-Jun	41.81 b	16.86 g	24.96 de	37.95	1.94	3.31	56.82
	13-Jul	47.74 a	21.23 e	26.51 d	42.25	1.99	2.00	53.78
	10-Aug	38.86 bc	19.92 ef	18.94 f	46.45	4.15	9.71	39.70
	Means	40.56 a	18.45 c	22.11 b				
Italia (incipient)	06-May	28.05 c	13.39 g	14.67 f	44.40	3.06	2.40	50.10
	10-Jun	40.74 a	18.95 e	21.79 d	43.10	3.07	2.00	51.80
	13-Jul	40.90 a	22.59 d	18.31 e	52.30	2.88	1.70	43.10
	10-Aug	34.84 b	22.90 d	11.93 h	58.90	6.41	6.20	28.50
	Means	36.16 a	19.46 b	16.68 c				
Vittoria (incipient)	06-May	34.89 c	18.92 f	15.97 g	51.50	2.44	2.30	43.80
	10-Jun	41.53 b	22.13 e	19.40 f	49.90	3.130	1.80	45.17
	13-Jul	49.55 a	27.49 d	22.06 e	52.90	2.24	1.45	43.41
	10-Aug	42.19 b	27.86 d	14.33 h	59.58	4.98	5.64	29.8
	Means	42.04 a	24.1 b	17.94 c				
Italia (evident)	06-May	27.10 d	12.52 i	14.58 h	51.00	3.87	2.40	42.74
	10-Jun	41.06 b	21.83 f	19.23 g	48.80	4.03	1.75	45.42
	13-Jul	49.24 a	35.47 c	13.77 h	67.53	4.47	1.00	27.00
	10-Aug	33.12 c	23.27 e	9.85 l	62.30	7.77	3.80	26.13
	Means	37.63 a	23.27 b	14.36 c				
Vittoria (evident)	06-May	45.29 a	32.99 c	12.30 g	70.30	2.16	1.50	26.04
	10-Jun	37.56 b	27.21 d	10.35 h	65.76	6.39	2.00	25.85
	13-Jul	49.08 a	30.54 c	18.54 f	59.16	2.73	1.30	36.81
	10-Aug	32.97 c	20.01 e	12.96 g	54.60	4.76	4.94	35.70
	Means	41.23 a	27.69 b	13.54 c				

The values of the "total" form may differ from the summation of "available" and "unavailable" forms due to the different analytical methodology.

correlated to Chl (Monge et al., 1993; Van Dijk and Bienfait, 1993; Zhang et al., 1995): in this its active fraction was significantly correlated to chlorophyll in both cultivars.

Manganese and copper percentage, both in their total and active form, exhibits a consistent trend to decrease throughout the season and is lower in non stressed plant (Tables 6 and 7). The higher percentage of such cations in sensitive plants can be explained as a plant defence strategy to balance the insufficient availability of Fe^{2+}, as recorded also in pea (Yi and Guerinot, 1996). This brought to a decrease of about 90% in the ratio Fe^{2+}/Cu^{2+} and about 85% in the ratio Fe^{2+}/Mn^{2+} in both sensitive cultivars (Figure 3): such a decrease can be taken as an indicator of iron availability (Lucena et al., 1990; Monge et al., 1993; Zhang, 1993).

Variations in Cu^{2+} and Mn^{2+} can depend on the need to contrast the unbalance in nutrients due to a reduction in active Fe^{2+} fraction, as reported for peach by Köseoglu

(1995) and Monge et al 1993. It has been recently reported that in the same substrate the uptaking of nutrients is different among the cultivars in dependence of their genetic characteristics, since plant ability to use available nutrients is conditioned by specific proteins managing the transport of bivalent elements (Welch et al., 1993).

The higher rate of Cu^{2+} mobilization in chlorotic scions may have caused an inhibition of ferrochelatoreductase activity (Yi and Guerinot, 1996) since it is implied in many factors such as competition for electrons, for chelating agents (Welch et al., 1993) or directly on the redox system (Welch et al., 1993). The competition between Cu^{2+} and Fe^{2+} has been confirmed by many studies with Fe-deficient solutions, where chlorosis was less evident if solutions were also Cu-deficient.

The stress to iron deficiency probably impacted the mobilization of all the cations interacting with nutrition (Mengel et al., 1995), with different responses in the two

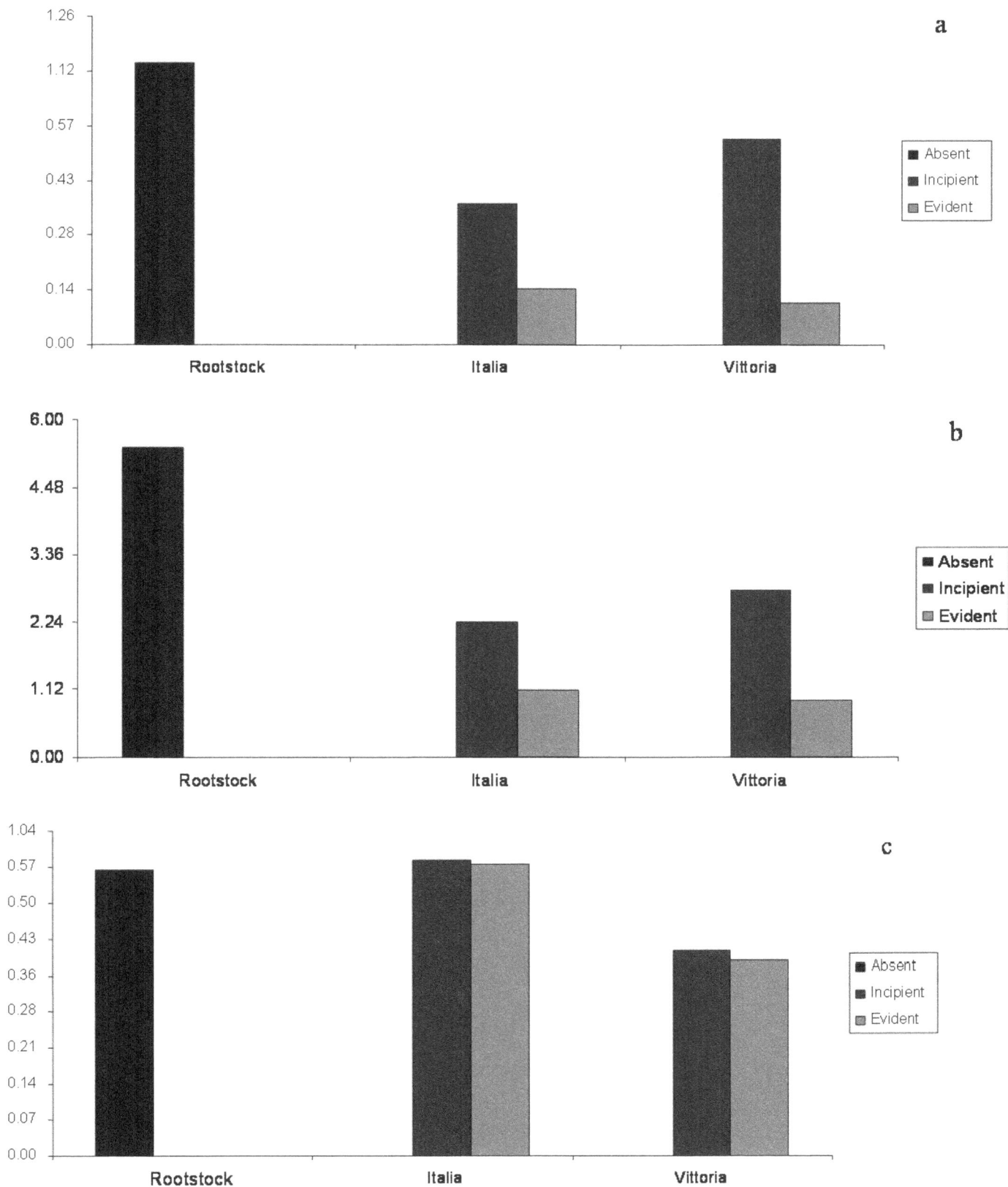

Figure 3. Ratios between the active forms of selected cations, Fe^{2+}/Cu^{2+} (a), Fe^{2+}/Mn^{2+} (b), K^+/Ca^{2+} (c) in the rootstock and the variously stressed scions.

Table 5. Content of the various forms of magnesium in the rootstock and the variously stressed scions.

| Parameters | | Magnesium (mg g^{-1} d.m.) | | | Magnesium (% on total) | | | |
| Cultivars and chlorosis intensity | Sampling date | Total | Available | Unavailable | Available | | Unavailable | |
					Extra cell and cytopl	Chloroplastic mobile	Strongly linked	Residual
Rootstock (absent)	06-May	9.42 d	5.55 g	3.88 i	48.00	9.81	4.82	37.37
	10-Jun	10.61 c	6.32 f	4.29 i	48.145	10.805	5.10	35.95
	13-Jul	12.28 b	7.37 e	4.97 h	50.27	9.40	4.85	35.48
	10-Aug	14.71 a	9.51 d	5.18 gh	53.38	10.99	5.00	30.64
	Means	11.76 a	7.21 b	4.58 c				
Italia (incipient)	06-May	6.39de	3.14d	3.25b	44.68	5.63	5.90	43.79
	10-Jun	6.70d	3.50d	3.20b	48.18	5.27	5.61	40.94
	13-Jul	7.38cd	3.70cd	3.68b	46.51	5.29	5.73	42.47
	10-Aug	8.64c	4.24c	4.40a	47.40	1.40	5.00	46.20
	Means	7.28 a	3.63 b	3.63 b				
Vittoria (incipient)	06-May	7.88 d	4.45 h	3.43 i	49.07	7.11	4.49	39.33
	10-Jun	9.09 c	4.86 gh	4.23 h	45.64	7.53	5.20	41.63
	13-Jul	10.33 b	5.97 f	4.36 h	50.81	6.68	4.85	37.66
	10-Aug	12.10 a	6.84 e	5.26 g	53.40	2.80	4.23	39.57
	Means	9.82 a	5.53 b	4.32 c				
Italia (evident)	06-May	4.46 c	1.86 gh	2.58 e	36.40	4.70	4.75	54.15
	10-Jun	4.96 bc	1.75 h	3.26 d	24.90	4.50	4.71	65.89
	13-Jul	5.26 b	1.55 i	2.75 e	22.40	5.10	4.78	67.72
	10-Aug	6.85 a	2.01 g	3.59 d	23.00	4.10	4.35	68.55
	Means	5.38 a	1.79 c	3.05 b				
Vittoria (evident)	06-May	4.49 d	1.88 f	2.63 f	37.00	5.00	4.78	53.22
	10-Jun	5.89 b	1.70 f	4.14 de	28.00	6.00	5.06	60.94
	13-Jul	5.58 bc	2.51 e	4.03 e	39.00	8.00	5.39	47.61
	10-Aug	7.27 a	3.25 d	5.26 c	40.00	7.00	4.90	48.1
	Means	5.81 a	2.32 c	4.02 b				

The values of the "total" form may differ from the summation of "available" and "unavailable" forms due to the different analytical methodology.

Table 6. Content of the various forms of manganese in the rootstock and the variously stressed scions.

| Parameters | | Manganese (mg g^{-1} d.m.) | | | Manganese (% on total) | | | |
| Cultivars and chlorosis intensity | Sampling date | Total | Available | Unavailable | Available | | Unavailable | |
					Extra cell and cytopl	Chloroplastic mobile	Strongly linked	Residual
Rootstock (absent)	06-May	0.33 a	0.27 c	0.07 f	63.28	17.25	7.45	12.02
	10-Jun	0.30 b	0.23 d	0.06 g	58.91	20.53	4.85	15.71
	13-Jul	0.27 c	0.22 d	0.04 h	66.87	20.02	5.09	8.02
	10-Aug	0.21 de	0.20 e	0.01 i	76.88	18.73	3.07	1.32
	Means	0.28 a	0.23 b	0.06 c				
Italia (incipient)	06-May	0.34 b	0.28 c	0.06 g	67.00	15.36	5.80	11.84
	10-Jun	0.39 a	0.29 c	0.10 f	54.40	20.27	5.20	20.13
	13-Jul	0.28 c	0.23 d	0.05 h	60.50	20.52	6.20	12.78
	10-Aug	0.20 e	0.19 e	0.02 i	68.60	21.77	6.90	2.73
	Means	0.30 a	0.25 b	0.06 c				

Table 6. Contd.

Cultivar	Sampling date							
Vittoria (incipient)	06-May	0.52 a	0.38 b	0.14 g	58.03	14.51	9.96	17.50
	10-Jun	0.35 b	0.28 de	0.06 h	61.20	19.88	4.50	14.42
	13-Jul	0.30 c	0.26 e	0.04 i	64.80	21.48	4.37	9.35
	10-Aug	0.22 e	0.21 f	0.01 l	71.40	22.78	3.42	2.40
	Means	0.35 c	0.28 b	0.06 c				
Italia (evident)	06-May	0.41 b	0.32 c	0.09 h	61.80	15.03	5.90	17.27
	10-Jun	0.45 a	0.32 c	0.13 g	50.50	20.21	7.00	22.29
	13-Jul	0.33 c	0.27 d	0.06 i	60.00	20.56	6.60	12.84
	10-Aug	0.24 e	0.22 f	0.02 l	65.12	25.88	5.48	3.52
	Means	0.36 a	0.28 b	0.07 c				
Vittoria (evident)	06-May	0.73 a	0.52 b	0.21 g	54.70	15.75	6.00	23.55
	10-Jun	0.52 b	0.37 c	0.15 h	53.10	18.50	4.12	24.28
	13-Jul	0.33 d	0.28 e	0.05 i	60.10	23.56	4.60	11.74
	10-Aug	0.24 f	0.22 fg	0.01 l	63.60	29.50	3.68	3.22
	Means	0.46 a	0.35 b	0.11 c				

The values of the "total" form may differ from the summation of "available" and "unavailable" forms due to the different analytical methodology.

Table 7. Content of the various forms of copper in the rootstock and the variously stressed scions.

Parameters		Copper (mg g^{-1} d.m.)			Copper (% on total)			
Cultivars and chlorosis intensity	Sampling date	Total	Available	Unavailable	Available		Unavailable	
					Extra cell and cytopl	Chloroplastic mobile	Strongly linked	Residual
Rootstock (absent)	06-May	1.06 b	0.97 b	0.09 d	66.80	27.55	3.32	2.33
	10-Jun	1.02 b	0.96 b	0.06 e	60.83	33.15	3.01	3.01
	13-Jul	0.96 b	0.91 c	0.06 e	47.875	47.265	2.45	2.41
	10-Aug	1.43 a	1.38 a	0.05 f	50.11	46.93	1.73	1.23
	Means	1.12 a	1.06 a	0.07 b				
Italia (incipient)	06-May	1.47 b	1.35 c	0.12 d	57.60	34.97	5.15	2.28
	10-Jun	1.46 b	1.34 c	0.11d	59.93	33.86	3.20	3.01
	13-Jul	1.43 bc	1.34 c	0.09e	52.54	40.52	3.09	3.85
	10-Aug	2.01 a	1.95 a	0.07f	51.10	44.12	2.21	2.57
	Means	1.59 a	1.50 a	0.10 b				
Vittoria (incipient)	06-May	1.43 b	1.33 b	0.10 e	57.60	34.97	5.15	2.28
	10-Jun	1.33 b	1.00d	0.13 f	59.93	33.86	3.20	3.01
	13-Jul	1.30 b	1.23 b	0.07 g	52.54	40.52	3.09	3.85
	10-Aug	1.84 a	1.78 a	0.05 h	51.10	44.12	2.21	2.57
	Means	1.48 a	1.3 b	0.09 c				
Italia (evident)	06-May	1.93 bc	1.79 c	0.13 e	54.17	38.05	4.80	2.98
	10-Jun	1.86 c	1.69 c	0.17 d	53.14	37.62	6.10	3.14
	13-Jul	2.17 b	2.03 b	0.14 e	54.01	39.07	3.94	2.98
	10-Aug	2.84 a	2.76 a	0.08 f	69.64	27.43	1.17	1.76
	Means	2.20 a	2.07 a	0.13 b				
Vittoria (evident)	06-May	2.14 ab	2.02 b	0.12 f	52.16	40.59	4.45	2.80
	10-Jun	2.31 a	2.15 ab	0.16 d	42.27	50.22	4.61	2.90
	13-Jul	1.67 c	1.53 c	0.14 e	49.3	41.92	4.78	4.00
	10-Aug	2.19 ab	2.08 ab	0.11 f	64.74	29.60	2.16	3.50
	Means	2.08 a	1.95 a	0.13 b				

The values of the "total" form may differ from the summation of "available" and "unavailable" forms due to the different analytical methodology.

cultivars Italia and Vittoria. Accordingly chlorotic scions resulted unable to take advantage of Fe^{2+} uptaken and made available by the tolerant rootstock. This in turn influenced the overall conditions of plants, which could not compensate the unbalances due to Fe^{2+} deficiency, mainly demonstrated by 1) the lower contents in Mg^{2+} and higher contents in K^+ and Ca^{2+} depending on the alterations in photosynthetic activity, and 2) the increased percentage of Mn^{2+} and Cu^{2+} in leaves: such elements, active in electron transfer and cofactors of enzymatic activities, can act synergically or antagonistically in Fe^{2+} mobilization in dependence of plant genetic characteristics.

REFERENCES

Arnon DI, Tsujimoto HY, Mcswain BD (1964). Ferredoxin in photosynthetic production of oxygen and phosphorylation by chloroplasts. Proceedings of the National Academy of Sciences of the United States of America 51: 1274-1282.

Bienfait HF, Scheffers MR (1992). Some properties of ferric citrate relevant to the iron nutrition of plants. Plant and Soil 143: 141-144.

Blatt MR (1990). Potassium channel currents in intact stomatal guard cells: rapid enhancement by abscisic acid. Planta. 180: 445-455.

Bradford MM (1976). A rapid and sensitive method for the quantitation of microgram quantities of protein utilizing the principle of protein-binding. Anal. Biochem. 72: 248-254.

Brancadoro L, Rabotti G, Scienza A, Zocchi G (1995). Mechanisms of Fe-efficiency in roots of Vitis spp. In response to iron deficiency stress. Plant Soil 171: 229-234.

Brown JC, Draper AD (1980). Differential response of "Blueberry " (Vaccinum) progenies to pH and subsequent use of iron. J. Am. Soc. Hort. Sci. 105: 20-24.

Cinelli F (1995). Physiological responses of clonal quince root-stocks to iron-deficiency induced by addition of bicarbonate to nutrient solution. J. Plant Nutr. 18: 77-89.

Grossman MJ, Hinton SM, Minak-Bernero V, Slaugheter C, Stiefel EI (1992). Unification of ferritin family of proteins. Biochem. 89: 2419-2423.

Gupton CL, Spiers JM (1992). Inheritance of the tolerance to mineral element-induced chlorosis in rabbiteye blueberry. Hort. Sci. 27: 148-151.

Köseoglu AT, Açikgöz V (1995). Determination of iron chlorosis with extractable iron analysis in peach leaves. J. Plant Nutr. 18: 153-161.

Lucena JJ, Garate A, Ramon AM, Manzanares M (1990). Iron nutrition of a hydroponic strawberry culture (Fragaria vesca L.) supplied with different Fe chelates. Plant Soil 123: 9-15.

Machold O (1968). Einflub der Ernahrunggsbedingungen auf den Zustand des Eisens in den Blattern, den Chlorophyllgehalt un die Katalase-sowie Peroxydaseaktivitat. Flora Abt. A 159.

McAinsh MR, Webb AAR, Taylor JE, Hetherington AM (1995). Stimulus-induced oscillations in guard cell cytosolic free calcium. The Plant Cell 7: 1207-1219.

Mench MJ, Fargues S (1994). Metal uptake by iron-efficient and inefficient oats. Plant Soil 165: 227-233.

Mengel K, Malissiovas N (1981). Bicarbonate as a factor inducing iron chlorosis in grapevine (Vitis vinifera). Vitis 20: 235-244.

Mengel K, Planker R, Hoffmann B (1994). Relationship between leaf apoplast pH and iron chlorosis of sunflower (Helianthus annuus L.). J. Plant. Nutr. 17: 1053-1065.

Monge E, Perez C, Pequerul P, Madero P, Val J (1993). Effect of iron chlorosis on mineral nutrition and lipid composition of thylakoid biomembrane in Prunus persica (L.) Bastch. Plant Soil 154: 97-102.

Nenova V, Stoyanov I (1995). Physiological and biochemical changes in young maize plants under iron deficiency: 2. Catalase, peroxidase and nitrate reductase activities in leaves. J. Plant Nutr. 18: 2081-2091.

Nikolic M, Römheld V (1999). Mechanism of Fe uptake by the leaf symplast: Is Fe inactivation in leaf a cause of Fe deficiency chlorosis? Plant Soil 215: 229-237.

Pal AR, Motiramani DP, Gupta SB, Bhargava BS (1990). Chlorosis in sugarcane: associated soil properties, leaf mineral composition and crop response to iron and manganese. Fert. Res. 22: 129-136.

Pich A, Scholz G, Stephan UW (1994). Iron-dependent changes of heavy metals, nicotianamine, and citrate in different plant organs and in the xylem exudate of two tomato genotypes. Nicotianamine as possible copper translocator. Plant Soil 165: 189-196.

Singh AL, Chaudhari V, Koradia VG, Zala PV (1995). Effect of excess irrigation and iron and sulphur fertilizers on the chlorosis, dry matter production, yield and nutrients uptake by groundnut in calcareous soil. Agrochimica 39: 184-198.

Szlek M, Miller GW, Welkie GW (1990). Potassium Effect on iron stress in tomato I. The effect on pH, Fe-reductase and chlorophyll. J. Plant Nutr. 13: 215-229.

Tagliavini M, Rombolà AD, Marangoni B (1995). Response to iron-deficiency stress of pea and quince genotypes. J. Plant Nutr. 18: 2465-2482.

Van Dijk HFG, Bienfait HF (1993). Iron-deficiency chlorosis in Scots pine growing on acid soils. Plant Soil 153: 255-263.

Ward JM, Schroeder JI (1994). Calcium-activated K^+ channels and calcium-induced calcium release by slow vacuolar ion channels in guard cell vacules implicated in the control of stomatal closure. The Plant Cell 6: 669-683.

Welch RM, Norvell WA, Schaefer SC, Shaff JE, Kochian LV (1993). Induction of iron (III) and copper (II) reduction in pea (Pisum sativum L.) roots by Fe and Cu status: Does the root-cell plasmalemma Fe(III)-chelate reductase perform a general role in regulating cation uptake? Planta 190: 555-561.

Yi Y, Guerinot ML (1996). Genetic evidence that induction of root Fe(III) chelate reductase activity is necessary for iron uptake under iron deficiency. Plant J. 10: 835–844.

Zhang C, Romheld V, Marschner H (1995). Distribution pattern of root-supplied [59]iron in iron-sufficient and iron-deficient bean plants. J. Plant Nutr. 18: 2049-2058.

Zhang FS (1993). Mobilisation of iron and manganese by plant-borne and synthetic metal chelators. Plant Soil pp. 155-156.

Zhen HH, Shen T, Korcak RF, Baligar VC (1994). Screening for iron-efficient species in the genus malus. J. Plant Nutr. 17: 579-592.

18

Tea *Artemisia annua* inhibits *Plasmodium falciparum* isolates collected in Pikine, Senegal

Gueye Papa El hadji Omar[1], Diallo Mouhamadou[1], Deme Awa Bineta[1], Badiane Aida Sadikh[1,2], Dior Diop Mare[1], Ahouidi Ambroise[1], Abdoul Aziz Ndiaye[2], Dieng Thérése[2], Lutgen Pierre[3], Mboup Souleymane[1] and Sarr Ousmane[1]

[1]Molecular Biology Unit, Bacteriology-Virology Laboratory at Aristide Le Dantec Hospital, 30 Pasteur Avenue, BP 7325 Dakar, Senegal.
[2]Department of Parasitology and Mycology, University Cheikh Anta Diop, Dakar, Senegal.
[3]Iwerliewen Fir Bedreete Volleker, Luxemburg NGO, Luxemburg.

Malaria is a major scourge of most countries in Africa which continues to defy science and technology. Several medicinal plants are traditionally used for the treatment of malaria and other protozoa infections. We aimed to assess by the Double-site Enzyme-linked Lactate dehydrogenase enzyme Immuno-detection (DELI) test for the first time in Senegal, *Plasmodium falciparum* isolates *in vitro* susceptibility to Tea *Artemisia annua* (TAA). In total, 40 field isolates have been tested and the mean IC_{50} was 0.095 µg/ml, while the IC_{50} for the 3D7 and W2 laboratory adapted strains were 0.14 and 0.39 µg/ml, respectively. Tea *A. annua* sensitivity was not obtained for three isolates because of lack of growth. The results suggest that tea *A. annua* has potent antiplasmodial activity against *P. falciparum* strains collected in Pikine, Senegal.

Key words: *Artemisia annua*, enzyme-linked lactate dehydrogenase enzyme immuno-detection (DELI), *Plasmodium falciparum*, IC_{50}.

INTRODUCTION

The combination of tools such as long-lasting insecticidal nets, artemisinin-based combination therapies, indoor residual spraying and intermittent preventive treatment in pregnancy has had a great public health impact in terms of reduction of malaria-associated death and morbidity especially in sub-Saharan Africa. Despite this gain, the number of estimated malaria cases remains still important with 225 million with 781,000 deaths (World Malaria, 2010). Many of them have died for reasons such as being out of reach of health centers or because of the cost of malaria treatment. This situation renders urgent the need to find ways to mitigate this problem. It is estimated that 80% of the world's population use herbal remedies to treat many illnesses and ailments (Zihiri et al., 2005). Additional reports shows that in the US, more than 158 million Americans spent US$ 17 billion on herbal medicines, while more than 70% of Germans recognized using herbal products in the management of many health conditions (US Report, 2002; Tuffs et al., 2002). In developing countries, the figure is almost the same with 80%

of the people using herbal remedies (Verma and Singh, 2008). In Africa also, medicinal plants are used for the treatment of many ailments (Wafo et al., 1999; Nkeh et al., 2001; Noumi and Yomi, 2001; Dieye et al., 2008). In this line, one acknowledges that plants such as the neem in Africa and artemisinin in China have been used for many years as drugs against falciparum malaria. The interest on artemisinin extracted from *Artemisia annua* has grown up since resistance to most of the affordable antimalarials has spread around the world.

A. annua L. is a member of the Asteraceace family and has been used by Chinese since ancient times in traditional medicine against fever and malaria. Artemisinin is a cadinane-type sesquiterpene lactone with an endoperoxide bridge which is presently the most potent and efficacious compound against the late-stage ring parasites and trophozoites of *Plasmodium falciparum* (Brisibe, 2009), the causative agent of malaria. The plant produces a portfolio of bioactive compounds including flavonoids, coumarins, steroids, phenolics, purines, lipids, aliphatic compounds, monoterpenoids, triterpenoids and sesquiterpenoids. Thus far, the most important of the sesquiterpenoids seems to be artemisinin, dihydroartemisinic acid, artemisinic acid and arteannuin B which are stored in the glandular trichomes found in the leaves and inflorescence (Ferreira and Janick, 1995). The action of artemisinin derivatives is different from that of the other antimalarials. Artemisinins have a very fast action and parasite clearance times are much shorter than with other malarial drugs. Whereas, most of antimalarials work at the late trophozoites and schizont stages of the malaria parasite, artemisinin derivates also act already at early trophozoites and ring stages.

Artemisinin acts on gametocyte development, resulting in decreased transmission in areas where artemisinin compounds are extensively used. Artemisinin is considered as the treatment of choice for malaria, and the WHO has called for an immediate halt to single-drug artemisinin preparations in favor of medications that combine artemisinin with another antimalarial, to reduce the risk of parasites developing resistance (WHO, 2006). Recently, the plant is introduced in Senegal and The Gambia and is grown in some areas where people have started to use it as herbal tea for malaria-like symptoms. To investigate whether the Tea *A. annua* has antiplasmodial effect, we studied its *in vitro* antimalarial activity on *P. falciparum* isolates collected from malaria patients at Pikine, Senegal. We used the DELI-microtest which is very sensitive allowing the measurement of drug response with very low parasite densities (0.005%) (Moreno et al., 2001).

MATERIALS AND METHODS

Study site and patients

The study samples were collected from malaria infected patients attending an outpatient clinic situated in Pikine (Figure 1) where

malaria transmission is highly seasonal with an entomologic inoculation rate ranging from 0 to 16.8 infective bites per person per year (Pages et al., 2008). Patients with peripheral blood smears positive for *P. falciparum*, with non-complicated malaria, aged 5 years or greater were invited to participate in the study as previously reported (Thomas et al., 2002; Sarr et al., 2005). They were excluded in case of complicated malaria, pregnancy and recent history of antimalarial treatment.

Artemisia annua dried leaves

Leaves of *A. annua* were graciously provided by Dr Pierre Lutgen and colleagues from Iwerliewen Fir Bedreete Volleker (IFBV), Luxembourg. They were dried under shade in Luxembourg before being sent to Dakar in paper bags weighting 20 g. They were then kept at room temperature until analyses. Chloroquine (CQ) was obtained from Sigma Chemical Co. (St Louis, Mo, USA).

Extraction method

We used the method described by Rath (Rath et al., 2004) which yields 70 to 76% of artemisinin. Briefly, 10 g of dried leaves of *A. annua* are weighed and put in a glass container. Then 1 L of warm water is added and the mixture is stirred up for 10 min and covered by a lid which is not of iron because the latter reacts easily with artemisinin and may decrease the output of the extraction. The solution is then filtered to collect the tea *A. annua*.

Inclusion and exclusion criteria

We recruited study participants in an outpatient clinic in Pikine. The principal criterion of inclusion was uncomplicated *P. falciparum* malaria diagnosed by malaria Rapid Diagnostic Test (RDT). Patients were excluded in the presence of another *Plasmodium* species or any signs of severe malaria.

Socio-demographic characteristics of the study population

In total, 40 patients were recruited and 33 of them were male. The mean age was 25.3 years (12 to 60) (Table 1). The mean temperature and parasite density were 37.9°C and 15,955 parasites/µL respectively.

Consent

In case of positive RDT, we proposed to any patient to participate in the study. A written consent form is translated into wolof for any patient or his legal guardian in case he/she can not read. The form is signed by the patient and the investigator. To preserve the confidentiality of the participants, we have used a code. The study was approved by the Ethics Committee of the Ministry of Health and the Medical Prevention.

Blood collection

For each patient, 5 to 10 ml of whole blood is collected onto ethylen diamine tetracetic acid (EDTA) vacutainer tubes which are kept at room temperature until they are conveyed to the laboratory for analyses.

Parasite culture

To validate our study, we tested our TAA, using field strains as well as two known laboratory strains 3D7, susceptible to CQ and W2, chloroquinoresistant. 3D7 is a clone from NF54 which was isolated

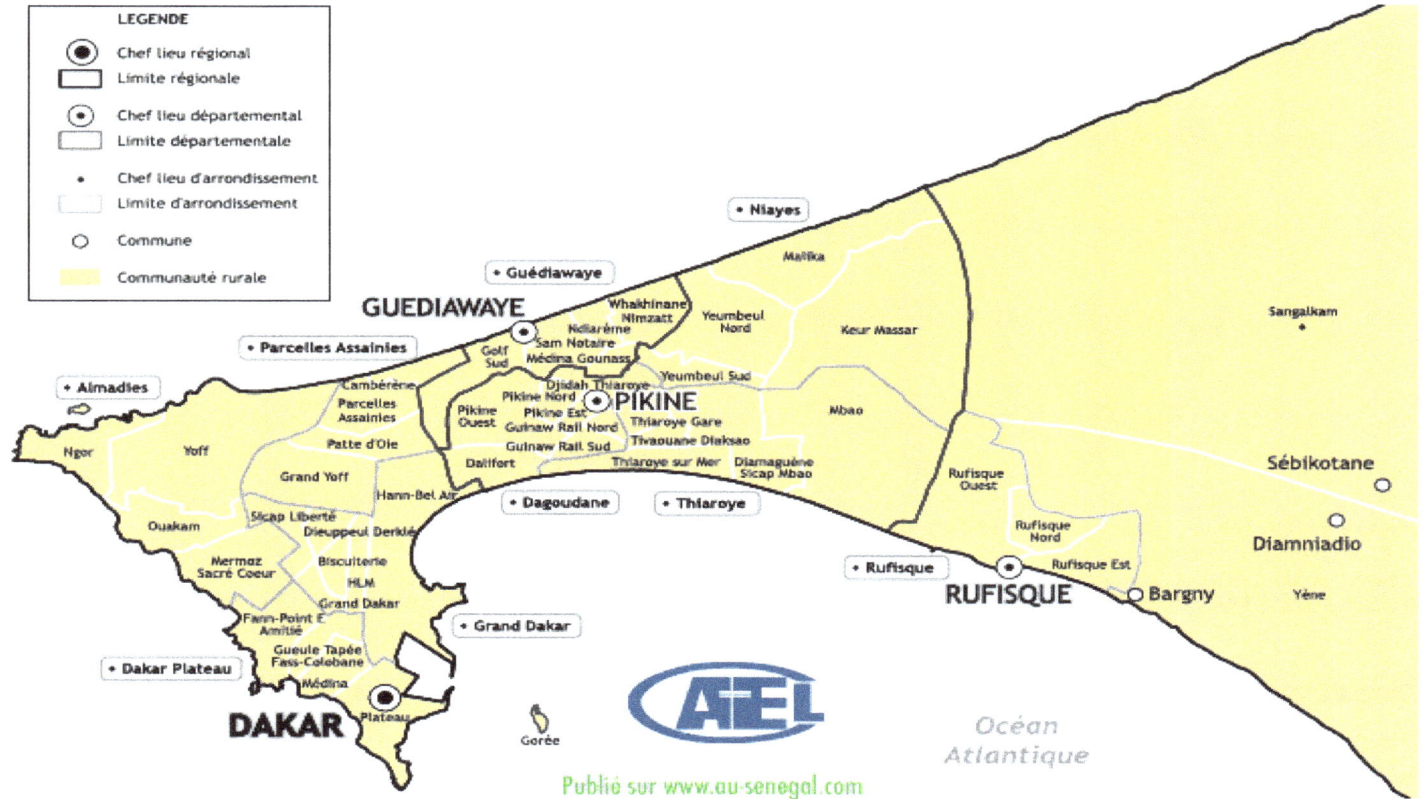

Figure 1. Administrative division of the region of Dakar, Senegal.

Table 1. Parasite density distribution based on the different age groups of our study population.

Different age groups (years)	Parasite density means (p/µL)
[12-15]	10739
[15-30]	18332.9
[30-60]	7390.1

from a patient who lived near Schipol Airport, Amsterdam, The Netherlands (Miller et al., 1993) while W2 was isolated from Indochina. Both strains are maintained in continuous culture in our laboratory.

When the isolates arrived in the laboratory, the parasitized blood sample is centrifuged at 2,000 rpm for 10 min. The plasma is collected in cryotubes and kept at –20°C. The pellet is washed twice with RPMI 1640 and the parasite load is adjusted between 0.5 and 1% with non-parasitized red blood cells. The last wash is done with supplemented RPMI 1640. The sample is put into 96-well microplates with supplemented RPMI 1640 along with tea A. annua at decreasing concentrations from 13.552 to 0.052 µg/ml. Two wells without tea A. annua are used as control. Each isolate is tested in duplicate. Parasites are allowed to grow during 48 h at 37°C in a candle jar after which, plates are taken out and frozen at –20°C until the DELI-test is performed.

The DELI-test

The test measures the amount of lactate dehydrogenase

(LDH) produced by the parasite. The level of pLDH is proportional to the growth of P. falciparum in vitro. This rate is measured by means of immuno-enzymatic technique in double sandwich as described by Druilhe (Druilhe et al., 2001). The technique uses two monoclenal antibodies Mab 17EA and Mab 19G7 directed against an epitop of pLDH.

Coating plates

The 96-wells plates were coated with Mab 17 at 1 µg/L diluted in phosphat buffered saline (PBS) pH 7.4 for 12 h at 4°C and then washed twice with PBS-BSA (bovine serum albumin) at 1%. Thereafter, 300 µl of PBS-BSA at 1% are distributed into the wells and left for 4 h at room temperature. Finally, the plates are emptied and sealed in aluminum foil for a maximum of 8 days.

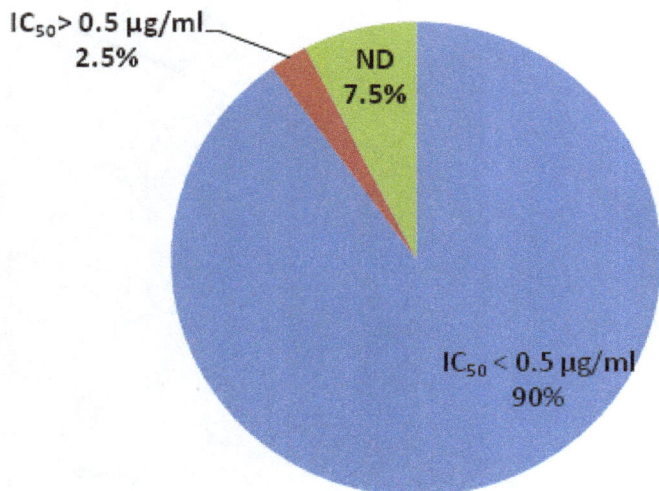

Figure 2. Representative curve of the different IC_{50} highlighting the distribution of isolates. In total, 7.5% assays were unsuccessful while 2.5% of the isolates were resistant, and 90% sensitive to the TAA.

The DELI-Test

The 96-well plates are thawed and frozen 3 times to lyze the red blood cells. One deposits 100 µL of supernatant of culture in the plates with dimensions and saturated. After incubation at 37°C for 1 h, they are washed 3 times with PBS-BSA with 1% and supplemented with biotinylated anti-pLDH Mab 19G7 antibody. After 1 h of incubation at 37°C, the plates are washed 3 times with PBS-BSA then a streptavidin-peroxidase solution is added to the wells followed by incubation for 30 min at room temperature. They are washed 3 times with PBS-BSA 1%. Peroxidase substrate, 3,3', 5,5'-tétraméthylbenzidine and hydrogen peroxide to 0.02% are added to the mixture. The enzymatic reaction is stopped after 5 min per addition of 100 µL 1M phosphoric acid per well. The intensity of the yellow color obtained is measured at 450 nm using a spectrophotometer (StatFax-2100, Awareness Technology Inc). The test is interpretable when the optical density of the well without tea A. annua lies between 0.4 and 1.5. The interpretation of the results is based on the optical density which is proportional to the quantity of pLDH. The results of the tests are expressed in 50% inhibitory concentration (IC_{50}), which is the tea A. annua concentration inhibiting 50% of the parasite growth. The IC_{50} are calculated starting from the curve dose/OD (optic density) by the software GraphPad PRISM 4.

RESULTS

Forty fresh samples collected from the study participants were tested and 37 P. falciparum isolates adapted to the in vitro growth. The cut-off value for resistance to TAA is 0.5 µg/ml. The geometric mean IC_{50} was 0.095 µg/ml.

The IC_{50} values were lower than 0.5 µg/ml for 36 samples and only one exhibited an $IC_{50} > 0.5$ µg/ml. The IC_{50} were undetermined for 3 isolates for unknown reasons. The 3D7 and W2 laboratory adapted strains exhibited IC_{50} of 0.146 and 0.394 µg/mL, respectively (Figure 2). The in vitro susceptibility of those isolates was also determined for CQ. The cut-off value for CQ resistance was set at 100 nM (Le Bras et al., 1990). The prevalence of isolates which exhibited resistance to CQ is 18.9%. Linear regression shows quite a weak association between TAA and CQ (R-squared = 0.0034). The geometric mean TAA IC_{50} in samples resistant and susceptible to CQ was 0.144 and 0.141 µg/ml, respectively (Table 2).

DISCUSSION

As the use of A. annua in Chinese population has been documented for many centuries against different ailments and malaria-like symptoms, we found it relevant to assess the in vitro antiplasmodial effect of TAA in P. falciparum strains. The known reference laboratory adapted strains 3D7 and W2 were tested in parallel as a control. The tea showed a strong antiplasmodial effect against the chloroquine sensitive 3D7 (0.14 µg/ml) and chloroquine resistant W2 (0.39 µg/ml) strains. The latter is considerably lower than the one obtained in Brazil, where a chloroquine resistant strain showed an IC_{50} of 6.1 µg/ml (De Mesquita et al., 2007). This difference is likely due to many parameters as the origin of the leaves of A. annua. The Brasilian leaves were grown locally and the local environment in which the plants are cultivated is of importance in the concentration of different chemical groups that are present in the leaves. However, the mean IC_{50} obtained in our study (0.095 µg/ml) is comparable to values obtained in Tanzania by Malebo and colleagues (Malebo et al., 2009) who showed a strong antiplasmodial activity of the tea A. annua with IC_{50} less than 5 µg/ml. The Senegalese and Tanzanian TAA have a strong antiplasmodial activity, though they are not as rich in artemisinin as the one from Brazil. This suggests that the antiplasmodial activity of the tea is caused by not only the artemisinin, but the combined effect of other substances including coumarine, scopoletines, ployphenols, flavonoids, camphors and microelements. This hypothesis is supported by other studies from Europe, where Artemisia absinthium and Artemisia abrotanum have demonstrated antiplasmodial properties, though they do not contain artemisinin (Cubukcu et al., 1990; Deans and Kenedy, 2002). Additionally, Valécha showed anti-plasmodial activity in plants of the genus Artemisia, though they also do not contain artemisinin (Valecha et al., 1994).

A. Afra extracts are effective against P. falciparum in vitro, and this activity is attribuate to a complex mixture of flavonoids and sesquiterpene lactones, rather than to a single compound (Kraft et al., 2003). Furthermore, the potential synergistic effects of artemisinin and flavonoids

Table 2. Correlation between parasite density and the IC_{50}. P represents the statiscal significance of the test (*p-value*) whereas N represents the number of analyzed isolates.

		Parasite density (parasites /µL)	IC_{50} (µg/ml)
Parasite density (parasites /µL)	Correlation coefficient	1	0.004
	p	-	0.981> 0.05
	N	40	37
IC_{50} (µg/ml)	Correlation coefficient	0.004	1
	p	0.981> 0.05	-
	N	37	37

were described in 1987 (Elford et al., 1987). *A. annua* produces at least 36 flavonoids (Liu et al., 1992). Five of these have been shown selectively to potentiate the *in vitro* activity of artemisinin against *P. falciparum* (Liu et al., 1992). A study shows also that flavonoids present in TAA have shown a variety of biological activities and may synergize the effects of artemisinin against malaria (Ferreira et al., 2010). *Coumarin*, in addition to its role in the immune system and on schizonts, is also known as a metal chelator, notably of iron, a chief element of malaria and other infectious diseases (Yang et al., 1992). In Bolivia, a serie of plants showed an IC_{50} of 9 µg/ml for *Amburana osarensis* against *P. falciparum* isolates essentially due to the coumarin (Bravo, 2003). *Curcumin*, likewise, is an excellent iron chelator, in addition to its numerous therapeutic properties (Means, 2009).

In conclusion, tea *A. annua* has shown strong *in vitro* antiplasmodial activities in *P. falciparum* isolates collected in Pikine. Additional investigations including *A. annua* of different origins are needed to determine which plants have the best antimalarial effects. In addition, more investigations should be carried out to assess the cytotoxicity levels when TAA is used daily by local populations to avoid adverse events and if necessary set a reporting system of those adverse reactions.

REFERENCES

Bravo JA, Sauvain M, Gimenez A, Massiots G, Deharo E, Lavand C (2003). A contribution to attenuation of health problems in Bolivia: Bioactive natural compounds from natives plants reported in traditional medicine. Revista Boliviana de Quimica 20:11.

Brisibe EA, Umoren E, Brisibe F, Magalhães PM, Ferreira JFS, Luthria D, Wu X, Prior RL (2009). Nutritional characterisation and antioxidant capacity of different tissues of Artemisia annua L. Food Chem. 115: 1240-1246.

Cubukcu B, Bray DH, Warhurst DC, Mericli AH, Ozhatay N, Sariyar G (1990). In vitro antimalarial activity of crude extracts and compounds from Artemisia abrotanum L. Phytother. Res. 40: 203-204

De Mesquita ML, Grellier P, Mambu L, De Paula JE, Espindola LS (2007). In vitro antiplasmodial activity of Brazilian Cerrado plants used as traditional remedies. J. Ethnopharmacol. 110: 165–170.

Deans SG, Kenedy AI (2002). Artemisia absinthium. In Artemisia, Wright, C.W., Ed. Taylor and Francis, London. pp. 79-89.

Dieye AM, Gueye I, Sy GY, Ndiaye-Sy A, Faye B (2008). Clinical trials in Dakar: survey on knowledge, attitudes and practices of key actors on the period from 2003 to 2007. Thérapie. 63(2): 89-96.

Druilhe P, Moreno A, Blanc C (2001). A colorimetric in vitro drug sensitivity assay for Plasmodium falciparum based on a highly sensitive double-site lactate deshydrogenase antigen capture enzyme-linked immunosorbent assay. Am. J. Trop. Med. Hyg. 64: 233-241

Elford BC, Roberts MF, Phillipson JD, Wilson RJM (1987). Potentiation of the antimalarial activity of quinghaosu by methoxylated flavones. Trans. R. Soc. Trop. Med. 61: 434-436

Ferreira JFS, Janick J (1995). Floral morphology of Artemisia annua with special reference to trichomes. Int. J. Plant Sci. 156: 807-815

Ferreira JFS, Luthria DL, Sasaki T, Heyerick A (2010). Flavonoids from Artemisia annua L. as Antioxidants and their potential Synergism with Artemisinin against Malaria and Cancer. Molécules 15: 3135-3170.

Kraft C, Jennett-Siems K, Siems K (2003). In vitro antiplasmodial evaluation of medicinal plants from Zimbabwe. Phytother. Res. 17: 123-128.

Le Bras J, Ringwald P (1990). Plasmodium falciparum chemoresistance. The situation in Africa in 1989. Med. Trop. 50: 11-16.

Liu KC-S, Yang S-L, Roberts MF, Elford BC, Phillipson JD (1992). Antimalarial activity of Artemisia annua flavonoids from whole plants and cell culture. Plant Cell Rep. 11: 637-640.

Malebo HM, Tanja W, Cal M, Swaleh SA, Omolo MO, Hassanali A, Sequin U, Hamburger M, Brun R, Ndiége IO (2009). Antiplasmodial, anti-trypanosomal, anti-leishmanial and cytotoxicity activity of selected Tanzanian medicinal plants. Tanzan. J. Health Res. 11(4): 226-234

Means RT (2009). Ironing out complementary medicine. Am. Soc. Hematologie 113: 270-271.

Miller LH, Roberts T, Shahabuddin M, McCutchan TF (1993). Analysis of sequence diversity in the Plasmodium falciparum merozoite surface protein-1 (MSP-1). Mol. Biochem. Parasitol. 59:1-14.

Moreno A, Brasseur P, Cuzin-Ouattara N, Blanc C, and Druilhe P (2001). Evaluation under field conditions of the colourimetric DELI-microtest for the assessment of Plasmodium falciparum drug resistance. Trans. R. Soc. Trop. Med. Hyg. 95:100-3.

Nkeh B, Chungag A, Njamen D, Dongmo A, Wandji J, Nguelefack TB, Wansi JD, Kamanyi A, and Fomum ZT (2001). Antiinflammatory and analgesic properties of the stem extracts of Drypetes molunduana Pax and Hoffm. (Euphorbiaceae) in rats. Pharm. Pharm. Lett. 11: 61-63.

Noumi E, Yomi A (2001). Medicinal plants used for intestinal diseases in Mbalmayo region, Central Province, Cameroon. Fitoterapia 72(3): 246-254.

Pages F, Texier G, Pradines B, Gadiaga L, Machault V, Jarjaval F, Penhoat K, Berger F, Trape JF, Rogier C, Sokhna C (2008). Malaria transmission in Dakar: a two-year survey Malar J. 7:178.

Rath K, Taxis K, Walz G, Gleiter GH, Li SM, Heide L (2004). Pharmacokinetic study of artemisinin after oral intake of a traditional preparation of Artemisia annua L. (annual wormwood). Am. J Trop. Med. Hyg. 70(2): 128-132.

Sarr O, Myrick A, Daily J, Diop BM, Dieng T, Ndir O, Sow PS, Mboup S, Wirth DF (2005). In vivo and in vitro analysis of chloroquine resistance in Plasmodium falciparum isolates from Senegal. Parasitol. Res. 97:136-40.

Thomas MT, Ndir O, Dieng T, Mboup S, Wypij D, Maguire JH, Wirth.

(2002). In vitro chloroquine susceptibility and PCR analysis of Pfcrt et Pfmdr1 polymorphisms in Plasmodium falciparum isolated from Senegal. Am. J. Trop. Med. Hyg. 66, 474-480.

Tuffs A (2002). Three out of four Germans have used complementary or natural remedies. BMJ 325: 990.

US Report (2002). Calls for tighter controls on complementary medicine. Br. Med. J. 324:870.

Valecha N, Biswas S, Badoni V, Bhandari KS, Sati OP (1994). Antimalarial activity of Artemisia japonica, Artemisia maritima and Artemisia nilegarica. Indian J. Pharmacol. 26: 144-146.

Verma S, Singh SP (2008). Current and future status of herbal medicines. Vet. World 1: 347-350.

Wafo P, Nyasse B, Fontaine C, and Sondengam BL(1999). Aporphine alkaloids from Enantia chlorantha. Fitoterapia 70: 157-160.

World Health Organization (WHO) (2006). National Policy on Traditional Medicine and Regulation of Herbal Medicine: Report of a WHO global survey.

World Malaria Report, World Health Organization, 2010.

Yang YZ, Ranz A, Pan HZ, Zhang ZN, Lin XB, Meshnick SR (1992). Daphnetin: a novel antimalarial agent with in vitro and in vivo activity. Am. J. Trop. Med. Hyg. 46: 15-20.

Zihiri GN, Mambu L, Guede-Guina F, Bodo B, Grellier P (2005). In vitro antiplasmodial and cytotoxicity of 33 West African plants used for treatment of malaria. J. Ethnopharmacol. 98: 281-285.

19

Comparison of peroxidase activities from *Allium sativum*, *Ipomoea batatas*, *Raphanus sativus* and *Sorghum bicolor* grown in Burkina Faso

Mamounata Diao, Oumou H. Kone, Nafissétou Ouedraogo, Romaric G. Bayili, Imael H. N. Bassole, and Mamoudou H. Dicko*

Laboratoire de Biochimie Alimentaire, Enzymologie, Biotechnologie Industrielle et Bioinformatique (BAEBIB), Département de Biochimie et Microbiologie, Université de Ouagadougou, Burkina Faso.

Current applications of peroxidase in various areas of biotechnology and clinical biochemistry show the interest for further screening for peroxidase. Thus, peroxidase activities were screened in higher plants such as *Allium sativum*, *Ipomoea batatas*, *Raphanus sativus* and *Sorghum bicolor* grown in a tropical environment. The enzymes were investigated for their specific activities and best physico-chemical conditions for activity and stabilities. Optima conditions with respect to pH, temperature and their heat inactivation were determined by monitoring the hydrogen peroxide-dependant oxidation of guaiacol. Results revealed that peroxidase specific activities in *R. sativus* were higher than the other three plant species. Optimum pHs of all screened peroxidase activities were in the acidic range (pH 4.5 to 6.5). Optimum temperatures were ranging from 30 to 40°C. Peroxidase from *R. sativus* was the most thermostable enzyme among the four plants. This suggests that *R. sativus* is a good source of plant peroxidase, which could be used for various applications.

Key words: *Allium sativum*, *Ipomoea batatas*, *Raphanus sativus*, *Sorghum bicolor*, peroxidase.

INTRODUCTION

Peroxidases (POXs) (E.C.1.11.1.7) are among the most ubiquitous enzymes in plant species. POXs are also found in some animal tissues and microorganisms where they are assigned to play a role of protection against toxic peroxides (Welinder, 1992). In plants they participate in the lignification process (Wakamatsu and Takahama, 1993) and in the mechanism of defense in physically damaged or infected tissues (Biles and Martin, 1993).

POXs are heme-containing enzymes that use H_2O_2 to oxidize a large diversity of hydrogen donors such as phenolic compounds, aromatic amines, ascorbic acid, auxin and certain inorganic ions (Vernwal et al., 2006). The family of plant POXs comprises yeast cytochrome c POXs, plant ascorbate POXs, fungal POXs and classical plant secretory POXs. The group of mammalian POXs includes myeloPOX, lactoperoxidase, thyroid POX and prosta-glandin H synthetase (Welinder, 1992). Reduction of hydrogen peroxides at the expense of electron dona-ting substrates makes POXs useful in a number of biotechnological applications (Regalado et al., 2004).

In the food industry for example, POXs have been widely used as an indicator of vegetables bleaching, due to their high thermal stability and wide distribution (Rodrigo et al., 1997). Plant POXs are used to produce dyes from natural phenolic compounds (Egorov, 1995). In analytical biochemistry, POXs are used as reagents in clinical diagnosis and in enzyme immunoassays (Agostini et al., 2002).

They can also be used for the treatment of containing phenols and aromatic amine (Kinsley and Nicell, 2000; Klibanov et al., 1980; Nakamoto and Machida, 1992; Diao et al., 2010). They found several applications in bio-bleaching processes, lignin degradation, fuel and chemi-cal production from wood pulp, production of dimeric

---

*Corresponding author. E-mail: mdicko@univ-ouaga.bf.

Abbreviations: BSA, Bovine serum albumin; HRP, horseradish peroxidase; POX, peroxidase.

alkaloids, biotransformation of organic compounds, etc. (Ryan et al., 1994). Although POXs are ubiquitous in the plant kingdom, until now the major source of commercially available POXs is from horseradish (*Armoracia rusticana*) and soybean (*Glycine max*). However, plants of other species may provide POXs whose characteristics are comparable or higher than those of horseradish (Dicko et al., 2006a, b).

The overall objective of this study is to find novel sources of plant POXs for biotechnological applications. Specifically, POXs from plants such as *Allium sativum, Ipomoea batatas, Raphanus sativus,* and *Sorghum bicolor,* grown in tropical climate may be predicted to display high thermostabilities and interesting catalytic properties.

MATERIALS AND METHODS

Chemicals and reagents

Guaiacol was purchased from Aldrich; bovine serum albumin was obtained from Sigma chemicals CO; hydrogen peroxide was purchased from Merck. Other chemicals and reagents employed were of analytical grade.

Plant materials

Enzymes were extracted from *A. sativum* bulbs, *I. batatas* tubers, *R. sativus* roots and *S. bicolor* germinated grains. To minimize stress related differences in POX biosynthesis, all the plant species were grown in the same farm and in the same natural environment in Ouagadougou (Burkina Faso), during the rainy-season 2007 to 2008.

Preparation of POX extracts

Enzyme extracts were prepared by mixing 250 mg of plant ground material with 1.2 mL of 50 mM Tris-HCl buffer pH 7.3 containing 0.5 M $CaCl_2$ and 5 mM β-mercapto-ethanol, at 4 °C for 1 h. The homogenate was centrifuged (14000 g, 4 °C, 45 min) and the resulting supernatant was used as crude extract of POX. Protein assay was performed by the linearized method of Bradford (Zor and Selinger, 1996) using the ratio of A_{620}/A_{450} versus protein concentration. Bovine serum albumin was used as standard protein.

POX assay

POX activity was measured with a spectrophotometric assay by monitoring the H_2O_2-dependent oxidation of guaiacol, at 25 °C. The reaction mixture consisted of 10 μL of 200-fold diluted crude enzyme extract, 20 μL of 100 mM guaiacol, 10 μL of 100 mM H_2O_2 and 160 μL of 50 mM sodium acetate pH 5.0. Control assays in which the enzyme extracts or substrates were replaced by buffer were performed. The reaction was monitored at 450 nm. One unit of POX activity (U) is defined as the amount of enzyme releasing 1 μmol of guaiacol radical/min under the assay conditions.

Determination the effect of pH on enzyme activities

The optimum pH was determined at 25 °C by measuring the activity of the enzyme in buffers of pH ranging from 3 to 8. The POXs activity was measured by performing the routine assay by changing the buffers at various pH values. The used buffers were those of McIlvaine, 50 mM citrate-phosphate.

Determination the effect of temperature on enzyme activities

The optimum temperature was determined at the optimum pH of each enzyme by measuring the activity of the enzyme in temperatures ranging from 30 to 90 °C. Thermal stability was studied by preincubating enzyme extracts at temperatures ranging from 30 to 90 °C for 10 min. After heating, samples were immediately cooled on ice during 10 min and the residual enzyme activity was then determined with the routine assay.

Statistical analysis

All spectrophotometric analyses were monitored with a MRX 96-well microplate reader on-line interfaced to a computer (Hewlett Packard). Kinetic data were determined in the linear phase of reaction traces using MRX revelation software version 1CXD-4239 (Dynex Technologies, Inc, USA). The reactions were monitored over 10 min. The initial slopes of the reaction traces caused by enzyme activities were corrected with the slopes of the blanks. All experiments were carried out in triplicate. Analysis of variance (ANOVA) and Student t-test (P= 0.05, considered as signification) were used to determine statistically significant differences between enzyme assays.

RESULTS

Comparison of POX specific activities

The results of POX specific activities from *A. sativum* bulb, *I. batatas* tuber, *R. sativus* roots and grain of *S. bicolor* are summarized in Table 1. These four plants display different levels of POX activities using guaiacol and H_2O_2 as substrates. POX specific activities among these plants ranged from 22.1 to 294.6 U/mg. The highest specific activity was found in plant tissues from *R. sativus*, followed by *S. bicolor* and *I. batatas. A. sativum* showed significantly lower activity than the other three species. POX specific activity in *R. sativus* was 13 fold higher than that of *A. sativum* and 9 and 4 fold higher than those of *I. batatas* and *S. bicolor*, respectively.

Comparison of optimum pHs

The effect of pH on the activities of POXs from the four POX are shown in Figure 1. All POXs of these plants have their optima pH for activity in acidic buffers (pH < 6). POXs from *S. bicolor* were most active at pH 3.5 to 4 but POXs from *I. batatas* were strongly inactivated at this pH. However, *I. batatas* POXs were most active at pH 6.0. The optimum pH of activity for POXs from *A. sativum* bulbs was around pH 5.5 to 6.5 and POXs from *R. sativus* tubers presented an optimum activity at pH 5.

Table 1. Comparison of peroxidase activities* among four plants.

Plant	Organs	Total protein (mg)	Total POX activity (U)	POX specific activity (U/mg)
A. sativum	Bulbs	24.3 ± 7	537.03 ± 14	22.1 ± 2
I. batatas	Tubers	36.9 ± 3	1154.97 ± 6	31.3 ± 2
R. sativus	Roots	14.7 ± 4	4348.3 ± 52	294.6 ± 13
S. bicolor	Grains (germinated)	28.4 ± 9	1948.24 ± 54	68.6 ± 6

*Enzymes were extracted with 50 mM Tris-HCl buffer, pH 7.3; containing 0.5 M $CaCl_2$ and 5 mM β-mercapto-ethanol. Enzyme activities were determined by monitoring the H_2O_2-dependant oxidation of guaiacol.

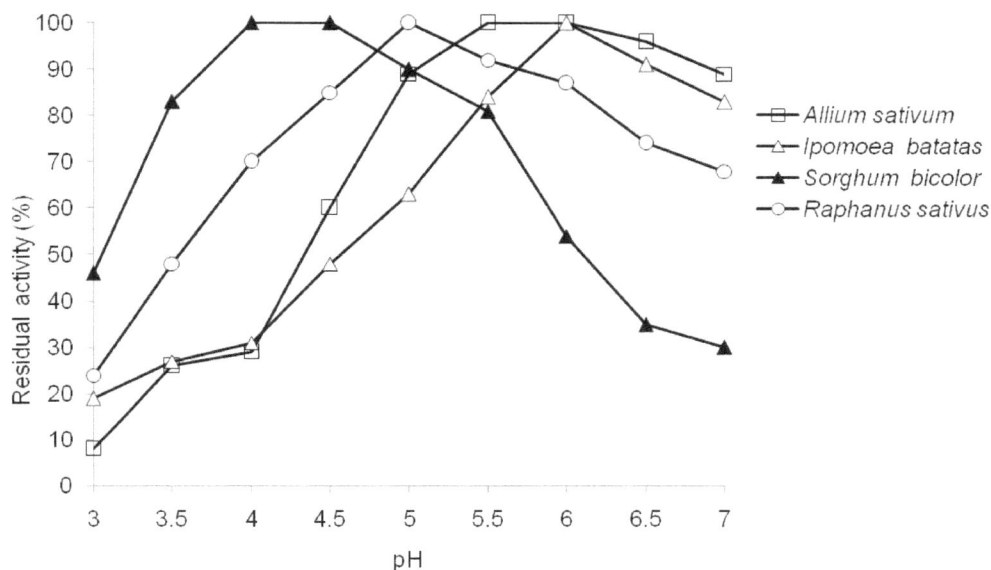

Figure 1. Comparison of the effect of pH on POX activities. Enzyme activities were determined by monitoring the H_2O_2-dependant oxidation of guaiacol in citrate-phosphate buffers (pH 3 to 7).

Comparison of temperature effect

The effect of temperature on the activities and stabilities of POXs from the four plant species was examined (Figures 2 and 3). POXs from sorghum were most active at 40°C, but they were completely inactive at 80°C. However, at this temperature, POXs of I. batatas and R. sativus conserved more than 30% of their activities and were then later completely inactivated at 100°C. These POXs also retained above half of their activities after 10 min of incubation at 70°C. POXs from I. batatas and R. sativus were most active at 30°C. However, POXs from A. sativum are most active at 40°C and were completely inactivated at 90°C.

DISCUSSION

In this contribution, POXs from four plants were compared for their specific activities, and catalytic properties. Based on their hydrogen peroxide-dependent oxidation of guaiacol, all the four plants had significant differences (P <0.05) in their content in POX activities. Among the four plant species, the highest POX activities were detected in R. sativus. However, these activities (294.6 ± 13 U/mg) were lower than those (413.5 U/mg) reported by Wang et al. (2004) on the black varieties of the same plant. POXs from A. sativum showed significantly lower activities than the other three species. POX specific activities in I. batatas (31 U/mg) are higher than data (15 U/mg) from the same plant reported by Santos et al. (2004). The environment may play a significant impact in these differences. POXs from these four plants had optimum pHs ranging from 4.0 to 6.5 with guaiacol as substrate.

Indeed using guaiacol as hydrogen donor, acidic optimum pHs have been reported for many plant POXs (Agostini et al., 1999; Mika and Lüthje, 2003). POXs from A. sativum and S. bicolor had a great range of pH activities. Other authors (Bhunia et al., 2002; Dicko et al., 2006a) showed that POX activities usually increased with decreasing pH.

Sakharov et al. (2003), using anionic POXs purified from African oil palm tree as biocatalysts, showed that

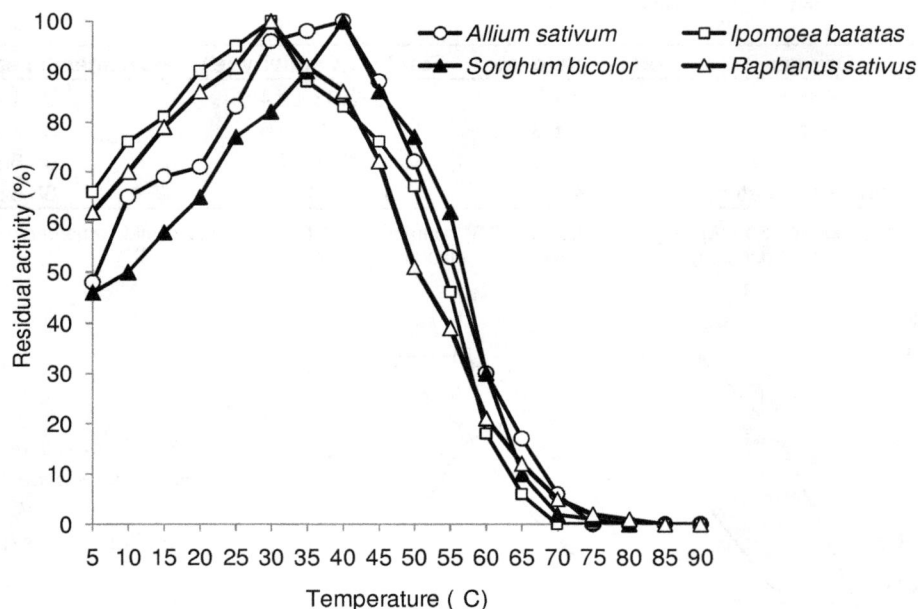

Figure 2. Comparison of the effect of temperature on POX activities. Reactions were performed in 50 mM sodium acetate buffer pH 5, with incubation temperatures ranging from 5 to 90°C. Enzyme activities were determined by monitoring the H_2O_2-dependant oxidation of guaiacol.

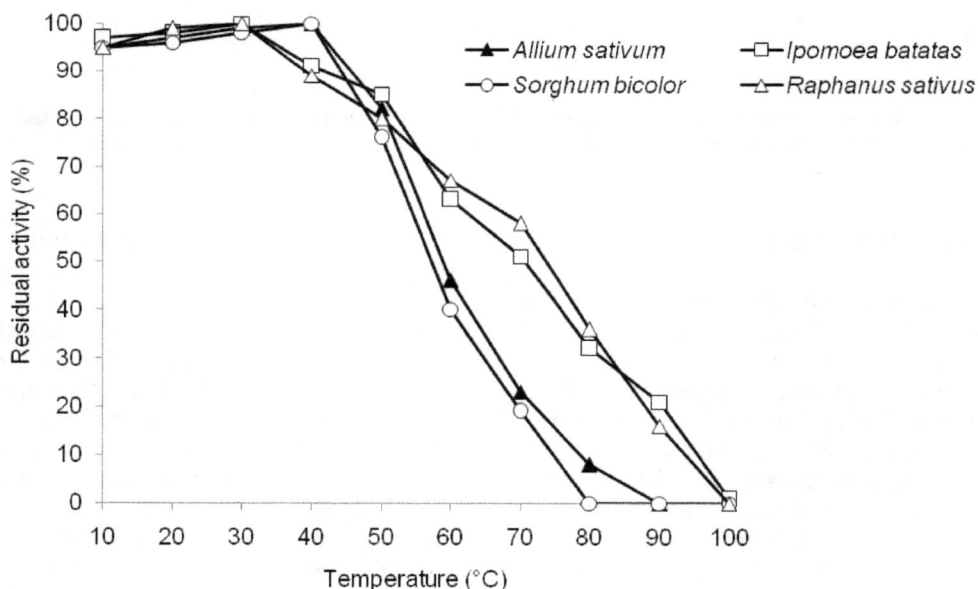

Figure 3. Comparison of thermal stability of POXs. Enzyme activities were determined by monitoring the H_2O_2-dependant oxidation of guaiacol.

polymerization of aniline was optimum at pH 3.5. Polyaniline is one of the most extensively investigated conducting polymers because of its high environmental stability and promising electronic properties. POX from *S. bicolor* displaying similar activities at the same range of pH may be efficient for the same applications. Optima temperatures of activities were at 30°C for POXs from *I. batatas*, *R. sativus* and 40°C for POX from *A. sativum* and *S. bicolor*. El Ichi et al. (2008) reported an optimum temperature of 30°C for POX from *A. sativum* bulbs cultivated in Tunisia similar to those of *I. batatas*, and *R. sativus*. Among the four plants, POXs from *R. sativus* are the most heat stable followed by POXs from *I. batatas*, *A. sativum* and *S. bicolor*.

Conclusion

It appeared that *A. sativum* bulbs, *I. batatas* tubers, *R. sativus* roots and *S. bicolor* grains have different level of peroxidase activities. Peroxidases from *R. sativus* exhibited the highest specific activities. The present contribution shows that *R. sativus* cultivated in tropical climate such as Burkina Faso may be an alternative source to horseradish for peroxidases. It may also display interesting catalytic properties as well as thermal resistance.

ACKNOWLEDGEMENTS

This research project was supported trough financial assistance from the Fondazione Lelio and Lisli Basso-Issoco, Italy, and Agence Universitaire de la Franco-phonie (AUF-GP3A), France.

REFERENCES

Agostini E, Millard De Forchetti SR, Tigier HA (1999). Characterization and application of an anionoic peroxidase from *Brassica napus* roots. Plant. Perox. Newslett., 13: 153-159.

Agostini E, Hernández-Ruiz J, Arnao MB, Milrad SR, Tigier HA, Acosta M (2002). A peroxidase isoenzyme secreted by turnip (*Brassica napus*) hairy-root cultures: inactivation by hydrogen peroxide and application in diagnostic kits. Biotechnol. Appl. Biochem., 35: 1-7.

Bhunia A, Durani S, Wangikar P (2002). Horseradish peroxidase catalyzed degradation of industrially important dyes. Biotechnol. Bioeng., 72: 562-567.

Biles CL, Martin RD (1993). Peroxidase, polyphenoloxidase, and shikimate dehydrogenase isozymes in relation to the tissue type, maturity and pathogen induction of watermelon seedlings. Plant Physiol. Biochem., 31: 499-506.

Diao M, Ouédraogo N, Baba-Moussa L, Sawadogo PW, Amani GN, Bassole IHN, Dicko MH (2010). Biodepollution of wastewater containing phenolic compounds from leather industry by plant peroxidase. Biodegradation. DOI 10.1007/s10532-010-9410-8, in press.

Dicko MH, Gruppen H, Hilhorst R, Voragen, AGJ, van Berkel WJH (2006a). Biochemical characterization of the major cationic sorghum peroxidase. FEBS J., 273: 2293-2307.

Dicko MH, Gruppen H, Zouzouho OC, Traore AS, van Berkel WJH, Voragen AGJ (2006b). Effects of germination on amylases and phenolics related enzymes in fifty sorghum varieties grouped according to food-end use properties. J. Sci. Food Agric., 86: 953-963.

Egorov AM (1995). Peroxidase biotechnology and application. International Workshop peroxidase Biotechnology and Application. Oral abstracts: part I. Moscow, Russia.

El Ichi S, Abdelghani A, Hadji I, Helali S, Limam F, Marzouki MN (2008). A newthermostable peroxidase from garlic (*Allium sativum*) bulb: its use in H₂O₂ biosensing. Biotechnol. Appl. Biochem., 51: 33-41.Mika A, Lüthje S (2003). Properties of guaiacol peroxidase activities isolated from corn root plasma membranes. Plant Physiol., 132: 1489-1498.

Kinsley C, Nicell JA (2000). Treatment of aqueous phenol with soybean peroxidase in the presence of polyethylene glycol. Biores. Technol., 73: 139-146.

Klibanov AM, Alberti BN, Morris ED, Felshin LM (1980). Enzymatic removal of toxic phenols and anilines from wastewaters. J. Appl. Biochem., 2: 414-421.

Nakamoto S, Machida N (1992). Phenol removal from aqueous solutions by peroxidase-catalyzed reactions using additives. Water Res., 26: 49-54.

Regalado C, Garcia-Almendárez BE, Duarte-Vázquez MA (2004). Biotechnological applications of peroxidases. Phytochem. Rev., 3: 243-256.

Rodrigo C, Rodrigo M, Alvarruiz A, Frigola A (1997). Inactivation and regeneration kinetics of horseradish peroxidase heated at high temperatures. J. Food Protect., 60: 961-966.

Ryan O, Smyth MR, Fágáin CO (1994). Horseradish peroxidase: the analyst's friend. Essays Biochem., 28: 129-146.

Sakharov IY, Vorobiev AC, Castillo LJJ (2003). Synthesis of polyelectrolyte complexes of polyaniline and sulfonated polystyrene by palm tree peroxidase. Enzyme Microb. Technol., 33: 661-667.

Santos de Araujo B, Santos de Oliveira JO, Machado SS, Pletsch M (2004). Comparative studies of the peroxidases from hairy roots of *Daucus carota*, *I. batatas*, and *Solanum aviculare*. Plant Sci., 167: 1151-1157.

Vernwal SK, Yadav RSS, Yadav KDS (2006). Purification of a peroxidase from Solanum melongena fruit juice. Indian J. Biochem. Biol., 43: 239-243.

Wang L, Kristensen BK, Burhenne K, Rasmussen SK, Chang G (2004). Purification and cloning of a Chinese red radish peroxidase that metabolises pelargonidin and forms a gene family in Brassicaceae. Genetic, 343: 323-335.

Wakamatsu K, Takahama U (1993). Changes in peroxidase activity and in peroxidase isozymes in carrot callus. Physiol. Plant., 88: 167-171.

Welinder KG (1992). Superfamily of plant, fungal and bacterial peroxidase. Curr. Opin. Struct. Biol., 2: 388-393.

Zor T, Selinger Z (1996). Linearization of the Bradford protein assay increases its sensitivity: theoretical and experimental studies. Anal. Biochem., 236: 302-308.

20

Effects of *Emilia praetermissa* leaf extract on the haematological and biochemical parameters of stress induced ulcerated Wistar rats

A. O. Ebunlomo[1], A. O. Odetola[1], O. Bamidele[2]*, J. N. Egwurugwu[3], S. Maduka[1] and J. Anopue[4]

[1]Department of Physiology, Nnamdi Azikiwe University, Nnewi, Anambra State, Nigeria.
[2]Department of Physiology, Bowen University, Iwo, Osun State, Nigeria.
[3]Department of Physiology, Imo State University, Owerri, Imo State, Nigeria.
[4]Department of Physiology, Madonna University, Elele, Rivers State, Nigeria.

Effects of *Emilia praetermissa* leaf extract on the haematological and biochemical parameters of cold-water stress induced ulceration in albino rats were the focus of this study. Blood samples collected from stress induced ulcerated Wistar rats by cold water immersion were used to evaluate the haematological and biochemical parameters. This study was conducted using twenty rats randomly divided into four groups, Group A which was control group and three test Groups B, C and D. All three test groups were subjected to cold water induced stress ulceration. Groups B and C were treated with normal saline (10 ml/kg body weight) and *E. praetermissa* (500 mg/kg body weight) respectively, for seven days after stress induced ulceration while the Group D was pretreated with *E. praetermissa* (500 mg/kg) for fourteen days before cold water stress induced ulceration. The results showed that Stress reduced all blood parameters tested except total white blood cell and platelet counts when compared with control while *E. praetermissa* increased all the blood parameters tested above that of control despite the induced stress. However, the extract significantly *(p < 0.05)* reduced biochemical parameters such as alanine aminotransaminase (ALT) and aspartate aminotransaminase (AST). The results therefore suggest that *E. pratermissa* has haemopoietic potential, it could stimulate blood cell formation and also powerful enough to suppress effects of stress on haematological parameters in stress-ulcerated Wistar rats.

Key words: *Emilia praetermissa*, cold water immersion, ulcer, haematological and biochemical parameters.

INTRODUCTION

Medicinal plants have been used as traditional treatment for numerous human diseases for thousands of years and in many parts of the world (Palombo, 2011).This interest in medicinal herbs has increased scientific scrutiny of the therapeutic potential and safety ('O'Hara et al., 1998). *Emilia pratermissa* which belongs to the family of Asteraceae is a useful plant of west tropical Africa used generally as food and medicine for general healing (Burkill, 1985). *E. praetermissa* has been established as an anti-ulcerogenic plant producing complete mucosal cytoprotection at a dose of 500 mg/kg (Tan et al., 1997).

Stress is a normal physical response to events that causes threat or upset an individual's balance in some way. It refers to any condition that arouses anxiety or fear. Anxiety can be defined as an emotion characterized by feeling of anticipated danger, tension and distress and by tendencies to avoid or escape (Qureshi et al., 2002). Stress can be helpful in small doses but long term exposure to stress can lead to serious health problems ranging from major damage to health to decrease level of productivity and quality of life. Physiological studies have

---

*Corresponding author. E-mail: femi_bayo_is@yahoo.co.uk.

shown that stress from any source can influence the endocrine, haemopoietic and immune systems. Cytokines and cortisol seem to play an important role in the communication between these systems (Maes et al., 1998; Benoit et al., 2001). Stress related mucosal disease is an acute erosive gastritis representing conditions ranging from stress related injury to stress ulcer (Spirt, 2004; Stollman and Metz, 2005). Stress related injury is superficial mucosa damage that present primarily as erosions whereas stress ulcers are deep focal mucosa damage penetrating the submucosa with high risk for gastrointestinal bleeding (Spirt, 2004).

The underlying cause of stress related mucosal disease is hypoperfusion of the mucosa in the upper gastrointestinal tract (Spirt and Stanley, 2006). Gastrointestinal microcirculation and the mucus layer normally maintain the integrity of the gastric mucosa by providing nourishment eliminating hydrogen ion, oxygen radicals and other toxic substances, increasing bicarbonate secretion to neutralize hydrogen ions (Spirt, 2004). Stress related mucosa damage occurs when the mucosal barrier is compromised and can no longer block the detrimental effects of hydrogen ions and oxygen radicals (Spirt and Stanley, 2006). Stress reduces blood parameters such as hematocrit, haemoglobin, erythrocyte count but increased leucocyte count (Gbore et al., 2006). Circulating levels of alanine transaminase (ALT) and aspartate transaminase (AST) increase under psychological and toxic stress; reflecting liver injury. This suggests that chronic stress relates to hepatic damage (Roland et al., 2009).

Cold-water restraint has been a useful procedure for the study of the underlying mechanisms involved in stress induced ulceration in which conscious animals have been noticed to develop more severe gastric damage compared to anesthetized animals, therefore the animal's conscious activity during cold-water immersion has been known to increase the severity of gastric mucosal damage (Fernandez, 2004). Since information on the effects of this plant on blood cells and biochemical parameters during stress has not been established, therefore the aim of this study was to investigate the effects of *E. praetermissa* leaf extract on the haematological and biochemical parameters of cold-water stress-induced ulceration in albino Wistar rats.

MATERIALS AND METHODS

Extract preparation

Leaves of *E. praetermissa* were collected from the Botanical garden of University of Port Harcourt, River State, Nigeria. The botanical identification and authentication were done by the Chief Herbarium Officer, University of Port Harcourt, River State, Nigeria. The leaves were dried, milled to fine powder in manual engine grinder (Modelcorene, A.5 lander YCIA S.A) and extracted using methanol. The extracted residue was dried in air oven of 50°C to obtain deep green homogenous substance which readily dissolves in distilled water.

Animal selection/handling

Twenty healthy albino Wistar rats were used for the study. They were kept in a spacious and well ventilated cage with suitable temperature, relative humidity, food and drinking water for 14 days to acclimatize before proceeding on the experiment. The animals with body weight ranging from 110 g to 220 g were randomly grouped into 4 groups. Group A served as control which received normal rat chow and water throughout the experiment. Group B represent animals treated with normal saline (10 ml/kg) for 7 days after ulcer induction. Group C animals were treated with extract of *E. praetermissa* (500 mg/kg) for 7 days after ulcer induction. The Group D animals were pre-treated with extract of *E. praetermissa* (500 mg/kg) for two weeks before ulcer induction.

Experimental procedure

Stress induced gastric ulcer was achieved using the cold water immersion method. After 48 h of food deprivation but free access to water, the animals were immersed in cold water of about 21°C and left to swim in the water tank of 17.5 cm long and having internal diameter of 5.2 cm (Landeira-Fernandez, 2004) for 3 h after which they were allowed to dry under 60 watts bulb lamp for another 3 h. Normal saline ((10ml/kg body weight)) was administered orally to animals in Group B while Group C animals were given the extract of *E. praetermissa* for 7 days respectively. The Group D animals were pretreated with the extract of *E. praetermissa* for 14 days before ulcer induction. All administration was through the oral route. 500 mg/kg of the crude extract of *E. praetermissa* was given to the experimental rats according to their weight. The rats were injected with thiopental which rendered them inactive for about 5 min putting them to sleep. The blood samples for analysis were collected through cardiac puncture and the samples were collected into two different containers. The first container was a plain bottle containing no anticoagulant with the sample used for biochemical determination while the second container containing EDTA anticoagulant was used for haematological parameters.

Blood analysis

The blood samples were analyzed to determine the haematological parameters such as: Packed cell volume (PCV), red blood cell (RBC) count, white blood cell (WBC) count, platelet count and haemoglobin concentration (Hb conc.) using an automated haematology ANALYZER KX-21N, made by sysmex Japan. The sysmex KX-21N is an automatic multi-pair blood cell counter for *in vitro* diagnostic use in clinical laboratory. It performs speedy and accurate analysis of blood parameters and detects the abnormal samples. The automated haematology analyzer reading correlated well with readings by the standard manual methods (Samuel et al., 2010). Mean corpuscular volume (MCV), mean corpuscular haemoglobin (MCH) and mean corpuscular haemoglobin concentration (MCHC) were calculated from values of RBC, PCV and HbC as follows:

MCV (fL) = PCV (%) x 10 / RBC count;
MCH (pg) = HbC (g/dl) x 10 / RBC count

And,

$MCHC$ (g/dl) = HbC (g/dl) x 100 / PCV (%).

Table 1. Haematological parameters in control and stress-induced ulceration treated with normal saline and *Emilia praetermissa* groups of rats (n=5).

Parameters	Group A	Group B	Group C	Group D
HbC (g/dl)	7.38 ± 0.13	4.90 ±1.33	9.62 ± 0.34*[a]	9.06 ± 0.78[a]
RBC (× 106 mm^3)	4.55 ± 0.09	4.06 ± 0.59	5.44 ± 0.43	5.90 ± 0.50
MCV (fl)	55.05 ± 0.78	55.8 ± 0.71	60.4 ± 0.60*	56.1 ± 1.90
MCH (pg)	16.18 ± 0.24	11.36 ± 1.41*	18.26 ± 2.89[a]	15.66 ± 2.89
HCT (%)	25.6 ± 0.50	22.8 ± 3.65	32.8 ± 2.49*[a]	33.0 ± 2.89*[a]
MCHC (g/dl)	29.4 ± 0.27	20.3 ± 2.33*	30.2 ± 1.76[a]	28.1 ± 2.72[a]
PLT(× 10^3 mm^3)	104.2 ± 3.79	192.4 ± 79.82*	194.2 ± 58.68*	229.2 ± 138.3*[a]
WBC(×10^3 -mm^3)	3.3 ± 0.25	5.14 ± 0.52	6.88 ± 1.48*	5.46 ± 1.01
LYMP (%)	84.1 ± 0.67	72.0 ± 4.54*	70.3 ± 2.07*	59.3 ± 1.18*[a]

*$P<0.05$ vs. control, Values are mean ± SEM; [a] $P<0.05$ Group B vs. Groups C and D. Values are mean ± SEM.

Biochemical analysis of the serum enzymes for ALT and AST was by the method of Reitman and Frankel (1957). Bilirubin concentration determination was by 2.5-dichlorphenyldiazonium method described by Schlebusch et al. (1988).

Statistical analysis

All data were presented as mean ± SEM. The one way ANOVA was used to analyze the data, followed by a post-hoc test (LSD). The results were considered significant at $P<0.05$.

RESULTS

Effect of stress-induced gastric ulceration on haematological parameters in albino rats

Table 1 shows the haematological parameters of the various groups after the duration of the experiment. The mean values for haemoglobin concentration (HbC), red blood cell count (RBC), hematocrit (HCT), mean corpuscular haemoglobin and mean corpuscular volume (MCV) were lower in Group B when compared with Group A and these differences were not statistically significant ($P>0.05$) while values for mean corpuscular haemoglobin (MCH) and mean corpuscular haemoglobin concentration (MCHC) and lymphocyte count were significantly lower in Group B when compared with Group A ($p<0.05$). White blood cell count and platelet counts were higher in Group B than Group A and the difference were statistically significant ($p<0.05$).

Effect of *E. praetermissa* on haematological parameters in stress induced gastric ulceration in albino rats

All values except the value for lymphocyte count were higher in Groups C and D when compared with Groups A and B respectively. The values were higher in Group D than in Group C. Significant differences were noticed in

haemoglobin concentration, mean corpuscular volume, heamatocrit, platelet and white blood cell count in Groups C and D when compared with Group A (Control). Haemoglobin concentration, haematocrit, mean corpuscular haemoglobin and mean corpuscular haemoglobin concentration in rats treated with *E. praetermissa* after ulcer induction (Group C) were significantly higher ($p<0.05$) than those in which ulcer was induced by stress only (Group B). The rats pretreated with *E. praetermissa* (Group D) had higher values for haemoglobin concentration, hematocrit and platelet count than those in Group B. The increases were statistically significant ($p<0.05$). Lymphocyte count in Group D was significantly reduced when compared with Group B.

Effect of stress-induced gastric ulceration on biochemical parameters in albino rats

Table 2 shows the concentration of alanine amino transferase (ALT), aspartate amino transferase (AST) and bilirubin in the blood of control and treated rats. There were slight increases in the concentration of alanine amino transferase (ALT), aspartate amino transferase (AST) and slight decrease in bilirubin concentration in Group B when compared with Group A. The differences observed were not statistically significant.

Effect of *E. praetermissa* on biochemical parameters in stress induced gastric ulceration in albino rats

There were significant decreases ($p<0.05$) in alanine amino transferase (ALT) and aspartate amino transferase (AST) in Groups C and D when compared with Group A (Control) and Group B (stress-ulcerated rats). Bilirubin concentration in Groups C and D appeared to be unchanged when compared with the control.

Table 2. Biochemical parameters in control and stress-induced ulceration treated with normal saline and *Emilia praetermissa* groups of rats (n=5).

Parameters	Group A	Group B	Group C	Group D
Aspartate amino transferase (µ/L)	196.2 ± 9.24	198.8 ± 2.03	126.8 ± 13.1*[a]	171.6 ± 26.71*
Alanine amino transferase (µ/L)	64 ± 4.9	67 ± 2.0	25.4 ± 3.26*[a]	53.6 ± 10.86*[a]
Bilirubin (µmol/L)	7.2 ± 0.23	5.96 ± 0.62	8.74 ± 0.77	8.76 ± 2.19

* $P < 0.05$ vs. control, Values are mean ± SEM; [a] $P < 0.05$ Group B vs. Groups C and D. Values are mean ± SEM.

DISCUSSION

The changes in haematological and biochemical parameters of stress-ulcerated animals treated with *E. praetermissa* extract were the focus of this study. Physiological studies have shown that stress from any source can influence on the endocrine, hemopoietic and immune systems. Cytokines and cortisol seem to play an important role in the communication between these systems (Maes et al., 1998 and Benoit et al., 2001) Previous studies have shown that stress increases erythrocytes, neutrophils and platelets counts, whereas lymphocytes, eosinophils and monocytes decrease in number (Dhabhar, 1995). The magnitude of stress-induced changes is significantly reduced in adrenalectomised animals. It is suggested that endocrine factors released during stress modulate leucocyte trafficking and result in the redistribution of leucocytes between the blood and other immune compartments (Dhabhar, 1995). The activation of sympathetic nervous system may also have a role to play. Lymphocytes and monocytes express receptors for several stress hormones, including norepinephrine and epinephrine. Thus stressful events could alter immune function. This alteration in immune function due to decrease in lymphocytes and basophils was found in the subjects of this study, confirming the stress related changes reported in the literature.

The results of this study showed that there were no significant changes in RBC and HCT in response to stress. This is similar to no significant change in red blood cell or haematocrit observed by Qureshi et al. (2002). The reason for this might be due to physiological factor pertaining to female animals such as low level of testosterone, No significant changes in MCV and Hb observed in this study contrast significant increase Hb and MCV reported by Maes et al. (1998). The differences might be due to the fact that we used majorly female animals which normally have low red blood cell count because of the reason stated above. The significant increase in Platelet count and decrease in lymphocyte count were consistent with the earlier findings reported by Qureshi et al. (2002) and Dhabhar et al. (1995). However, nearly all the values of haematological parameters (HbC, RBC, HCT, MCV, MCH, MCHC and Platelet) determined except for lymphocyte increased in the groups treated with *E. praetermissa* before and after

stress induction. Owing to the fact that it is a vegetable eaten as food (Burkill, 1985), the increase might be due to the presence of blood forming vitamins (Bamidele et al, 2010).

The slight increases in ALT and AST in stress-ulcerated rats were closely related to the findings of earlier researchers (Roland et al., 2009). The leaf extract of *E. praetermissa* decreased the activities of liver enzymes ALT and AST in stress-induced rats. Liver enzymes (ALT and AST) are released into the blood whenever liver cells are damaged and enzyme activity in the plasma is increased (Edwards et al., 1995). The fact that the enzyme activities were reduced showed that the extract protects against hepatic damage. The hepatoprotective functions of the plant might be due the presence of flavonoids, which are antioxidants that mop up super oxide anions. No significant change in bilirubin concentration observed in this present work can be attributed to any change in the RBC or HbC in the stress ulcerated animals (destruction of red blood cells produce bilirubin).

Conclusion

The results suggest that *E. praetermissa* improves haemopoiesis even in the presence of stress induced ulceration in albino Wistar rats. It also protects against hepatic damage. Therefore *E. Praetermissa* consumption as vegetable as well as anti-ulcerogenic agent may be encouraged since it brings about reduction on the effect of stress on haematological parameters.

REFERENCES

Bamidele O, Akinnuga AM, Olorunfemi JO, Odetola OA, Oparaji CK and Ezeigbo, N (2010). Effects of aqueous extract of *Basella alba* leaves on haematological and biochemical parameters in albino rats. Afr. J. Biotechnol. 9(41): 6952-6955.

Benoit D, Esa L, Ralph G (2001). The driving test as a stress model: Effects on blood picture, serum cortisol and the production of interleukins in man. Life Sci. 68(14):1641-1647.

Burkill HM (1985). The useful plants of west tropical Africa. 2nd Edition. Vol. 1. Families A – D Royal Botany Garden. p. 960.

Dhabhar FS, Miller AH, McEwen BS, Spencer RL (1995). Effects of stress on immune cell distribution. Dynamics and hormonal mechanisms. J. Immunol. 154 (10):5511-5527.

Edwards CRW, Bouchier IAD, Haslet C, Chilvers ER (1995). Davidson's Principles and Practice of Medicine, 17th edn. Churchill Livingstone, pp. 488-490.

Fernandez JL (2004). Analysis of the cold restraint procedure in gastric ulceration and body temperature. Physiol. Behav. 82(5):827-833.

Gbore FA, Oginni O, Adewole AM Aladetan JO (2006). The effect of transportation and handling stress on haematology and plasma biochemistry in fingerlings of *Ciarias gariepinus* and *Tilapia zilli*. World J. Agric. Sci. 2(2):208-212.

Landeira-Fernandez J (2004). Analysis of the cold-water restraint procedure in gastric ulceration and body temperature. Physiol. Behav. 82:827-833

Maes M, Vander Planken M, Van Gastel A (1998). Influence of academic stress on haematological measurements in subjectively healthy volunteers. Psychiatry Res. 80:201-212.

O'Hara M, Keifer D, Farrel K, Kemper K (1998). A review of 12 commonly used medicinal herbs. Arch. Family Med. 7:527-536.

Palombo EA (2011). Traditional medicinal plant extracts and natural products with activity against oral bacteria: potential application in the prevention and treatment of oral diseases. Evidenced-based Complementary and Alternative Medicine. 2011: Article ID 680354, p. 15.doi:10.1093/ecam/nep067

Quareshi F, Alam J, Khan MA, Sheraz G (2002). Effect of examination stress on blood parameters of students in a pakistani medical college. J. Ayub. Med. Coll. Abbottabad, 14(1):20-22.

Reitman S, Frankel AS (1957). A colorimetric method of determination of serum glutamic oxaloacetic and glutamic pyruvic transaminases. Am. J. Clin. Pathol. 28:53-63.

Roland Von K, Chiara CA, Stefan B, Marie LG, Hugo S, Jean PS (2009). Association between posttraumatic stress disorder following myocardial infarction and liver enzyme levels: A prospective study. J. Digest. Dis. Sci. 0163 -2116.

Samuel OI, Thomas N, Ernest OU, Imelda N, Elvis NS, Ifeyinwa E (2010). Comparison of haematological parameters determined by the Sysmex KX-21N automated haematology analyzer and the manual counts. BMC Clin. Pathol. 10:3.

Schlebusch H, Liappis N, Niesen M (1988). Determination of bilirubin in capillary plasma by a direct photometric method (DPM, bilirubinometer) and the chemical determination of bilirubin in the bilirubin determination (2,5-dichlorophenyldiazonium method) in serum of venous blood samples. Klin Padiatr. 200(4):330-334.

Spirt MJ (2004). Stress related mucosal disease: risk factors and prophylactic therapy. Clin. Ther. 26(2):197-213.

Spirt MJ, Stanley S (2006). Update on stress ulcer prophylaxis in critically ill patients. Crit. Care Nurse 26(1):18-20, 22-28.

Stollman N, Metz DC (2005). Pathophysiology and prophylaxis of stress ulcer in intensive care unit patients. J. Crit. Care 20(1):35-45.

Tan PV, Njimi CK, Ayafort JF (1997). Screening of some African Medicinal plants for antiulcerogenic activity. Phytother. Res. 11:45-47.

21

Extraction and application of dye extracted from eriophyid leaf galls of *Quercus leucotrichophora*-A Himalayan bluejack oak

Preeti Mishra* and Vidya Patni

Department of Botany, Plant Pathology, Tissue Culture and Biotechnology Laboratory, University of Rajasthan, JLN Marg, Jaipur-302004, India.

Oaks (*Quercus* sp.) are susceptible to a wide diversity of gall-forming insects. When the insect pupates and leaves, the gall can be used as a dyestuff. Kalamkari unit's method (India) was followed for the extraction of dye. A dark brown dye was obtained from the leaf gall of *Quercus leucotrichophora* caused by *Eriophyes* sp. Three different types of cloths and three different types of yarns were used in the experiment to observe the strength of dye. Cotton-Jute (C-3) sample showed dark brown color with myrobalan and ferrous sulphate, yellowish, cream and purplish color with potassium dichromate, stannous chloride and aluminium sulphate respectively.

Key words: Extraction, jute, Kalamkari, myrobalan.

INTRODUCTION

Galls are typical plant growths that provide nourishment, shelter and protection to the inducer or its progeny. They are in a sense new plant organs because it is the plant that produces the gall in response to a specific stimulus provided by the invading insect (Shorthouse et al., 2005). With about 2000 different galls, the Indian subcontinent displays a rich variety of gall flora. Gall inducing parasites of peninsular India are endemic, whereas those in the temperate Himalayan slopes and in the Indo-Gangetic plains show affinity to central Asian and European gall inducing elements (Raman, 2007; Ramani and Kant, 1989; Mani, 2000). The number and diversity of gall-inducing insects are great (Espirito-Santo and Fernandes, 2007). Oaks (*Quercus* sp.) are susceptible to a wide diversity of gall-forming insects. When the insect pupates and leaves, the gall can be used as a dyestuff. Galls are used as powder or decoction. Oak galls have been used extensively over a period of centuries in the manufacture of certain inks and tanning of leather.

Gallnuts (Aleppo galls) of *Q. infectoria* are a tannin mordant and a golden-tan dye with alum mordant, pale grayed-tan with copper mordant, light brown with tin mordant. Thus obtained Iron-gall ink was the most important ink in western history. To make iron-gall ink, galls from oak trees were crushed to obtain gallotannic acid. The gallotannic acid was mixed with water. This ink did not rub off documents. Unlike paper, parchment was not absorbent, so carbon-based ink easily rubbed away (Shirley, 2005). Leaf galls of Manjuphal tree (*Q. leucotrichophora*) can also be used for the manufacture of dyestuff. The status for detection of the dye in Indian oak species is yet to be set. Not any remarkable work has been done in the particular area. Pigmentary molecules containing aromatic ring structure coupled with a side chain are usually required for resonance and thus to impart colour (Vanker, 2000; Purohit et al., 2007, Kamel et al., 2005).

The present study deals with eriophyid leaf galls on *Q. leucotrichophora*. The study area (Uttranchal, central Himalayas) is located between 79°23' and 79°42' E, and

---

*Corresponding author. E-mail: drpreetipathak@gmail.com.

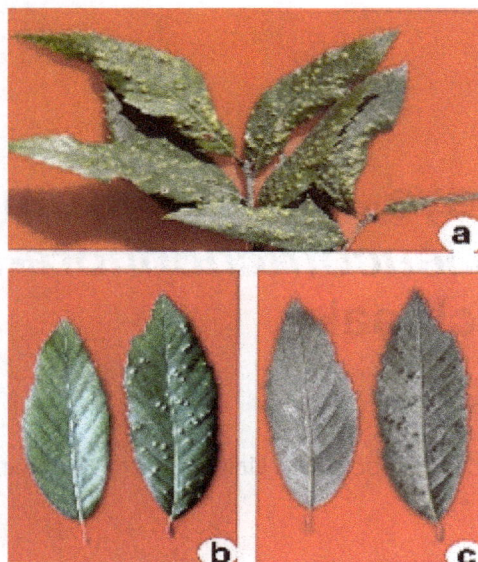

Figure 1. Morphology of eriophyid galls, a-Pustule like galls on adaxial surfaces of leaves of *Quercus leucotrichophora* caused by *Eriophyes* sp.; b and c - Comparison between affected and normal leaf surfaces (adaxial and abaxial) of *Quercus leucotrichophora*.

29°20' and 29°30' N. The altitude ranges between 1300 and 2600 meter above sea level. The art of making vegetable dyes is one of the oldest known to man and dated back to the dawn of civilization. In India, it was widely used for colouring of fabrics and other materials (Siva, 2007). Though, the very earliest dyes were discovered by accident using berries and fruits, with the experimentation and gradual development the vegetable dyes have resulted into a highly refined art. India's expertise in vegetable dyes dates back to ancient times. As a defense response gallotannin, gallic acid, tannin acid etc start accumulating in the gall tissues.

Oak galls contain large amounts of gallic and tannic acid, which are widely used in the manufacture of medicines, insecticides and permanent inks.

MATERIALS AND METHOD

The dye was extracted from the leaf galls of *Q. leucotrichophora* caused by *Eriophyes* sp. These galls are found in abundance on the leaf\ves and after drying of leaf, the whole galled leaf can be used for dyeing. For the extraction of dye Kalamkari unit's method was followed (The Art and Craft of National Dyes by Shakuntala Ramani) with some modifications.

Procedure

The galled leaves were dried in shade, crushed and powdered in

grinder. Gall powder thus prepared was taken in distilled water as solvent for extraction. This pasty mass was kept for 10 to 15 days to get colour of dye. This extract was filtered and used for dyeing. Cloth used for dyeing was boiled in NaOH solution to remove the starch from the cloth, then washed with cold distilled water. This cloth was then transferred in mordant (Myrobalan) then it was put in the gall extract for some time. Then the cloth was treated with (colour fixative) and dried in sunlight. Different types of cotton cloths (5 X 5 cm) and yarns were experimented for dyeing. Three different types of cloths and three different types of yarns were used in the experiment to observe the strength of dye. Cotton-synthetic mix (C-1), Cotton pure (C-2) and Cotton- Jute mix (C-3) were taken. Yarns of silk (Y-1), cotton (Y-2) and wool (Y-3) were taken into account.

Similarly effect of various mordants on color of dye extracted from the galls was also studied on the cloth found best in the above experiment. This was achieved by incorporating different mordants like:

i) Stannous chloride
ii) Aluminium sulphate
iii) Ferrous sulphate and
iv) Potassium dichromate separately, each at a concentration of 3% of the gall extract (5 ml).

Cloth pieces were individually soaked with the mixture of extract mordant solution. After soaking, the cloth was dried in sunlight for 2 h. The sun-dried cloth was further evaluated for its colour, lightness and was fastness. Wash fastness was tested by washing with soap water (10% w/v) and heat resistance was tested by keeping the cloth at various temperatures namely: 50, 60, 70°C for 30 min in the oven.

RESULTS AND DISCUSSION

Leaf galls of *Q. leucotrichophora* are induced by a mite *Eriophyes* belonging to family Eriophydae, well known for inducing galls on different plants. Galls on leaf affect the plant in many ways. Heavily infected plants can be found throughout, between the months of August to December. The leaf galls caused by *Eriophyes* were epiphyllous, hemispherical and pustule like. These were solitary pouch-gall, about 5 mm in diameter, generally about 20 or many galls were observed on a leaf. The galls were smooth, yellowish green above and below with brown erineum (Figure 1 a to c). Structural and physiological significance of different plant galls have been critically reviewed by Ramani and Kant (1989), Schonrogge et al. (2000), Stone et al. (2002), Price (2003), Raman et al. (2005), Raman (2007) and Diamond et al. (2008).

A dark brown dye was obtained from the leaf of gall *Q. leucotrichophora* caused by *Eriophyes* sp. Effects of Myrobalan and dye colour are presented in Figures 2, 3 and 4 and Tables 1 and 2. Cotton-synthetic mix (C-1) showed light brown coloration, while Cotton-Jute (C-3) and pure cotton (C-2) cloth samples showed best coloration (dark brown and brown) by repeated dipping in gall extract (Figures 2 a to c). Different types of yarns, silk yarn (Y-1), Cotton yarn (Y-2) and wool yarn (Y-3) showed bright brown, brown and camel color respectively (Figure

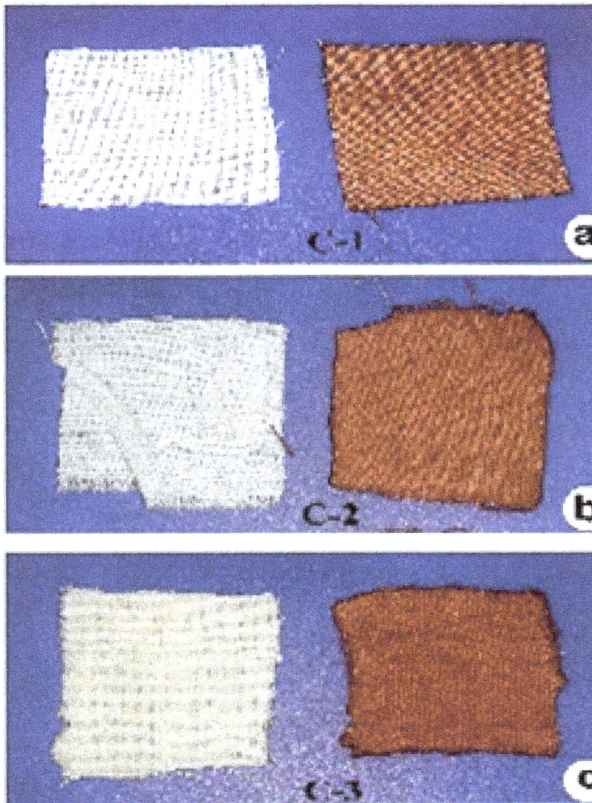

Figure 2. Application of dye extracted from eriophyid leaf galls of *Quercus leucotrichophora* + myrobalan on different type of cloth types, a-Untreated white cotton-synthetic mix (C-1) cloth and dyed light brown colored cloth; b- Untreated white cotton pure (C-2) cloth and dyed brown colored cloth; c- Untreated white cotton-jute mix (C-3) cloth and dyed dark brown colored cloth.

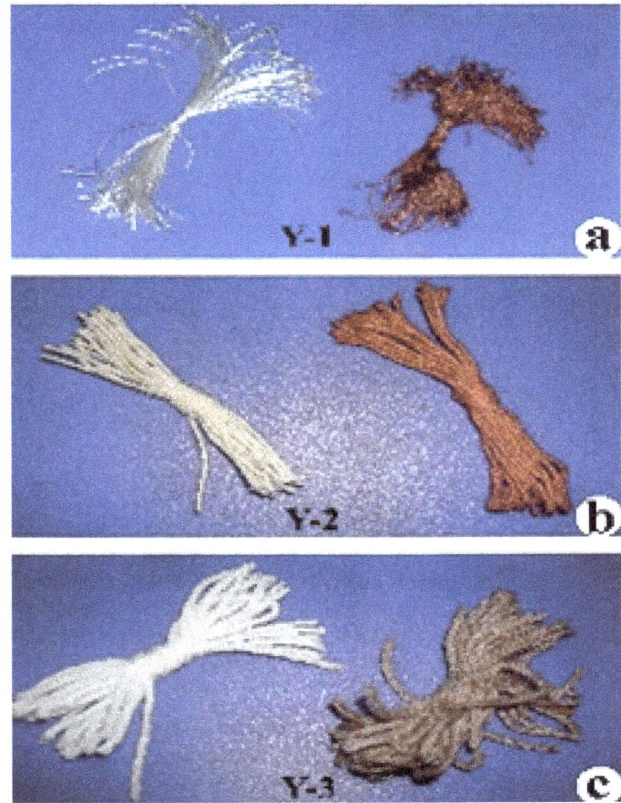

Figure 3. Application of dye extracted from eriophyid leaf galls of *Quercus leucotrichophora* + myrobalan on different type of yarns, a- Untreated white silk yarn (Y-1) and dyed bright brown colored yarn; b- Untreated white cotton yarn (Y-2) and dyed brown colored yarn; c- Untreated wool yarn (Y-3) and dyed grayish colored yarn.

3 a to c) with the treatment of myrobalan and dye extracted. When Cotton Jute (C-3) was subjected to various mordants, it showed different type of shades according to mordants. Effects of different mordants on the cloth with dye and color fastness are presented in Figure 4 and Table 2.

Cotton Jute cloth showed best coloration (Figure 2 c) after treating with dye hence it was further selected for dyeing with mixture of different mordant and extracted dye. Cotton-Jute (C-3) sample showed dark brown color with Myrobalan and Ferrous sulphate, yellowish, cream and purplish color with potassium dichromate, stannous chloride and aluminium sulphate respectively (Figure 4). The application of natural dyes in textile industry in the form of:

a) Dyeing of yarns, which are then woven into cloth, carpet or any other usable form.

b) Dyeing of cloths woven earlier.
c) Block printing, where the textile materials are printed with the help of printing blocks and,
d) Kalamkari where the "Kalam" or pen is used to draw beautiful designs on the cloth (Gopi, 2004).

Eriophyid leaf gall extract of *Q. leucotrichophora* yielded large amount of pigments. Three types of cloths and three types of yarns were tested against dye extracted from leaf galls. Galls of some *Quercus* spp. From India (Himalayan region) are used for dyeing of silk and wool from ancient times. It gives brown and camel colours with alum and iron mordants. The consequent solar drying, soap washing and heating did not alter the colour shade developed during the dyeing process that is colour, light and wash fastness. Similar findings were reported in *Zizyphus* gall (Gopi, 2004). Shirley (2005) also reported some natural dyes form oak galls. Oak galls contain large amounts of gallic and tannic acid, which are widely used

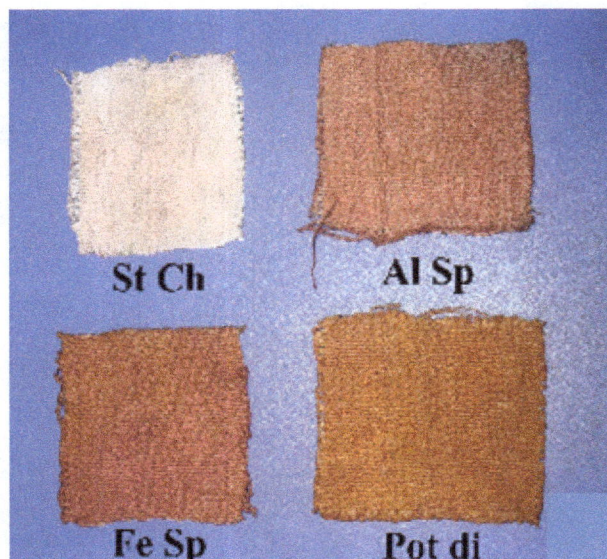

Figure 4. Effect of different mordants on the cotton-jute cloth
(C-3) with dye obtained. Cream color with stannous chloride,
pink brown with aluminium sulphate, dark brown with ferrous
sulphate, yellowish brown with potassium dichromate.

Table 1. Effect of Myrobalan and dye extracted from eriophyid leaf galls *of Quercus leucotrichophora* on different types of cloth and yarns.

S.No.	Sample	Color obtained with the treatment of myrobalan+dye extract
1.	Cotton-synthetic mix (C-1)	Light brown
2.	Cotton pure (C-2)	Brown
3.	Cotton- Jute mix (C-3)	Dark brown
4.	Silk yarn (Y-1)	Bright-brown
5.	Cotton yarn (Y-2)	Brow
6.	Wool yarn (Y-3)	Grayish

Table 2. Effect of different mordants with dye extracted from eriophyid leaf galls of *Quercus leucotrichophora* on the cotton-jute cloth.

S.No.	Mordants	Color obtained	Color fastness
1.	Myrobalan	Dark brown	Slight
2.	Stanous chloride	Cream	Slight
3.	Alluminium sulphate	Purplish	Slight
4.	Ferrous sulphate	Dark brown	Slight
5.	Potassium dichromate	Yellowish brown	Slight

in the manufacture of medicines, insecticides and permanent inks. The Aleppo oak gall of Asia Minor, produced by a cynipid wasp, contains about 65% tannic acid.

For centuries the best permanent inks were made from these galls (Elmer, 1999).

REFERENCES

Diamond SE, Blair CP, Abrahmson WG (2008). Testing the nutrition hypothesis for the adaptive nature of insect galls: does a non-adapted herbivore perform better in galls? Ecol. Entomol. Doi: 10.1111/j.1365-2311.2007.00979.x.

Elmer E (1999). "Iron Gall Ink." Paper Conservation Department. Rotterdam. www.knaw.nl/ecpa/ink/html/make.html.

Gopi M (2004). Biotechnlology and industrial application of a mite gall on *Zizyphus jujuba*. XXVII Conference of Indian Botanical Society, Oct. 29-31, S.K. University, Anantapur.

Kamel MM, El-Shishtawy RM, Yussef BM, Mashaly H (2005). Dyes Pigments, 65: 103-110.

Mani M S (2000). Plant Galls of India. Oxford. IBH. New Delhi. pp. 490.

Price PW (2003). Macroevolutionary theory on macroecological patterns, Cambridge University Press, Cambridge, pp. 291

Purohit A, Mallick S, Nayak A, Das NB, Nanda B, Sahoo S (2007). Developing multiple natural dyes from flower parts of Gulmohur. Curr. Sci., 92(12): 1681-1682.

Raman A (2007). Insect-induced plant gall of India: unresolved questions. Curr. Sci., 92(6): 748-757.

Raman A, Schaefer CW, Withers TM (2005). Galls and gall-inducing arthropods: an overview of their biology, ecology and evolution. In:

Raman, A., Schaefer, C.W. and Withers, T.M. eds. Biology, ecology, and evolution of gall-inducing arthropods Enfield, CT, USA: Science Publishers, pp. 1–33.

Ramani V, Kant U (1989). Phenolics and enzymes involved in phenol metabolism of gall and normal tissues of *Prosopis cineraria* (Linn.) Druce *in vitro* and *in vivo*. Proc. Indian Nat. Sci. Acad., B55 (5,6): 417-420.

Schonrogge K, Harper LJ, Lichtenstein CP (2000). The protein content of tissues in cynipid galls (Hymenoptera: Cynipidae): similarities between cynipid galls and seeds. Plant Cell Environ., 2: 215-222.

Siva R (2007). Status of natural dyes and dye yielding plants in India. Curr. Sci., 92(7): 916-925.

Stone GN, Schonrogge K, Atkinson RJ, Bellido D, Pujade-Villar J (2002). The population biology of oak gall wasps (Hymenoptera: Cynipidae). Ann. Rev. Entomol., 47: 633-668.

22

Fermentation in cassava *(Manihot esculenta* Crantz) pulp juice improves nutritive value of cassava peel

Naa Ayikailey Adamafio*, Maxwell Sakyiamah and Josephyne Tettey

Department of Biochemistry, University of Ghana, P. O. Box LG 54, Legon, Ghana.

A major challenge in using cassava peel as feed for animals is the presence of cyanogenic glycosides and the low concentration of protein. The present study investigated the possibility of upgrading cassava peels using fermented cassava pulp juice. Cassava pulp juice was squeezed out of grated cassava pulp and fermented for 3 days at ambient temperature. The microorganisms in the fermented pulp juice were identified as *Aspergillus niger*, *Aspergillus flavus* and *Lactobacillus spp.* Non-sterile cassava peels were sun-dried, milled and inoculated with fermented cassava pulp juice over a 7-day period. Controls were treated with either sterile distilled water, autoclaved inoculum or phosphate buffer (pH 5) over the same period. After 7 days, the cyanogenic glycoside content of the peels, determined by the silver nitrate titration method, had decreased to 12.3% (p < 0.05) of the value for untreated peels while the cyanogenic glycoside content of the controls was 38.8 - 42.9%. Proximate analysis of 7-day inoculum-treated and untreated cassava peels showed that the protein content of the treated peels had increased 10-fold and significant decreases in starch and fat content were recorded. The fibre content remained unchanged. The present findings show that microorganisms present in fermented cassava pulp juice are capable of enhancing the nutritional value of cassava peels by increasing the protein content and reducing the cyanogenic glycoside content to levels safe for consumption by livestock.

Key words: Cassava peel, pulp juice, cyanogenic glycosides, microorganisms.

INTRODUCTION

Substantial quantities of cassava (Manihot esculenta Crantz) peel, that could provide carbohydrates for livestock (Ezekiel et al., 2010; Ukanwoko et al., 2009), are generated in Ghana annually from the processing of cassava into starch, chips and gari. Maximum utilization of this bioresource in an integrated agricultural system would alleviate the major challenge of inadequate dry-season feed for livestock in Ghana and generate additional income for cassava processors and farmers. Amino acid-derived β-glycosides of α-hydroxynitriles, termed cyanogenic glycosides, are produced by a variety of plants as defense biomolecules (Gleadow et al., 2008; Bak et al., 2006). The cassava plant produces two toxic cyanogenic glycosides, linamarin (2-β-D-glucopyranosyloxyl isobutyronitrile) and lotaustralin

(methylbutyronitrile), a large proportion of which is present in the peel (Cardoso et al., 2005). The enzymes linamarase and hydroxynitrile lyase which catalyze the degradation cyanogenic glycosides to release hydrogen cyanide (HCN), are sequestered in different tissues of the cassava plant and released when the tissue is disrupted (Kimaryo et al., 2000). Chronic ingestion of fresh or processed cassava peel-based diets containing sub-lethal dietary cyanide has reportedly caused impaired thyroid function and growth, neonatal deaths and lower birth rates in animals (Fatufe et al., 2007; Ernesto et al., 2000). Another limitation to the use of cassava peels as animal feed is its low protein content (Ezekiel et al., 2010).

Sun drying, the commonest method used in the treatment of cassava peels for livestock feeding by subsistence farmers in Ghana, is only partially effective in reducing cyanogenic glycoside content (Tewe, 1989). Various methods of processing, some more effective than others, have been described (Perera, 2010; Kuti and Konoru, 2006). However, it is important to identify

---

*Corresponding author. E-mail: adamafio@gmail.com.

Abbreviation: HCN, hydrogen cyanide.

methods that are highly effective, require no sophisticated equipment and can readily be adopted by subsistence farmers.

Although the use of pure cultures of microorganisms such as *Aspergillus spp, Saccharomyces spp* and *Lactobacillus spp*, or combinations of these is reported to cause a substantial decrease in cyanogenic glycoside content (Oboh, 2006; Oboh and Akindahunsi, 2003), its application by subsistence farmers in Ghana may be undermined by the cost and techniques involved. An alternative approach may be the use of fermented cassava pulp juice, which contains cyanohydrophilic microorganisms and is readily available as a waste product of starch and gari producing industries. The present study was, therefore, aimed at investigating the effectiveness of naturally fermented cassava pulp juice in the remediation of cassava peels.

MATERIALS AND METHODS

Cassava roots were purchased from Madina market in the Ga East District of Accra, Ghana. The samples were washed with water and peeled to obtain both parenchyma and peels.

Treatment of peel and cyanogen determination

Peeled cassava parenchyma was grated and pressed in a fine cloth to obtain pulp juice. The juice was allowed to ferment for 3 days at ambient temperature (26 - 32°C). This served as the inoculum. The peels were cut into small segments, sun-dried for 3 days and milled into fine powder. A 5 ml aliquot of the inoculum was added to 3.0 g of milled cassava peels and incubated, without aeration or shaking in capped plastic containers, at 32°C for 1 to 7 days. Control samples were treated with equal volumes of either 0.2 M phosphate buffer (pH 5.3) or autoclaved inoculum. After termination of the fermentation process, samples were dried at 40°C to constant weight and the cyanogenic glycoside content was determined using the silver nitrate method (AOAC, 1965).

Identification of microorganisms

Nutrient agar plates were prepared from 1.3% nutrient broth and 1.5% agar-agar, inoculated with 0.5 ml of the inoculum and incubated at 37°C for 24 h. The isolates were subcultured to obtain distinct colonies. These were subsequently used to inoculate 6.5% Sabouraud Dextrose Agar plates which were then incubated at 37°C for 3 days. Light microscopy was performed to identify fungi on the basis of their colony colours and hyphal characteristics (Thom and Raper, 1945; Burnett, 1972). In order to identify bacteria, peptone agar plates (2.5% peptone water broth and 1.5% agar-agar) were inoculated aseptically with the distinct colonies from the nutrient agar plates and incubated at 37°C for 24 h. The bacteria were identified on the basis of colony and cell morphology, gram stain reaction and standard biochemical tests (Harley and Prescott, 1990; Buchnan and Gibbons, 1974).

Nutrient composition of cassava peels

Protein, non-structural carbohydrate composition and content were determined using the macro-Kjedahl method (AOAC, 1970), the Luff-Schoorl method (Kirk and Sawyer, 1991) and a gravimetric method (AOAC, 1970), respectively. Moisture content was also measured (AOAC, 1970) and values of nutrients were expressed on dry weight basis.

Data analysis

Statistical analysis was performed using Microsoft Excel and SPSS. Analysis of variance was carried out using Statgraphics-plus Software Programme, Version 3.0. Differences were considered significant if p was less than 0.05.

RESULTS

As shown in Figure 1, the cyanogenic glycoside content of all categories of cassava peels diminished during the 7 day incubation period. However, cassava peels treated with fermented cassava pulp fluid showed the most significant decrease in cyanogenic glycosides (p < 0.05).The level dropped to 53.8 mg/kg after the fermented pulp fluid treatment from 351 mg/kg in the dried untreated peels. This represents 12.3% HCN relative to the untreated peels which was set as 100%. The buffer and sterile inoculum treatment gave values of 43.1% and 34.3% respectively after the 7th day of fermentation.

The inoculum-treated peels displayed the highest reduction in HCN content. The buffer-treated control samples showed the least reduction in HCN.

A general increase in protein content was observed for all samples during the treatment period (Figure 2) The most dramatic increase (10.8 fold) in protein occurred in the inoculum-treated peels after the 7th day (p < 0.05). No significant alteration in fibre content was recorded (Figure 3). However, the amount of non-structural carbohydrates, comprising starch and sugars, reduced after treatment with both the inoculum and autoclaved inoculum (Figure 4).

The protein content of the inoculum-treated peels increased considerably (p < 0.05). There was no significant change in the fibre content of the inoculum-treated peels. The non-structural carbohydrate con-tent of both autoclaved inoculum-treated and inoculum-treated peels reduced significantly(p < 0.05).

Both fungi and bacteria were present in the inoculum. The fungi were identified as *A. niger* and *A. flavus*. Gram positive motile cocci and rod-shaped (bacilli) bacteria were present in the fermented cassava pulp juice. Biochemical tests showed that the bacteria included lactose fermenters and non gas producers. They neither metabolized indole nor sulphur.

DISCUSSION

The objective of the present study was to determine whether or not fermented cassava pulp juice would be an effective agent in the bioremediation of cassava peels. A reduction in the pH of the cassava pulp juice from 7.5 to

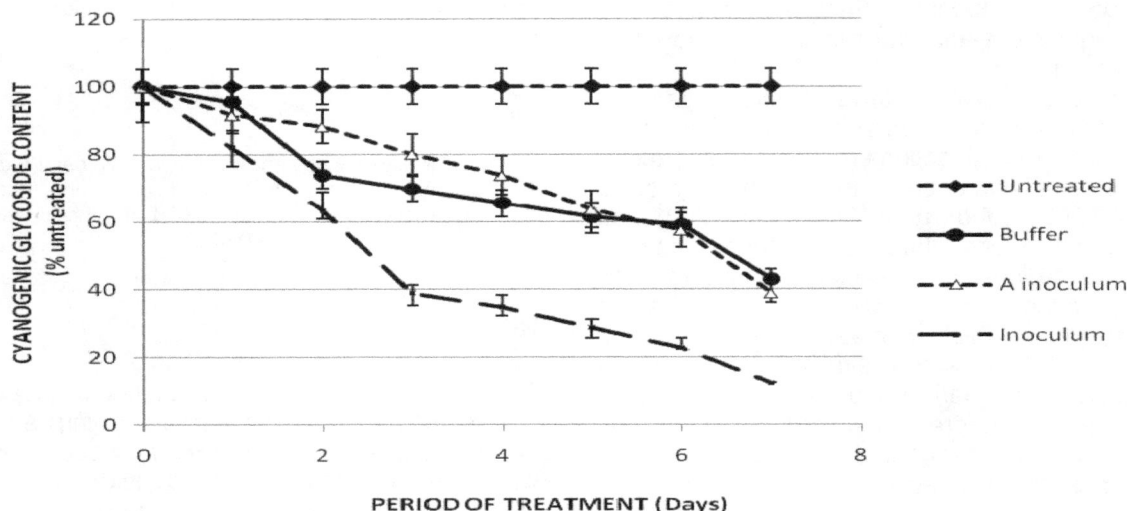

Figure 1. Effect of microbial treatment on cyanogenic glycoside content of cassava peel. Dried milled cassava peels were treated with fermented cassava pulp juice, autoclaved inoculum (A-inoculum) or phosphate buffer for 7 days and the amount of cyanogenic glycosides was determined by the silver nitrate method. The values are presented as mean ± SEM.

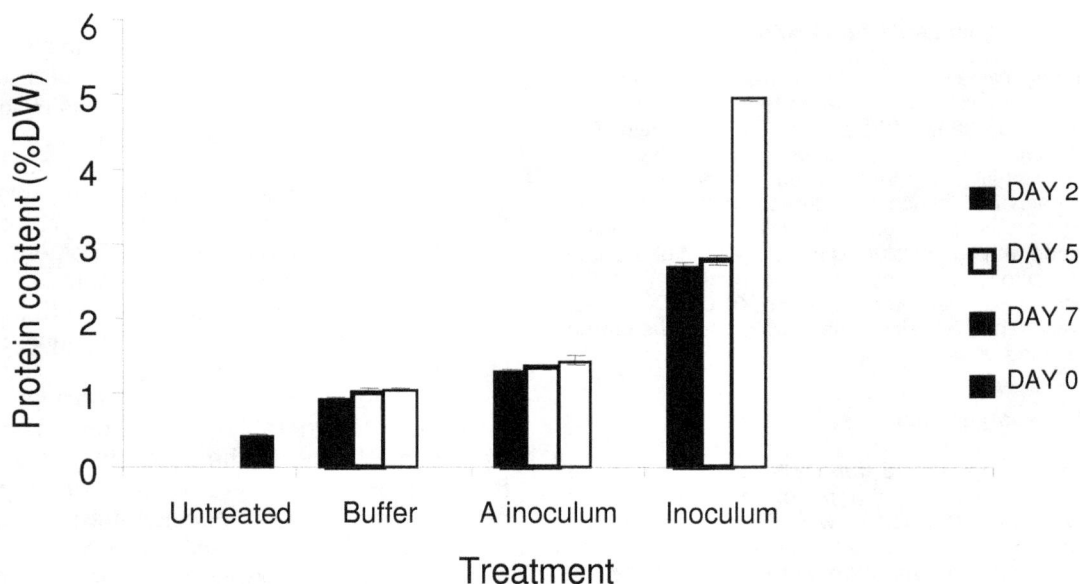

Figure 2. Effect of microbial treatment on the protein content of cassava peels. Dried milled cassava peels were treated with fermented cassava pulp juice, autoclaved inoculum (A-inoculum) or phosphate buffer for 7 days and protein content of the peels was determined using the Kjedahl method.

5.3 after 3 days of fermentation prior to the treatment of cassava peels was observed. This necessitated the inclusion of buffer (pH 5.3) treatment to discount the possibility that any alterations in cyanogen content were caused by acidification.

Treatment of the cassava peels with fermented cassava pulp juice resulted in a progressive decrease in cyanogen content during the 7 days period by 86.2% compared with 58.8 and 64.4% for buffer-treated and autoclaved inoculum-treated peels respectively (Figure 2). The observed reduction in cyanogen content of the buffer-treated and autoclaved inoculum-treated peels is attributable to degradation by indigenous cyanophilic microorganisms associated with the non-sterile cassava

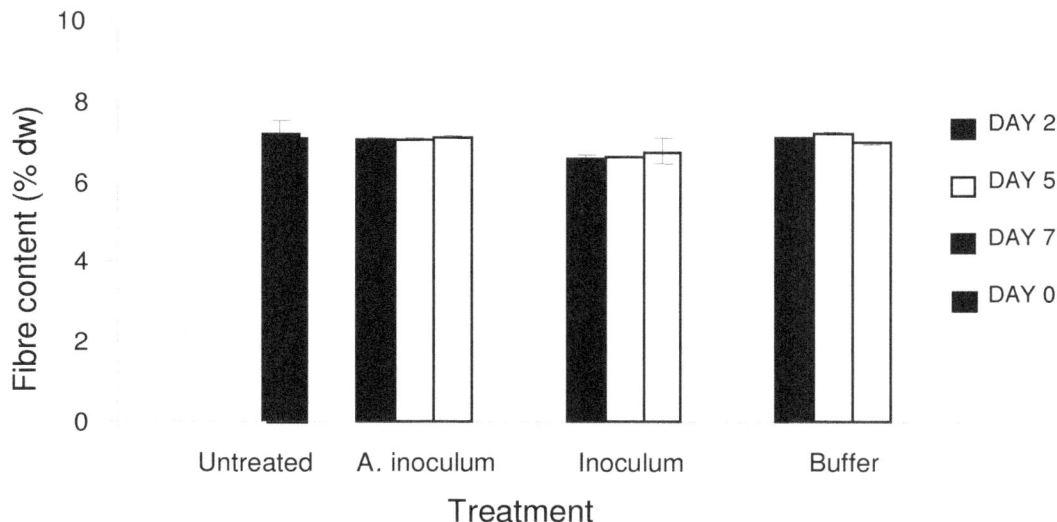

Figure 3. The effect of microbial treatment on fibre content of cassava peels. Dried milled cassava peels were treated with fermented cassava pulp juice, autoclaved inoculum (A-inoculum) or phosphate buffer for 7 days and crude fibre was measured using a gravimetirc method.

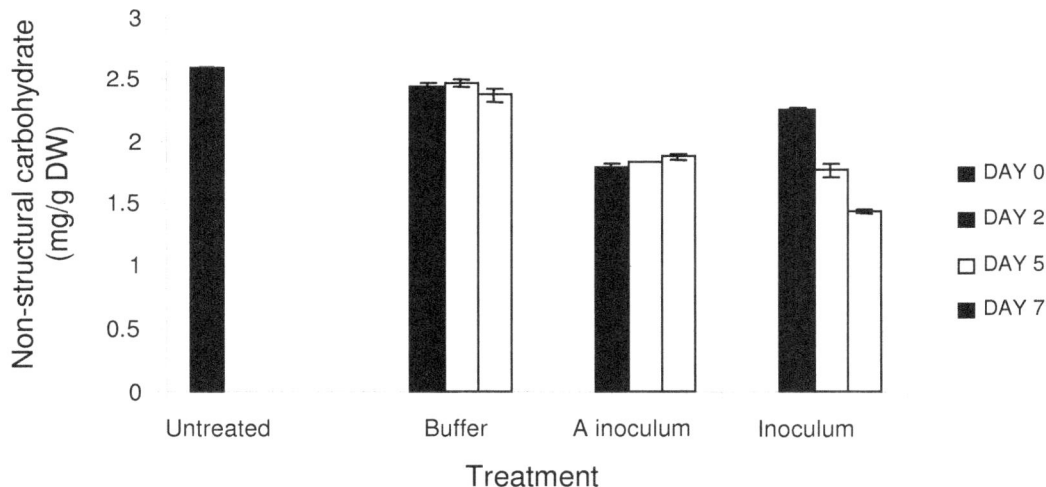

Figure 4. Non-structural carbohydrate content of cassava peels after microbial treatment. Dried milled cassava peels were treated with fermented cassava pulp juice, autoclaved inoculum (A-inoculum) or phosphate buffer for 7 days and starch and sugar content of the peels was determined using the Luff-Schoorl method.

peels. This would account, in part, for the decrease in the cyanogen content of the inoculum-treated peels. However, a comparison of the magnitude of reduction in treated and control peels indicates that cyanogenic glycoside degrading microorganisms were present in the fermented cassava pulp juice and contributed significantly to the level of reduction observed. This is consistent with the observation by Amoa-Awua et al. (1997) that all yeasts and moulds identified in traditional cassava dough inocula exhibited linamarase activity and

were therefore capable of degrading cyanogenic glycosides.

The extent of reduction of cyanogens in the treated cassava peel compares favourably with the findings of studies conducted using pure strains of *Saccharomyces cerevisae* and *Lactobacillus spp.* (Oboh, 2006; Amoa-Awua and Jakobsen, 1995). Furthermore, the present findings confirm the observation by other researchers (Tweyongyere and Katongole, 2002; Wickham, 2001) that microbial treatment achieves a far greater degree of cyanide detoxification than methods such as drying, steaming and boiling, which reduce cyanogen content by approximately 27, 16 and 47%, respectively. Cyanophilic microorganisms have been shown to possess the enzymes linamarase, hydroxynitrile lyase and cyanide hydratase that catalyze the sequential degradation of cyanogenic glycosides into HCN which is subsequently converted into formamide which they use as both a nitrogen and carbon source.

Analysis of the nutrient composition of inoculum-treated and control cassava peels showed a considerable increase in protein content of the latter by day 7 (Figure 3), probably reflecting a rise in microbial mass as well as an increase in the concentration of extracellular microbial enzymes. In contrast to the effect on protein content, there was no significant alteration in fibre content. This is advantageous since it indicates that the cellulose and hemicellulose components of fibre, vital sources of energy for ruminants, were conserved. In contrast, non-structural carbohydrate content decreased somewhat (Figure 4). Presumably, the microorganisms oxidized carbohydrates and lipids to obtain energy. The enhanced protein content of the inoculum-treated peels should be beneficial from the standpoint of animal nutrition. The observation that the quantity of non-structural carbohydrates, comprising starch and sugars, reduced after treatment with both the inoculum and autoclaved inoculum suggests that amylolytic microorganisms associated with the non-sterile peels were largely responsible for this reduction.

Since a natural process of fermentation was employed in the present study, the spectrum of indigenous microorganisms present should be fairly similar to that present in naturally-fermented cassava dough, *Agbelima*. A mixed population of microorganisms, mainly lactic acid bacteria, *Bacillus* species, moulds and yeasts are involved in the fermentation of cassava dough (Amoa-Awua et al., 1996). Since this product has been consumed for centuries, in Ghana, Togo and Benin, without any deleterious effects, it is highly improbable that the consumption of inoculum-treated cassava peels by livestock would lead to adverse consequences. Nonetheless, the identification of *A. flavus* in the inoculum used in the present study necessitates future studies on mycotoxin content, to ascertain the wholesomeness of fermented cassava pulp juice-treated cassava peels as an animal feed resource. Future studies would also investigate whether the inclusion of GRAS (Generally Recognized As Safe) microorganisms in the fermentation process would increase the velocity of the fermentation process appreciably.

Conclusion

In conclusion, the findings of the present study show that liquid waste discarded by cassava processing industries, can be fermented and effectively utilized to achieve remediation of cassava peels, another waste product. The fermented pulp juice contains cyanophilic microorganisms which are capable of reducing the levels of cyanogenic glycosides in cassava peels to non-toxic levels and also improving the nutritional value of the peels by increasing the protein content of the peels appreciably. The cassava pulp juice can be readily obtained from starch and gari industries and distributed to subsistence farmers for the purpose of improving the nutritional value of cassava peels after sun-drying. This would be an effective and simple means of detoxifying and enriching cassava peels as a feed resource for livestock production.

ACKNOWLEDGEMENTS

The authors gratefully acknowledge the assistance provided by Mr. K. Bako of the Department of Botany and Ms. R. Nyarko of the Department of Biochemistry in the identification of microorganisms.

REFERENCES

AOAC (1965). Official Methods of Analysis (10th edn). Horwitz W. (ed.) Association of Official Analytical Chemists, Washington, USA p. 341.

AOAC (1970). Official Methods of Analysis (11th edn). Association of Official Analytical Chemists, Washington, USA p. 438.

Amoa-Awua WKA, Jakobsen M (1995). The role of *Bacillus* sp in the fermentation of cassava. J. Appl. Bacteriol., 79: 250-256.

Amoa-Awua WKA, Appoh FE, Jakobsen M (1996). Lactic acid fermentation of cassava dough into agbelima. Int. J. Food Microbiol. 31: 87-98.

Bak S, Paquette S, Morant M, Morant A, Saito S, Bjarnholt N, Zagrobelny M, Jørgensen K, Osmani S, Simonsen H, Pérez R, Heeswijck T, Jørgensen B, Møller B. (2006). Cyanogenic glycosides: A case study for evolution and application of cytochromes p450. Phytochem. Rev. 5: 309-329.

Buchnan RE, Gibbons NE (1974). Bergey's Manual of Determinative Bacteriology (8th edn). Waverly Press Inc. Baltimore, USA pp. 270-575.

Burnet HL (1972). Illustrated Genera of Imperfect Fungi, Burgess Publishing Company, Minneapolis, Minnesota, USA pp. 34-45.

Cardoso AP, Mirione E, Ernesto M, Massaza F, Cliff J, Haque MR, Bradbury HJ (2005). Processing of cassava roots to remove cyanogens. J. Food Composition and Analysis 18: 451-460.

Ernesto M, Cardoso AP, Nicala D, Mirione E, Massaza F, Cliff J, Haque MR, Bradbury JH (2002). Persistent konzo and cyanogens toxicity from cassava in northern Mozambique. Acta.Tropica. 82: 357-362.

Ezekiel OO, Aworh OC, Blaschek HP, Ezeji TC. (2010). Protein enrichment of cassava peel by submerged fermentation with *Trichoderma viride* (ATCC 36316). Afr. J. Biotechnol. 9: 187-194.

Fatufe AA, Akanbi IO, Saba GA, Olowofeso O, Tewe OO (2007). Growth performance and nutrient digestibility of growing pigs fed a mixture of palm kernel meal and cassava peel meal. Livestock Research for Rural Development. Volume 19, Article #180. Retrieved September 26, 2008, from http://www.cipav.org.co/lrrd/lrrd19/12/fatu19180.htm

Gleadow RM, Haburjak J, Dunn JE, Conn ME, Conn EE (2008). Frequency and distribution of cyanogenic glycosides in *Eucalyptus* L'Hérit. Phytochem. 69: 1870-1874.

Harley PJ, Prescott ML (1990). Laboratory Exercises in Microbiology, W.C. Brown Publishers, Toronto, Canada.

Kimaryo VM, Massawe GA, Olasupo NA, Holzapfal WN (2000). The use of a starter culture in the fermentation of cassava for the production of "kivunde", a traditional Tanzanian food product. Int. J. Food Microbiol. 56: 179-190.

Kirk RS, Sawyer R (1991). Sugars and preserves. In: Pearson's Composition and Analysis of Foods, Longman Scientific and Technical, Essex pp. 183-235.

Kuti JO, Konoru HB (2006). Cyanogenic glycosides content in two edible leaves of tree spinach (Cnidoscolus sp) J. Food Composition and Analysis 19: 556-561.

Oboh G (2006). Nutrient enrichment of cassava peels using a mixed culture of Saccharomyces cerevisae and Lactobacillus sp solid media fermentation techniques. Elect. J. Biotechnol. 9: 203-206.

Oboh G, Akindahunsi AA (2003). Biochemical changes in cassava products(flour and gari) subjected to Saccharomyces cerevisae solid media fermentation. Elect. J. Biotechnol. 82: 559-602.

Perera CO (2010). Removal of Cyanogenic Glycoside from Cassava during Controlled Drying. Drying Technol. 28: 68-72.

Tewe OO (1989). Indices of cassava safety for livestock feeding. In: M. Bokanga,. Essers AJA, Poulter N, Rosling H, Tewe O. (eds). Int. Soc. Hort. Sci. pp. 241-249.

Thom C, Raper KB (1945). A Manual of the Aspergilli, The Williams and Wilking Company, Baltimore, USA pp. 25-68.

Tweyongyere R, Katongole I (2002). Cyanogenic potential of cassava peels and their detoxification for utilization as livestock feed. Vet. Human Toxicol. 44: 366-369.

Ukanwoko AI, Ahamefule FO, Ukachukwu SN (2009). Nutrient Intake and Digestibility of West African Dwarf Bucks Fed Cassava Peel-Cassava Leaf Meal Based Diets in South Eastern Nigeria. Pak. J. Nutr. 8: 983-987.

Wickham LD (2001). FAO report Small-scale processing of starchy staplescaricomcountries.http//journals2.iranscience.net800/www.fao.org/ag/ags/Agsi/starchfinal/hs_257Ewickhamrev3proof.htmCASSA retrieved on 1104-2006.

23

Activities of some enzymes, enzyme inhibitors and antinutritional factors from the seeds of sponge gourd (*Luffa aegyptiaca* M.)

Gloria N. Elemo[1], Babajide O. Elemo[2] and Ochuko L. Erukainure[1]*

[1]Department of Food and Analytical Services, Federal Institute of Industrial Research, Lagos, Nigeria.
[2]Biochemistry Department, Lagos State University, Lagos, Nigeria.

Activities of some of the enzymes; enzyme inhibitors and antinutritional factors of the seeds of *Luffa aegyptiaca* were investigated. Enzymes were extracted and their activities determined using standard procedures. The activities of trypsin and amylase inhibitors were also determined using standard procedures. While antinutritional factors were detemined using the AOAC method as well as standard procedures. The activies of β-amylase and phytase were low. High activities were observed in peroxidase and urease. While lipoxygenase exhbited the highest activity. Trypsin inhibitor had a higher activity than amylase inhibitor. The concentrations of phytic acid, lectins and tannnin levels were much higher compared to that of saponins. The highest concentration was observed in sapogenin. The observed amylase inhibitor activity may be responsible for the use of the seeds in the management of diabetes mellitus, however further work is needed in this area. Further work is also recommended on the separation of the sapogenin into its different components.

Key words: Lipoxygenase, β-amylase, trypsin, sapogenin.

INTRODUCTION

Luffa aegyptiaca M. belongs to the family of cucurbitaceae. Its origin can be traced to tropical Asia (Siqueira et al., 2010). It is a crawling plant that grows in the wild and on abandoned building structures and fence walls in towns and villages in Nigeria (Dairo, 2008). It is a climbing annual wild vine with lobed cucumber-like leaves that are dark green in colouration with rough surface. The plants with yellow flowers bear fruits that are cucumber shaped but larger in size and contain fibrous sponge in which the hard black seeds are enmeshed (Dairo, 2008). It is a lignocellulosic material composed of 60% cellulose, 30% hemicellulose, and 10% lignin (Mazali and Alves, 2005). It has been discovered that the consumption of sponge gourds can supply some antioxidant constituents to human body (Oboh and Aluyor, 2009). In oriental medicine, *L. aegyptiaca* has effect on the treatment of fever, enteritis and swell etc. The extracts from vines alive are used as an ingredient in cosmetics and medicine (Lee and Yoo, 2006). Immature fruit is used as vegetables, which is good for diabetes (Bal et al., 2004). One of the main uses of sponge gourd is as sponges, but they are also used as filler in the production of composites materials, materials of absorption in water treatments stations during the step of ion exchange (Tanobe et al., 2005). Its use in the cosmetic industry for the production of various bath and cosmetics products has been reported (Davis and Decourley, 1993).

Seeds are well known rich sources of minerals but the bioavailability of these minerals is usually low due to the presence of antinutrients and enzyme inhibitors (Valencia et al., 1999). These antinutrients and enzyme inhibitors interfere with absorption of nutrients from foodstuff thus affecting their metabolism. Enzymes are also present in seeds which aid digestion when consumed, as well as serve as nutrition source during germination.

There exist little or no data on the enzyme activities,

*Corresponding author. E-mail: loreks@yahoo.co.uk

enzyme inhibitors and antinutritional factors of the seeds of sponge gourd. Therefore this present study aims at ascertaining some of the enzymes; enzyme inhibitors and antinutritional factors of the seeds of *L. aegyptiaca*, in order to provide data, either for practical use or for basic research needs. This study will contribute to the knowledge of the seeds of this vegetable that could improve its uses.

MATERIALS AND METHODS

Plant material

L. aegyptiaca was collected from the bush at Ojo community in Lagos, Nigeria. The coats (covering) were removed and the seeds grounded into smooth powder. The powdered sample was defatted in a soxhlet extractor using hexane as solvent.

Enzyme extraction and determination

The defatted flour was homogenized with distilled water at 0 to 4°C for 10 min followed by centrifugation to provide a clear supernatant as an enzyme extract. Activities of β-amylase (Swain and Dekkar, 1966), Urease (Hofman, 1963) and lipase (Marchis-Mauron et al., 1959) were determined and expressed as μmol of respective product liberated per minute.

Lipooxygenase activity was assayed at 234 nm by the spectrophotometric method of Ben-Azis et al. (1970). Unit lipooxygenase activity corresponds to a change in absorbance of 0.001 min^{-1}.

Peroxidase activity was determined spectrophotometrically at 420 nm using o-phenylenediamine and hydrogen peroxide (Gregory, 1966). A change in absorbance of 1×10^{-3} min^{-1} corresponds to unit activity.

Phytase was assayed by measuring the rate of increase in inorganic phosphorus using the ascorbic acid method (Watanabe and Olsen, 1965).

Determination of enzyme inhibitors

For trypsin inhibitor, the defatted powdered sample was extracted using dilute alkali and the activity was determined as described by Kakade et al. (1969). Amylase inhibitor activity was determined from the aqueous extracts of the defatted powdered sample against a standard enzyme preparation of known activity at pH 6.0 and 37°C and expressed as for trypsin inhibitor (Marshall and Lauda, 2006).

Determination of antinutritional factors

Phytic acid was extracted from the defatted powdered sample with tricholoroacetic acid and estimated as ferric phytate by the method of Wheeler and Farrel (1971).

Lectin activities were quantified by using the hemagglutination method described by Aregheore et al. (1998). Lectin content was expressed in terms of the dilution value, 2^n where n represents the number of dilutions performed affecting agglutination.

Tannin was estimated by the standard AOAC (1997) method. Saponin and sapogenin content of the sponge gourd seeds (dehulled) were determined by the method of Gestetner et al. (1966). Defatted soybean flour was used as a standard.

Table 1. Activities of some enzymes in the dehulled seed of *L. aegyptiaca*.

Enzyme	Dehulled seeds (Units/mg protein)
β-amylase	0.04
Urease	24.30
Lipase	0.11
Lipooxygenase	29.00
Peroxidase	14.30
Phytase	0.03

Values are mean of triplicates.

RESULTS AND DISCUSSION

This paper reports the enzyme activities of *L. aegyptiaca* seeds and its inhibitors. Living cells perform a multitude of chemical reactions very rapidly because of the participation of enzymes. Plants in their natural states contain enzyme that they made for their own use. However, changes in the activities of these enzymes have been shown to accompany seed deterioration, which precedes loss of seed viability (Sharma and Kumar, 2010).

The activities of the enzymes of *L. aegyptiaca* are shown in Table 1. β-amylase activity was observed to be quite low. This can be attributed to the dormancy of the seed. It is known that amylase activities are usually enhanced during germination (Awoyinka and Adebawo, 2008). Zhang and Wang (2002) attributed this rise in β-amylase activities to increasing expression of the enzyme proteins. Zhang and Wang (2002) further reported that the observed rise in the activity of β-amylase is largely associated with the decline in starch concentration during fruit development. Seed ageing may also be responsible for the observed low β-amylase activity as aging is commonly accompanied by loss of enzyme activity (Ganguli and Sen-Mandi, 1993).

Urease activity was quite high as depicted in Table 1. Urease acts on urea liberating ammonia (Mertzler, 2001). Urea is usually obtained from the hydrolysis of arginine to ornithine and urea, during fermentation of seeds. Urease may play a necessary role in nitrogen metabolism (Eskew et al., 1983) as green plants have been reported to recycle nitrogen via urea and the Ni^{2+}-dependent urease (Mertzler, 2001). The observed high urease activity is of advantage to the seeds of *L. aegyptiaca* as reduction in the activities of urease leads to the accumulation of urea, which causes necrotic spots on leaves. This will further have consequences on the deficiency and metabolism of ureides, amino acids, and other organic acids (Malavolta and Moraes, 2007).

The observed high level of lipoxygenase activity shown in Table 1 may be due to tissue wounding. Lipoxygenase (LOX) is one of the most widely studied enzyme in plants and animal kingdom which is found in more than 60

Table 2. Some enzyme inhibitors in the dehulled seeds of *L. aegyptiaca*.

Enzyme inhibitor	Dehulled seed (Units/mg protein)
*Trypsin inhibitor	24.00
Amylase Inhibitor	19.60

* Extract had 48.6% inhibition for × 2.5 dilution. Values are in triplicates.

Table 3. Some antinutritional factors of the dehulled seeds of *L. aegyptiaca*

Antinutritional factor	Level
Phytic acid (mg/g)	5.16
Lectins (expressed as number of agglutination)	3.00
Tannins (g/100 g seed)	3.13
*Saponins (g/100 g seed)	0.11
§Sapogenin (mg/g seed)	83.40

*The estimated saponin value of soybean was found to be 0.487%. § Soybean sapogenin was 349 mg/g. Values are in triplicates.

species (Baysal and Demirdoven, 2007). Reports from recent studies show that lipoxygenases are involved in the mobilization of storage lipids contained in oil bodies (Heldt, 2005). It catalyses the bioxygenation of poly-unsaturated fatty acids (PUFA) containing a cis,cis-1,4-pentadiene unit to form conjugated hydroperoxydienoic acids (Baysal and Demirdoven, 2007). The observed high level of lipoxygenase activity may be an indication of a fairly high acid value of oil extracted from the seed. It is also crucial to the defence strategies of the seeds as there is ample evidence that lipoxygenase is a crucial element of plants defence strategies (Baysal and Demirdoven, 2007).

The objective of investigating phytase activity was to study the enzyme of *L. aegyptiaca* that hydrolyses phytic acid with a view to utilize this enzymatic activity for reducing the phytic acid content of the seeds. The activity was quite low at 0.3 units/mg protein (Table 1). However there is need to characterize this enzyme in order to know the conditions of its optimum activity.

Lipase hydrolyses triglycerides to glycerol and fatty acid (Pahoja and Sethar, 2002). Its presence in a seed could affect the storage stability of the oil extracted. Its activity is increased during germination of oilseeds as it hydrolyzes triacylglycerol to glycerol and fatty acids which are converted to sugars to support the growth of young sprouts (Hills and Beevers, 1987). Lipase activity and peroxidases levels of the seeds of *L. aegyptiaca* were quite high compared to other oil seeds. This observation however, contradicts reports that lipase activity is absent in dormant seeds (Enujiugha et al., 2004). Oilseed lipases have great potential for commercial exploitation as industrial enzymes, especially those oilseeds that are presently considered under-utilized, among which are the *L. aegyptiaca* seeds.

The trypsin inhibitor of the seed of *L. aegyptiaca* was quite high at 24.00 TIU/mg protein (Table 2). However this level is quite low when compared to literature values of legumes as chick pea (46.5 TIU/mg protein), lentils (40.0 TIU/mg protein), pinto bean (75.5 TIU/mg protein) and tepay bean (61.7 TIU/mg protein). Trypsin inhibitors form irreversible complexes with trypsin and inhibit its activity (Liener and Kakade, 1980). Heat treatment however has been shown to reduce trypsin inhibitor activity.

The aqueous extract of *L. aegyptiaca* seeds exhibited amylase inhibition activity of about 20 unit/mg protein (Table 2). Amylase inhibition activities of plant seeds can

be attributed to defence mechanism against pest such as weevils (Huesing et al., 1991). Interference with the digestion and utilization of dietary starches by naturally occurring amylase inhibitor is of great nutritional and practical concern (Puztai et al., 1995). Purified preparations of amylase inhibitor perfused into the duodenum of humans at clinically acceptable doses have been shown to inhibit in traluminal amylase activity (Layer et al., 1985). This activity has been argued to be good in the management of non insulin-dependent diabetes mellitus as ingestion of the inhibitor with dietary starch significantly reduced postprandial increases in glucose and insulin levels in both normal subjects and patients with diabetes mellitus (Boivin et al., 1988). This may be responsible for the use of the seeds in the management of the diabetes mellitus. Further work in this area is therefore recommended.

Although, oilseed meals have high protein levels and favourable essential amino acid (EAA) profiles they are known to contain a variety of growth inhibiting anti-nutritional factors (NRC, 1993; Francis et al., 2001). This acts as advantages to the plant such as resistance to some insects and soil nematodes.

The nutritional importance of phytic acid lies in its ability to chelate several mineral elements especially calcium, magnesium, iron and zinc, thereby reducing availability in the intestinal track (Hertramf and Piedad-Pascual, 2000). Phytic acid was quite high in the seeds of *L. aegyptiaca* (Table 3). This level however corresponds to reported high level of phosphorus in the seeds; since phytic acid is the major storage form of phosphorus. The observed high content of phytic acid may pose a problem of inavailibilty of micronutrients when consumed.

The level of saponin was not significantly ($p > 0.05$) high compared to that observed in soybean. Saponins are toxic to cold-blooded animals and are generally known for their bitter taste, foaming in aqueous solutions, and their ability to haemolyse red blood cells (Birk and Peri, 1980). Saponin has health benefits as well. The presence of saponin may account for the reported therapeutic actions of *L. aegyptiaca* seeds.

The sapogenin level was observed to be significantly lower ($p > 0.05$) than that of the soybean (Table 3). More work is needed in separating the sapogenin into its different components.

The *L. aegyptiaca* seed extract was observed to affect agglutination only at low dilution values as indicated by its low lectin level (Table 3).

Conclusion

Results of this study reveal the presence of enzymes with lipoxygenase having the highest activity. Sapogenin had the highest level among the antinutritional factors investigated. The observed amylase inhibitor activity may be responsible for the use of this seeds in the management of diabetes mellitus, however further work is needed in this area. Further work is also recommended on the separation of the sapogenin into its different components.

REFERENCES

Association of Official Analytical Chemist (AOAC) (1997). Official methods of analysis of the Association of Analytical Chemists. Washington DC.

Aregheore EM, Makkar HPS, Becker K (1998). Assessment of Lectin Activity in a Toxic and a Non-toxic Variety of *Jatropha curcas* using Latex Agglutination and Haemagglutination Methods and Inactivation of Lectin by Heat Treatments. J. Sci. Food Agric., 77, 349-352.

Awoyinka OA, Adebawo OO (2008). Influence of malting time on α-and β-Amylases secretions in Nigerian amylolytic maize cultivars. Afr. J. Agric. Res., 3(1): 007-012.

Bal KE, Bal Y, Lallam A (2004). Gross morphology and absorption capacity of cell-fibers from the fibrous vascular system of Loofah (*Luffa cylindrica*). Textile Res. J., 74: 241-247.

Baysal T, Demirdoven A (2007). Lipoxygenase in fruits and vegetables: a review. Enzyme Microb. Technol., 40(4), 491-496.

Ben-Aziz A, Grossmein S, Ascarlli I, Budowski P (1970). Linoleate oxidation induced by lipoxygenase and proteins. Anal. Biochem., 36: 88-100.

Birk Y, Peri I (1980). Saponins. In: Liener I.E. (Ed.). Toxic constituents of plant foodstuffs, 2nd edn. Academic Press. pp. 161-182. New York.

Boivin M, Flourie B, Rizza RA, Go VLW, DiMagno EP (1988) Gastrointestinal and metabolic effects of amylase inhibition in diabetics. Gastroenterol., 94: 387-394.

Davis JM, Decourley CD (1993). New crops, Janick J and Simon JE (eds.), John Wiley and Sons, New York. Pp 560-561.

Enujiugha VN,Thani FA, Sanni TM, Abigor RD (2004). Lipase activity in dormant seeds of the African oil bean (*Pentaclethra macrophylla* Benth). Food Chem. 88: 405–410

Eskew DL, Welch RM, Cary EE (1983). Nickel: An Essential Micronutrient for Legumes and Possibly All Higher Plants, Sciences, 222, 621 – 623.

Francis G, Makkar HPS, Becker K (2001). Antinutritional factors present in plant-derived alternate fish feed ingredients and their effects in fish. Aquaculture, 199(3-4): 197-227.

Ganguli S, Sen-Mandi S (1993). Effects of ageing on amylase activity and scutellar cell structure during imbibition in wheat seed. Ann. Bot., 91: 411–416.

Gestetner B, Brik Y, Bondi A, Tencer Y (1966). Soya bean saponins VII. A metrhod for the determination of sapogenin and saponin contents in soya beans. Phytochem., 5: 803-806.

Gregory RPF (1966). A Rapid Assay for Peroxidase Activity. Biochem. J., 101: 582-583.

Heldt H (2005). Plant Biochemistry. 3rd ed. Elsevier Inc.

Hertramf JW, Piedad-Pascual F (2000). Handbook on ingredients for aquaculture feeds. Kluwer Academic Publishers. The Netherlands.

Hills MJ, Beevers H (1987). An antibody to the castor bean glyoxysomal lipase (62 KDa) also binds to a 62 KDa proteins in extracts from young oil seed plants. Plant Physiol., 85: 1084-1089.

Hofman E (1963). Urease. In: Methods of enzymatic analysis, Bergmeyer H. (ed.). Academic Press. Pp. 913-916. New York.

Huesing JE, Shade RE, Chrispeels MJ, Murdock LL (1991). α-Amylase Inhibitor, Not Phytohemagglutinin, Explains Resistance of Common Bean Seeds to Cowpea Weevil. Plant Physiol., 96: 993-996.

Kakade ML, Simons L, Lienar IE (1969). Evaluation of natural versus synthetic substrates for measuring the antitryptic activity of soya beans samples. Cereal Chem., 46: 518-528.

Layer P, Carlson GL, DiMagno EP (1985). Partially purified white bean amylase inhibitor reduces starch digestion in vitro and inactivates intraduodenal amylase in humans. Gastroenterol., 88: 1895-1902.

Lee S, Yoo JG (2006). (WO/2006/019205) method for preparing transformed luffa cylindrica Roem (World Intellectual property organization) http://www.wipo.int/pctdb/en/wo.jsp?IA=KR2004002745&DISPLAY=STATUS.

Liener IE, Kakade ML (1980). Protease inhibitors. In: Anonymous toxic constituents of plant foodstuffs. Academic press. pp. 7-71. New York, USA.

Marchis-Mauron G, Sardi L, Desneulle P (1959). Purification of hog pancreatic lipase. Arch. Biochem. Biophys., 83: 309-319.

Malavolta E, Moraes MF (2007). Nickel – from Toxic to Essential Nutrient. Better Crop. , 91(3): 26-27.

Marshall JJ, Lauda CM (2006). Assay of α-Amylase Inhibitor Activity in Legumes. Starch – Stärke, 27(80): 274 – 278.

Mazali IO, Alves OL (2005). Morphosynthesis: high fidelity inorganic replica of the fibrous network of loofa sponge *(Luffa cylindrica)*. Anais da Academia Brasileira de Ciências, 77(1): 25-31.

Mertzler DE (2001). Biochemistry: the chemical reaction of living cells. 2nd ed. Vol. 1 & 2. Elsevier Inc. pp. 836-903.

National Research Council (NRC) (1993). Nutrient Requirements of Fish. Washington D.C.: National Academy Press.

Pahoja VM, Sethar MA (2002). A review of enzymatic properties of lipase in plants, animals and microorganisms. Pak. J. Appl. Sci., 2(4): 474-484.

Puztai A, Grant G, Duguid T, Brown DS, Peumans WJ, Van Damme EJM, Bardocz S (1995). Inhibition of starch digestion by a-amylase inhibitor reduces the efficiency of utilization of dietary proteins and lipids and retards the growth of rats. J. Nutr., 125: 1554-1562.

Sharma RML, Kumar V (2010). Viability and enzyme activity of ageing seeds of bamboo (*Dendrocalamus strictus* (Roxb.) Nees) in relation to exogenous plant growth regulators. Curr. Sci., 99(11): 1590 – 1593.

Siqueira G, Bras J, Dufresne A (2010). Luffa cylindrica as a lignocellulosis source of fiber, microfibrillated cellulose, and cellulose nanocrystals. BioResources, 5(2): 727-740.

Swain RR, Dekkar EE (1966). Seed germination studies. I. Purification and properties of β amylase from the cotyledons of germinating seeds. Biochem. Biophys. Acta, 122: 75-86.

Tanobe VOA, Sydenstricker THD, Munaro M, Amico SC (2005). A comprehensive characterization of chemically treated Brazillian sponge-gourds (*Luffa cylindrica*). Polym. Test., 24(4): 474-482.

Valencia S, Svanberg U, Sanberg AS, Ruals J (1999). Processing of quinoa (*Chenopodium quinoa, Willd*). Effects on *in vitro* iron availability and phytate hydrolysis. Intl. J. Food Sci. Nutr., 50: 203-208.

Watanabe FS, Olsen SR (1965). Test of an ascorbic acid method for determining phosphorus in water and NaHCO$_3$ extracts from the soil. Soil Sci. Soc. Am. Proc., 29: 677.

Wheeler EL, Farrel RE (1971). A method of phytic acid determination in wheat and wheat fractions. Cereals Chem., 48: 312.

Zhang D, Wang Y (2002). β-amylase in developing apple fruits: activities, amounts and subcellular localization. Sci. China (Series C), 45(4): 429-440.

24

Corncob is the optimum carbon resource for the growth of *Tremella aurantialba*

Zhi-Cai Zhang*, Mingxia Chen, Xin Li and Wangli Shen

School of Food Science and Bio-Technology, Jiangsu University, Zhenjiang 212013, P. R. China.

The minimum medium without carbon source was used to investigate the effects of five kinds of hexoses, four kinds of pentose, four kinds of bi-saccharide and three kinds of polysaccharides on *Tremella aurantialba* growth. The result demonstrated that (1) The fittest carbon source of *T. aurantialba* was the five-carbon aldose in all the tested monosaccharides; (2) Polysaccharides linked with β-1,4-glycosyl bond is fitter for the growth of *T. aurantialba* than that linked with α-1,4-glycosyl bond (3) Corncob, which contained abundant poly-xyloses linked with β-1,4-glycosyl bond, is the fittest for growth of *T. aurantialba*. The reaction system of α-1,4-glycosidase was employed to study the reason that the polysaccharides linked in β-1,4-glycosyl bond is fitter for growth of *T. aurantialba*. The result showed that *T. aurantialba* broth can inhibit the activity of α-1,4-glycosidase produced by *T. aurantialba*.

Key words: *Tremella aurantialba*, glycosyl bond, carbon source, aldose, ketose.

INTRODUCTION

Tremella aurantialba, extensively used in Chinese medicine (inhabits in wood). During the last 15 - 20 years, studies on *Tremella mensenterica* polysaccharides from fruit body, mycelium, and the crude extract of broth have been carried out in China and Japan. It is reported (Xiong and Yu, 2000) that the composition of *T. mensenterica* decoction remarkably relaxes the tracheal smooth muscle and is a kind of anti-asthmatic. The report also shows that orally taking or subcutaneous injection of crude extract of *T. aurantialba* to mice can increase macrophages, enhance phagocytic function, adjust the effects on specific immunocompetence of mice, and increase mice nonspecific immunocompetence (Li et al., 2000). The fermented products from *T. aurantialba* can remarkably prolong the clotting time of the thrombogen and increase the volume of meninges' blood flow (Liu et al., 2003). The polysaccharides from fruit bodies of *T. aurantialba* exhibited remarkable hypoglycemic activity in normal mice and two diabetic mouse models; streptozotocin-induced diabetes and genetic diabetes, following intraperitoneal administration (Kiho et al., 1995).

The polysaccharides from mycelia of *T. aurantialba* were

reported to prevent and cure diet-induced hyperlipidemia (Wang et al., 2002). Zhang et al. (2004) reported that the polysaccharides of mycelia could decrease the blood sugar content. It is favour to increase the yield of mycelia in the fermentation of *T. aurantialba*. Via comparing with different kinds of hextoses, pentoses, bi-saccharides and polysaccharides, the different kinds of carbon source was found to affect significantly the growth of *T. aurantialba*. According to their structure's discrepancy, it was deduced that corncob is the fittest for the growth of *T. aurantialba*. The paper presented reported on the deducing process.

MATERIALS AND METHODS

Chemicals

α-D-glucosidase and 4-Nitrophenyl-α-D-glucopyranoside used in this experiment were purchased from Sigma. Albumin bovine was made in National Medical Chemical Reagent Co. Ltd, AB-8 macropores resin was purchased from Chemical plant of Nankai University. All other chemicals used in the test were of analytical grade.

Submerged incubation and fermentation

T. aurantialba, kindly presented by Weikjing Qu, a professor of East China Normal University, and grown in PDA medium at 28°C, was used in this study. The fungus was grown in 35 g solid seed

*Corresponding author. E-mail: Zhangzhicai2001@yahoo.com.cn.

medium containing 15 g bran and 20 ml water. Solid seed was incubated at 25°C for 7 days. During the culturing course, the flask was shaken once per day from the third to the fifth day.

The solid seed aforementioned was inoculated in 100 ml seed medium containing 2% sucrose, 1% corn powder, 0.5 % peptone. The medium was sterilized under 0.1 MPa for 30 min. The seed was incubated at 27°C in a rotary shaker (Shanghai Pharmaceutical Industrial Academe, China) at 150 rpm for 2 days. 10 ml seed culture was inoculated in 100 ml fermentation medium and fermentation was carried out under the same condition for 7 days. The fermentation medium contained 4% carbon source, 0.4% peptone, 1% corn powder, 1% bran, 0.15% KH_2PO_4, and 0.075% $MgSO_4$.

Isolation of mycelia

The broth was filtered using 40-mesh stainless sieve and the mycelium was obtained. The mycelium was washed using water, dried to constant weight and then weighed.

Inhibitory ratio on α-1,4-glucosidase broth of *T. aurantialba*

The inhibitory activity on α-1,4-glucosidase was determined using 4-Nitrophenyl-α-D-glucopyranoside as the substrate. The reaction mixture contained 0.3 ml broth of *T. aurantialba*, 50 μl enzyme solution in which 1 mg/ml bovine serum protein and 0.53 μ/ml α-D-glucosidase were dissolved, and 50 μl 0.116 mol/L 4-nitrophenyl-α-D-glucopyranoside in 1.7 ml 0.2 mol/L of pH 6.8 phosphate buffer solution. After incubation for 20 min at 37°C, the reaction was stopped by adding 10 ml 0.1M Na_2CO_3. Water instead of sample solution was used as blank. UV absorbance was measured at the wavelength of 400 nm. The ratio of inhibition was calculated as shown in the following equation:

$$I = \frac{X - Y}{X} \times 100\%$$

Where I is inhibition rate, X = absorbency of the blank, and Y = absorbency of the sample.

Statistical analysis

The analyses were made at least in quintic replication and results presented were expressed as mean ± S.D. Statistical analysis was performed using one way analysis of variance (ANOVA) followed by Duncan's Multiple Range Test (DMRT). P-values < 0.05 were considered significant.

RESULTS AND DISCUSSION

Effect of different hexose on growth of *T. aurantialba*

Five kinds of hexose were employed in the present test. They were glucose, mannose, altrose, galactose and fructose. The result is as shown in Table 1. Table 1 demonstrated that all tested aldoses had nearly the same dry weight of mycelia (glucose: 2.01 ± 0.25; mannose: 1.96 ± 0.35; altrose: 1.85 ± 0.29; galactose: 1.94 ± 0.27) and were not significantly different from each other.

But when compared with ketoses (fructose), their dry weights of mycelia were significantly higher than that of

Table 1. Effect of different hextose on dry weight of *T. aurantialba*.

Kind of hextose	The dry weight of mycelium (g)
Glucose	2.01 ± 0.25
Mannose	1.96 ± 0.35
Altrose	1.85 ± 0.29
Galactose	1.94 ± 0.27
Fructose	1.53 ± 0.18**

Values represent the dry weight of mycelia in 100mL medium and were expressed as mean ±S.D. *p ≤ 0.05 (compared to glucose, mannose, altrose and galactose). **p ≤ 0.01 (compared to glucose, mannose, altrose and galactose).

fructose (1.53 ± 0.18) (p < 0.01). The results proved that the more optimum carbon of *T. aurantialba* was aldoses and not ketose.

Effect of different pentose on growth of *T. aurantialba*

Four kinds of pentoses were employed in the test. They were lyxose, xylose, ribose and xylulose. The result is shown in Table 2. Table 2 demonstrated that all tested aldose had nearly the same dry weight of mycelia (lyxose: 2.41 ± 0.23; xylose: 2.53 ± 0.17; ribose: 2.48 ± 0.27) and there were no significantly difference between them. But when compared with ketose (xylulose), their dry mycelia weights were significant higher than that of xylulose (1.98 ± 0.26) (p < 0.01). The results proved that the more optimum carbon of *T. aurantialba* was aldoses and not ketose.

Pentose is more optimum carbon source within mono-saccharide

When Table 1 was compared with Table 2, the conclusion was obviously gotten: (1) that dry weights using the five-carbon aldose as carbon source were higher than the use of six-carbon aldose as source; (2) that dry weight using the five-carbon ketose as carbon source was higher than the use of six-carbon ketose as source; (3) the five-carbon aldose was the most optimum carbon source within all tested mono-saccharide.

Effect of different bi-saccharides on growth of *T. aurantialba*

Four kinds of bi-saccharides employed in the test were maltose, sucrose, lactose and cellobiose. The result is as shown in Table 2. Table 3 demonstrated that maltose, lactose and cellose had nearly the same dry weight of mycelia (2.32 ± 0.18; 2.38 ± 0.19 and 2.41 ± 0.23, respectively) and were not significantly different from each other. But the dry weights of mycelia using sucrose

Table 2. Effect of different pentose on dry weight of *T. aurantialba*.

Kind of hextose	The dry weight of mycelium (g)
Lyxose	2.41 ± 0.23
Xylose	2.53 ± 0.17
Ribose	2.48 ± 0.27
Xylulose	1.98 ± 0.26**

Values represent the dry weight of mycelia in 100mL medium and were expressed as mean ± S.D. *p ≤ 0.05 (compared to lyxcse, xylose and ribose). **p ≤ 0.01 (compared to lyxose, xylose and ribose).

Table 3. Effect of different bi-saccharides on growth of *T. aurantialba*.

Kind of bi-saccharide	The dry weight of mycelium (g)
Maltose	2.32 ± 0.18
Sucrose	2.06 ± 0.17**
Lactose	2.38 ± 0.19
Cellobiose	2.41 ± 0.23

*p ≤ 0.05 (compared to maltose, lactose and cellobiose). **p ≤ 0.01 (compared to maltose, lactose and cellobiose).

as carbon source (2.06 ± 0.17) were less than the use of other tested bi-saccharides as carbon source, which was highly significant (p < 0.01). The molecule of sucrose linked one molecule of aldose with ketose, and other three kind of molecular were made of two molecule of aldose. The result proved that the more optimum carbon of *T. aurantialba* was aldoses and not ketose again.

Effect of different poly-saccharides on growth of *T. aurantialba*

Three kinds of polysaccharides were used in the test. They were starch, dextrin and cellose. The result is shown in Table 4. Table 4 demonstrated that the dry weight of *T. aurantialba* using starch and dextrin as carbon source (1.47 ± 0.35 and 1.41 ± 0.28, respectively) were less than that using cellose as carbon source (1.83 ± 0.25) (p < 0.01). The result demonstrated that the polysaccharides linked with β-1,4-glycosyl bond is fitter for growth of *T. aurantialba* than that linked with α-1,4-glycoside bond

Corncob was the optimum carbon for growth of *T. aurantialba*

Effect of diverse carbon sources on *T. aurantialba* growth showed significant difference. Among tested carbon sources, five-carbon aldose is fitter for the growth of *T.*

Table 4. Effect of different polysaccharides on growth of *T. aurantialba*.

Kind of polysaccharide	The dry weight of mycelium (g)
Starch	1.47 ± 0.35
Dextrin	1.41 ± 0.28
Cellose	1.83 ± 0.25

*p ≤ 0.05 (compared to starch and dextrin). **p ≤ 0.01 (compared to starch and dextrin).

aurantialba than hextose or five-carbon ketose and the polysaccharides linked with β-1,4-glycosyl bond is fitter for *T. aurantialba* growth than that linked with α-1,4-glycosyl bond.

Glucose is metabolized via the Embden-Meyerhof-Parnas pathway (EMP) and xylose via the hexose phosphate shunt. In the test, the five-carbon aldose as carbon source produces more mycelia. The result provides the proof that the ingredients with the inhibitory activity on the Embden-Meyerhof-Parnas pathway (EMP) exist in the *T. aurantialba* broth or there is lack of the key enzyme of the Embden-Meyerhof-Parnas pathway in *T. aurantialba*.

Polysaccharides linked with α-1,4-glycosyl bond was fitter for *T. aurantialba* growth than that linked with β-1,4-glycosyl bond in theory. It is possible that the ingredients with the inhibitory activity on α-1,4-glycosidase exist in the broth of *T. aurantialba*.

The analysis of inhibitory effect on α-1,4-glucosidase broth of *T. aurantialba* demonstrated that 0.3 ml broth resulted to 47.5% inhibitory ratio. The analysis of remnant dextrin, starch and cellulose in the broth showed that unused starch or dextrin was not less than 2.3%, however, the unused cellose was less than 0.5%. These tests proved that the broth of *T. aurantialba* can inhibit α-1,4-glycosidase. It is the inhibitory activity that decreases the utilization ration of starch and dextrin in the fermentation of *T. aurantialba*.

In conclusion, the five-carbon aldose is the fittest carbon source of *T. aurantialba*. During the growing and metabolizing process, *T. aurantialba* can produce the ingredients that can inhibit α-1,4-glycosidase and decrease the utilization ratio of starch and dextrin, which result in cellose being fitter for the growth of *T. aurantialba* than starch and dextrin. Corncob contained more than 80% poly-xylose linked with β-1,4-glycosyl bond, so, it is reasonable to deduce that the corncob is the fittest carbon for the growth of *T. aurantialba*.

ACKNOWLEDGEMENT

This work was supported by a grant from Research Foundation for Advanced Talents of Jiangsu University (1821360005).

REFERENCES

Kiho T, Morimoto H, Sakushima M, Usui S, Ukai S (1995). Polysaccharides in fungi.XXXV. Anti-diabetic activity of an acidic polysaccharides from the fruiting bodies of *Tremella aurantialba*. Biol. Pharm. Bull., 18: 1627-9.

Li XH, Yang LL, Zhu SF (2000). *Tremella aurantialba* on immunnocompetence in mice. J. Shanxi Med. Univ. (in Chinese) 3: 206-207.

Liu CH, Xie H, Su BN, Han JX, Liu YP (2003). Antithrombus effect on the fermented products of mycelium from *Tremella aurantialba*. Nat. Prod. R. & D. (in Chinese) 4: 289-292.

Wang H, Qu WJ, Chu SD, Li MJ, Tian CP (2002). Studies on the preventive and therapeutic effects of the polysaccharides of Tremella aurantialba mycelia on diet-hyperlipidemia in mice. Acta Nutrimenta. Sinica 24: 431-433.

Xiong YK, Yu B (2000). Experimental studies on anti-asthmatic effect of composite *Tremella mesenterica* decoction. J. Zhejiang College of TCM (in Chinese) 1: 86-87.

Zhang W, Qu WJ, Zhang XL, Deng YX, Zhu, SD (2004). the anti-hyperglycemic activity of polysaccharides froom *Tremella aurantialba* mycelium. Acta Nutrimenta. sinica. 26: 300-303.

25

Characterization of the polysaccharide material that is isolated from the fruit of *Cordia abyssinica*

Benhura, M. A. N. and Chidewe, C.*

Department of Biochemistry, University of Zimbabwe, P.O. Box Mp 167, Mt Pleasant , Harare, Zimbabwe.

Treatment of aqueous extracts of the fruit of *Cordia abyssinica*, containing 0.25% sodium chloride with three volumes of ethanol produced an acidic polysaccharide with a 2% yield, on a fresh weight basis. Upon precipitation of the polysaccharide with acid, the yield of the polysaccharide decreased to 1.2% and the polymer showed some level of degradation. With protein content between 2.6 and 4.6% for the acid and ethanol precipitates respectively, the polymer contained 0.29% hydroxyproline. Treatment of the ethanol precipitated polysaccharide with pronase E resulted in a decrease in viscosity of polysaccharide solutions and high performance size exclusion (HPSEC) chromatograms of the protease treated samples showed peaks that had slightly shifted to low molecular weight. Uronic acid content, determined using the m-hydroxydiphenyl method was 9%. Some uronic acid residues along the polymer chain were methyl esterified, with the methoxyl content being 38%. With an ash content of 17% the polymer had a mineral ion content of Ca, 0.3%, Mg, 0.3%, Na, 0.2% and K, 4.8%. The optical rotation of a 0.25% solution was -50°.

Key words: *Cordia abyssinica*, polysaccharide, yield, viscosity, hydroxyproline, protein content.

INTRODUCTION

Polysaccharides or polyglycans are polymers of monosaccharide residues that are joined together by glycosidic bonds, which are formed by the elimination of elements of water, between the hemiacetal hydroxyl group of one residue and a primary or secondary hydroxyl group of an adjacent residue (Laere et al., 2000). The monomer species may be simple monosaccharides or sugar derivatives such as N-acetylaminosugars, uronic acids or ester sulphate sugars. Uronic acids are constituents of hemicellulose, pectin, gums, mucilages and other plant polysaccharides. Uronic acids occur widely in nature and much of the carbohydrate materials in plants contain this important component (Ridley et al., 2001). Typically, in the polyuronide molecule the neutral monosaccharide and uronic acids are joined by glycosidic linkages to form complex acidic polymers. Many polyuronides contain methyl groups that are linked through ether bonds to the uronic acid. In hemicelluloses and gums, the content of uronic acids is low but in pectic acids, uronic acid units

may constitute essentially the entire polysaccharide chain (Figure 1).

The ability of polysaccharides to produce high viscosity in water at low concentrations is a major property of polysaccharides that gives them valuable and widespread use in the food and non-food industries. Polysaccharide gums have mainly been used for thickening, modification of texture, gelling, formation of protective films, and for stabilization of emulsions, foams and suspensions (Kossori et al., 2000, Euston and Hirst, 2000). *Cordia abyssinica*, a member of the family Boraginaceae, is a small to medium sized tree that grows to about 9 m in height. The tree is fast growing and occurs in medium to low altitudes in woodland and bush. The tree is found in warm moist riverine areas, often along riverbanks (Van Wyk and Van Wyk, 1998). *C. abyssinica* grows in north-eastern Africa, extending southwards to Angola, Mozambique, Zimbabwe and the Limpopo province of South Africa (Palmer and Pitman, 1972). In Zimbabwe, the tree is commonly found in the South-Eastern parts of the country, such as Masvingo province and near the boarder with Mozambique.

The fruit of *C. abyssinica* is a drupe, about 10 to 30 mm in diameter, which has a globose shape and a sharp tip. Green when unripe the fruit turns yellow to orange on

---
*Corresponding author. E-mail: cchidewe@science.uz.ac.zw.

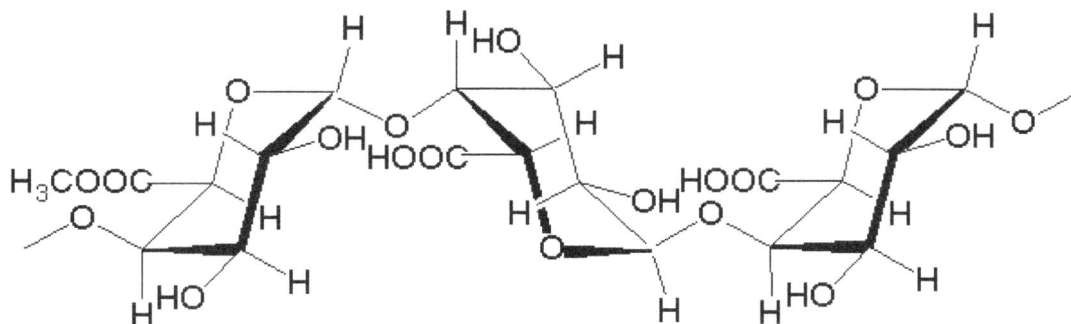

Figure 1. Structure of pectin showing uronic acid residues and methylated groups (http://www.kjemi.uio.no/Polymerkjemi/Research/pectin.jpeg).

ripening, which occurs between December and April. The shell encloses a sweet mucilaginous flesh, which is highly viscous and sticky. Rural school children often use the mucilage from the fruits of *C. abyssinica* as glue. Although edible, the fruit of *C. abyssinica* is not normally consumed by humans but is eaten by wild animals. The fruit of *C. abyssinica* was chosen as a suitable candidate for study because of the unique adhesive properties that the mucilage of the fruit possesses and high viscosity of the solutions formed when the polymer is dissolved in water. The polysaccharide from *C. abyssinica* has potential for application as a thickener (Benhura and Katayi-Chidewe, 2000), emulsion stabiliser (Benhura and Chidewe, 2004), and as a binding agent in the food industry and as an effective adhesive in the non-food industry. The polysaccharide that was isolated from *C. abyssininica* is made up of the sugars, mannose, glucose, galactose, arabinose, xylose, rhamnose, fucose and an unidentified methyl sugar (Benhura and Chidewe, 2002).

The objectives of the study were to determine some of the physical properties of the polysaccharide of *C. abyssinica*.

EXPERIMENTAL

Collection of the fruit

Mature but unripe fruit of *C. abyssinica* were picked from trees in Bikita, South Eastern Zimbabwe. The fruit was collected when in season during the period between December and April. Fruits were collected as available from trees occurring alongside streams in the same area. Harvested fruit, with their calyces on, were stored at room temperature and processed within 72 h from the time of collection.

Extraction of pulp from the fruit

The pulp was extracted by squashing the fruit by hand to release the stones, in which was most of the fruit pulp. The stones, in a strong plastic or stainless steel container, were vigorously agitated with a robust wooden rod during which process the pulp separated as a thick sticky mass. The separated stones were removed and

the pulp, where necessary, was stored frozen at -20°C until required.

Precipitation of polysaccharide using 0.25 M sodium chloride and ethanol

Water was added to the sticky freshly prepared or thawed pulp in order to make a workable mixture, which was centrifuged in a BHG Hermle ZK 401 centrifuge at 6000 rpm for 30 min to remove insoluble material. To the supernatant, solid sodium chloride was added to make a 0.25 M solution. In routine preparation of the polysaccharide, four volumes of ethanol were then added to the supernatant to precipitate the polysaccharide. The polysaccharide was dried in a pre-heated oven set at 100°C or freeze dried in a Christ-Alpha 2 to 4 freeze drier. The dry gum was stored at room temperature until required.

Precipitation of polysaccharide using 0.2 M HCl

Sodium carbonate (1 M, 120 ml) was added to the sticky pulp (400 ml) of *C. abyssinica*. When the sample dissolved, it was diluted to a final volume of 1200 ml with water so that the final concentration of sodium carbonate in solution was 0.1 M. The mixture was centrifuged in a BHG Hemle ZK 401 centrifuge at 6000 rpm for 30 min, to remove insoluble material. To precipitate the polysaccharide, 0.2 M HCl (500 ml) was added to the supernatant with stirring. The precipitated polysaccharide was washed five times with water and freeze- dried in a Christ Alpha 2 to 4 freeze-drier.

Determination of yield of polysaccharide

A known number of fruits were de-capped, weighed and the pulp was extracted from the fruit. After the polysaccharide was precipitated using sodium chloride-ethanol or HCl the freeze-dried mass of the polysaccharide was determined. Yield was expressed as percentage of the mass of the dry precipitate against the mass of the whole fresh de-capped fruit (James, 1995).

Determination of moisture, ash and mineral ion content of the polysaccharide

The dry polysaccharide (2 g) of *C. abyssinica* was weighed into a previously ignited, cooled and weighed porcelain crucible and the sample heated to constant weight in a pre-heated oven at 100°C. To determine ash content, the sample, dried at 100°C was first charred at 200°C for two hours in order to prevent the foaming that is likely to occur as a result of too rapid rise in temperature (James, 1995). The charred mass was ashed at 550°C in a Phoenix MRB2-017-8 furnace. To determine the mineral ion content, the ash was dissolved in 5 ml of concentrated HCl and the mixture boiled for 5 min on a hot plate in a fume cupboard, with acid being added as necessary in order to maintain constant volume. The mixture was transferred to a beaker and the crucible washed with distilled water pouring the washings into the beaker containing the sample. The volume was adjusted to about 40 ml and the mixture boiled for 10 min. The mixture was cooled and filtered through glass wool into a 100 ml volumetric flask and the beaker was rinsed into the volumetric flask.

The solution, cooled and made up to 100 ml, was used for the determination of the individual mineral ions including, sodium, calcium and magnesium using a Perkin Elmer 500 atomic absorption spectrophotometer and potassium was determined using a Corning 400 flame ionisation photometer (Rojas et al., 2004).

Determination of specific optical rotation of the polysaccharide

The optical rotation for solutions of polysaccharide at concentrations up to 0.5% was measured at room temperature on an Otago Polax-D polarimeter using the D-line of polarised sodium light and a 100 mm cell (Saka and Msonthi, 1994).

Determination of uronic acid content of the polysaccharide

Uronic acids were determined using the p-hydroxydiphenyl-sulphuric acid method with galacturonic acid as the standard (Chaplin and Kennedy, 1986).

Determination of the methyl ester content

Sodium hydroxide (0.75 M, 0.25 ml) was added to aliquots (0.5 ml) of the polysaccharide (50 to 200 µg) and the tubes gently swirled. After 30 min at room temperature, the samples and methanol standards (2 to 40 µg) were acidified with 2.75 M H_2SO_4 (0.25 ml) and cooled in an ice-water bath, for permanganate oxidation. To aliquots (1 ml) of the saponified polysaccharide or methanol standards aqueous potassium permanganate (2% w/v, 0.2 ml) was added, taking care not to splash liquid onto the sides of the tube. The mixture was agitated by swirling gently and the tubes held in an ice bath for 15 min. Sodium arsenate (0.05 M, 0.2 ml), followed by water (0.6 ml) was added, and the thoroughly mixed solution left for 1 h at room temperature. After addition of pentane-2,4-dione (2 ml) and thorough mixing, the tubes were closed with marbles, heated at 60°C for 15 min, and cooled to room temperature.

Absorbance at 415 nm was determined in a Spectronic 20 Genesys spectrophotometer, using a blank of water (1 ml) treated identically to the samples.

Determination of the protein content of the polysaccharide

Crude protein content of the polysaccharide preparations was determined using the Kjeldahl method with the nitrogen content being multiplied by a factor of 6.25 (Rodriguez et al., 2004).

Treatment of the polysaccharide with protease

Pronase E (52.5 mg), from Sigma, was dissolved in phosphate buffer (13.82 ml, pH 7.5). The solution of enzyme (300 µl) was mixed with solution of the polysaccharide (1%, 17.7 ml) and

incubated at 37°C. In the control tube, an equal volume of the phosphate buffer was used instead of the solution of enzyme. The action of the enzyme was followed by measuring the viscosity at 37°C using a Cannon Fenske routine viscometer # 350, from PSL Ltd, England. In order to determine the nature of interaction between the protein and the polysaccharide, the native and protease treated solutions of the polymer were analysed by HPSEC using a Zorbax GF 250 size exclusion column and water as the mobile phase at a flow rate of 2 ml/min. A 1% solution of sample (50 µl) was injected into the column.

Determination of hydroxyproline residues in the polysaccharide chains

Solutions (100 µg/ml) were prepared by dissolving 0.05 g of the standard amino acids in water (400 ml). Concentrated HCl (11 M, 20 ml) was added to the solutions to prevent microbial degradation and the solutions were made up to 500 ml with distilled water. Working solutions of standards, concentrations of up to 20 µg/ml were prepared by diluting the 100 µg/ml standard solutions with water. Polysaccharide that had been prepared by precipitation with acid or ethanol were used for determination of hydroxyproline, with the ethanol precipitate being dissolved in water and the acid precipitate was dissolved in acetate buffer, pH 5.5, to make 0.5% solutions. Copper sulphate solution (0.05 M, 1 ml) was added to samples and standards (1 ml), in duplicate, in rimless Pyrex tubes each tube followed by 2.5 M sodium hydroxide (1 ml), and the tubes were agitated by gentle swirling. The tubes were placed in a water bath at 40°C for 5 min after which hydrogen peroxide (6%, 1 ml) was added with immediate mixing by swirling of the tubes while still in the bath. The tubes were left in the bath for a further 10 min with occasional swirling. Tubes were cooled to room temperature with tap water and sulphuric acid (1.5 M, 4 ml) was added followed by 5% p-dimethylaminobenzaldehyde (1 ml).

The contents of the tubes were mixed on a vortex mixer after each addition. The tubes were capped with marbles and placed in a water bath at 70°C for 16 min. After this time, the solutions were left to cool to room temperature and the mixtures were agitated thoroughly on a vortex mixer and the absorbance at 555 was measured using a Genesys Spectronic 20 spectrophotometer from Spectronic instruments, USA.

RESULTS AND DISCUSSION

Precipitation of the polysaccharide of *C. abyssinica* using sodium chloride and ethanol, and acid

Initial attempts to precipitate the polysaccharide out of aqueous extracts using ethanol alone were not successful. When sodium chloride was added to the extract before adding ethanol, a white fibrous precipitate that could be spooled onto a glass rod was produced. On addition of acid to extracts of *C. abyssinica* fruit, a white particulate precipitate was obtained. On analysis of the acid precipitate by HPSEC using water as the mobile phase, the elution pattern was similar to that of the ethanol precipitates, but the peaks had shifted to lower molecular weight (Chidewe, 2004). It is possible that precipitation of the polysaccharide using acid had resulted in degradation of the polymer to some extent. Such degradation would give rise to polymers of reduced molecular weight. The degradation of *C. abyssinica* polysaccharide would be similar to that observed for other polysaccharides.

Thomas and Coworkers (2003) have pointed out that when precipitating pectin using acid it is difficult to avoid some degradation of the polymer that takes place.

Characterisation of some properties of the polysaccharide of *C. abyssinica*

More polysaccharide material was recovered when ethanol was used for precipitation than when acid was used. The difference in yield, expressed as percent fresh weight, could arise if acid did not precipitate neutral polysaccharides that would be precipitated using salt and ethanol.

At less than 1%, the ash content of the polymer precipitated with acid was much lower than that for the polymer precipitated using salt and ethanol at 17.4%. It can be concluded that the polysaccharide of *C. abyssinica* was associated with metal ions. As shown in Table 1, potassium was the most abundant ion, with levels of Ca, Mg and Na being ten times lower. When ethanol was added to the extracts, the metal ions would have been precipitated together with the ionised polysaccharide. Addition of acid would have replaced metal ions associated with the polymer with hydrogen ions, leading to a reduced metal ion content upon precipitation. The ash content can be taken as a measure of the ions or salts that were associated with the polymer (James, 1995).

There is no obvious explanation for the observation that the protein content of polysaccharide precipitated with acid was just over half of that precipitated with ethanol. The uronic acid content of 9% for both the acid and ethanol precipitates is consistent with the acidic nature of the polysaccharide.

Some of the uronic acid groups in the polysaccharide were methyl esterified as indicated by the methoxyl content of 38% for both the acid and ethanol precipitates. The methoxyl groups would be expected to have an effect on the functional properties of the polysaccharide such as gel formation, with different gel forming mechanisms being observed for high and low methoxy pectins (Barnavon et al., 2001).

Effect of protease treatment of the polysaccharide of *C. abyssinica*

When solutions of the polysaccharide that had been precipitated with ethanol were treated with pronase E, a decrease in viscosity was observed as shown in Figure 2. The decrease in viscosity would result from hydrolysis of protein portions occurring in the polysaccharide, by pronase E. Polysaccharides that have been isolated are

often associated with proteins (Sims and Furneaux, 2003). The protein could be free protein that was co purified with the polysaccharides during isolation or protein that was covalently bound. Proteins that are non-covalently bound may be removed by physical methods such as gel chromatography, density gradient centrifugation or treatment with dissociating agents. The HPSEC profile for the native and protease treated samples was similar but the peaks in the protease treated sample shifted to low molecular weight, as shown in Figure 3. On the basis of these results it was not possible to conclude whether the protein digested by protease was free or covalently bound to the polysaccharide.

Determination of hydroxyproline residues in the polysaccharide chains

The hydroxyproline content obtained for both the acid (0.28 ± 0.01) and ethanol (0.29 ± 0.01%) precipitates, was low compared to the hydroxyproline content reported for arabinogalactan-peptide preparations isolated from wheat endosperm which contained 15 to 20% hydroxyproline and a protein content of 6 to 8% (Strahm et al., 1981). The relatively low content of hydroxyproline could be related to low levels of integrated protein occurring in the polysaccharide of *C. abyssinica*. The proteoglycans isolated from various plant tissues have been shown to contain arabinose linked covalently to the hydroxyl group of hydroxyproline. For example, gum arabic is believed to be a member of the arabinogalactan-protein group of proteoglycans with 25% hydroxyproline content (Osman et al., 1993). The gum exudate from *Acacia robusta* has been found to contain protein (18%) bound to arabino-galactan (Churms and Stephen, 1984).

From the characteristic presence of hydroxyproline in the polysaccharide from *A. robusta* it has been suggested that hydroxyproline occurs in the polysaccharide- protein linkages. The hydroxyproline o-arabinosyl linkage in cells provides cross-links in the polysaccharide network (Vidal et al., 2003).

Conclusion

The polysaccharide of *C. abyssinica* is an acidic polymer that is associated with protein, some of which appears to be covalently bound. Further work will be done to try and determine the nature and types of linkages between the monosaccharide constituents of the polysaccharide and ultimately determine the actual structure of polysaccharide using techniques such as mass spectrometry and nuclear magnetic resonance (NMR) spectroscopy.

ACKNOWLEDGEMENTS

We would like to acknowledge funding from Swedish

Table 1. Some properties of the polysaccharide isolated from the fruit of *C. abyssinica*. The uncertainties shown are standard deviations for at least three determinations. ND indicates that the measurement was not made.

Parameter	Sample Ethanol precipitate (%)	Acid precipitate (%)
Yield (fresh weight)	2.0 ± 0.4	1.2 ± 0.5
Moisture	10.3 ± 0.6	9.1 ± 0.1
Ash	17.4 ± 0.6	0.7 ± 0.5
Specific optical rotation	-50.0° ± 0.0	-50.0° + 0.5
Protein	4.6 ± 0.6	2.6 ± 0.4
Uronic acids	9.2 + 0.4	8.7 ± 0.8
Methoxyl content	38.3 ± 0.5	38.0 ± 0.9
Mineral ion content		
Sodium	0.2 ± 0.1	ND
Calcium	0.3 ± 0.1	ND
Magnesium	0.3 ± 0.1	ND
Potassium	4.8 ± 0.1	ND

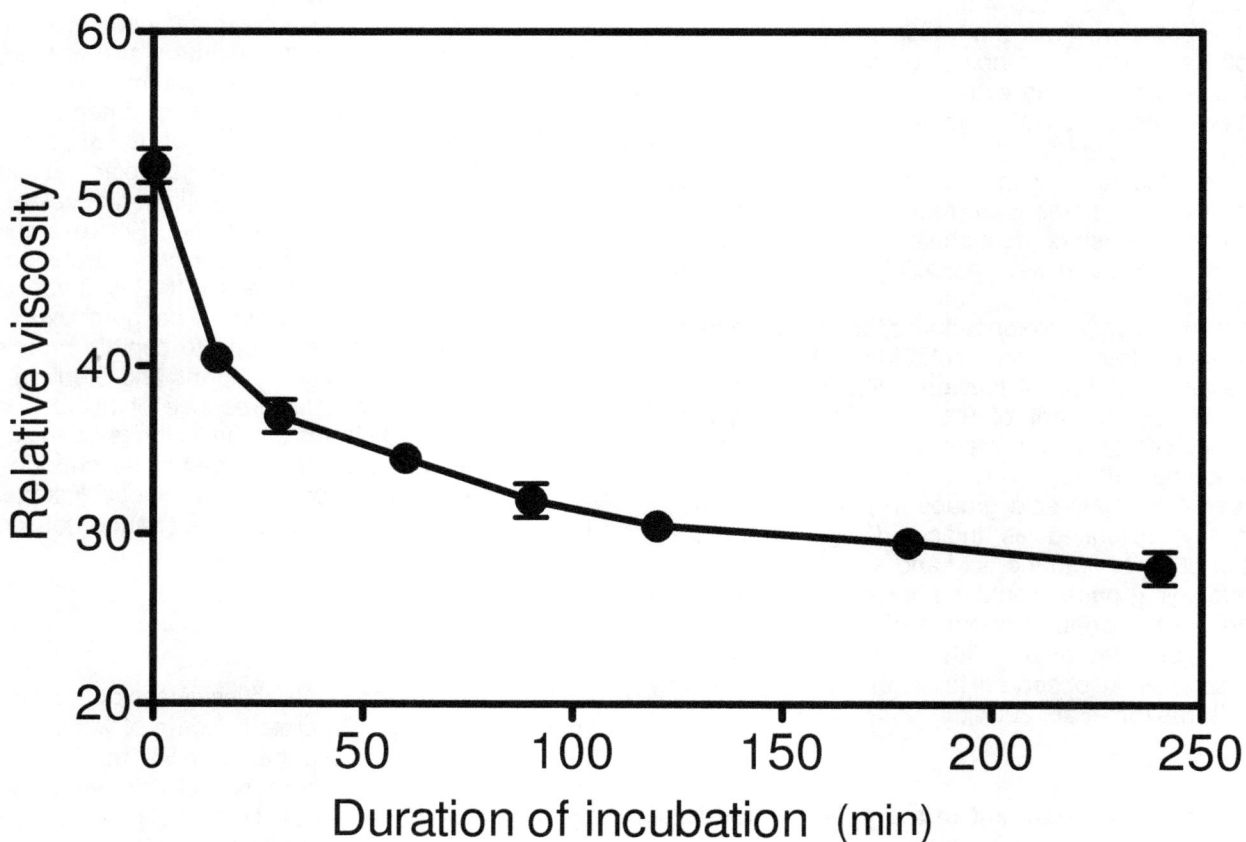

Figure 2. Change in viscosity of the polysaccharide of *C.abyssinica* during treatment with pronase E at 37℃. The uncertainties shown are standard deviations for at least three determinations. The points were standard deviation seem to be missing mean that SD was negligible.

Agency for Research Co-operation with developing countries (SAREC) and the research board of the

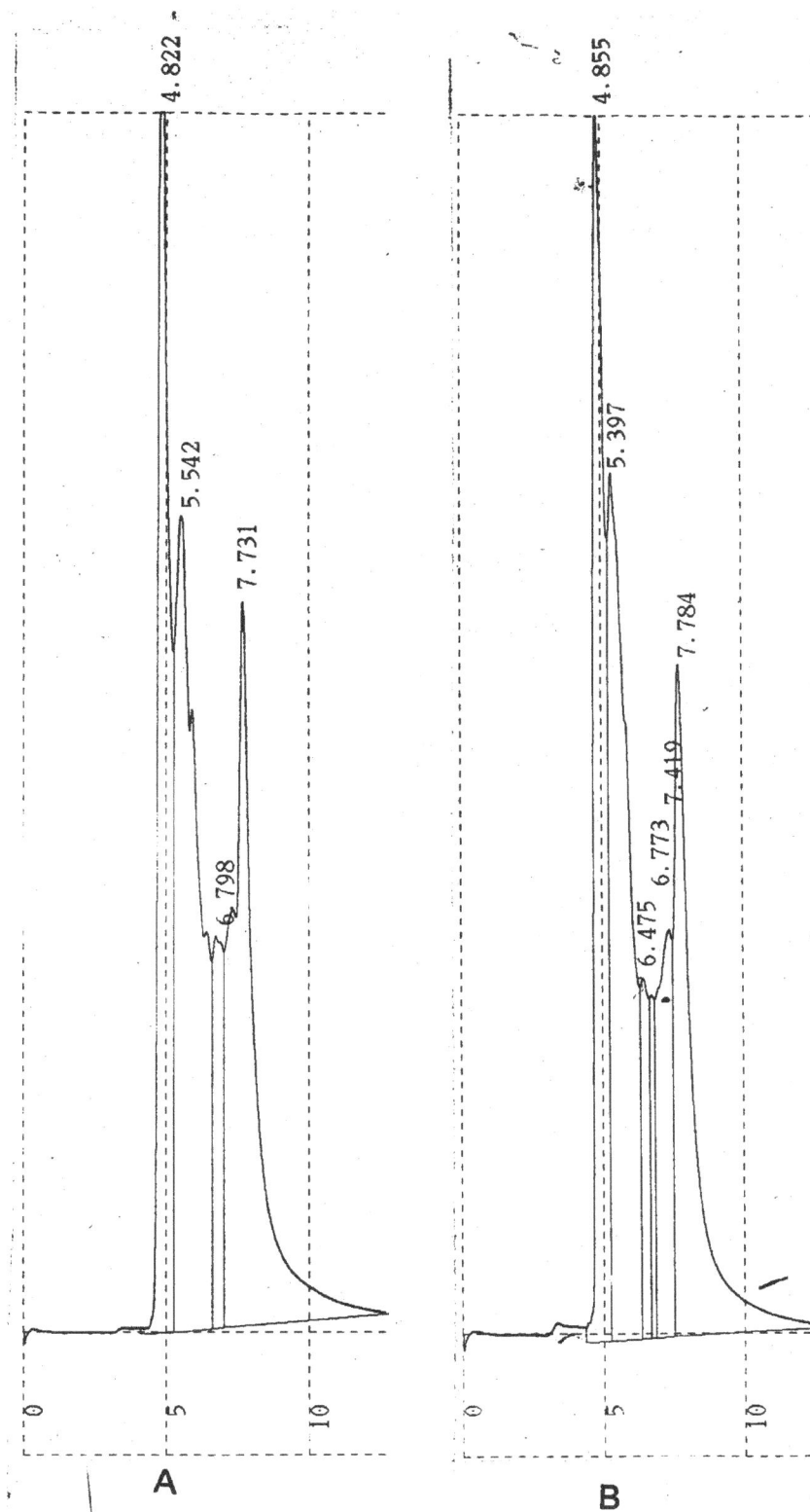

Figure 3. HPSEC profiles of the native (A) and protease treated (B) polysaccharides of *C. abyssinica*. HPSEC was done using a Zorbax GF 250 column and water as the mobile phase at a flow rate of 2 ml/min. Summary of retention times of substances was as follows: A (4.822, 5.542, 6.798 and 7.731) and B (4.855, 5.397, 6.475, 6.773, 7.419 and 7.784).

University of Zimbabwe.

REFERENCES

Barnavon L, Doco T, Terrier N, Ageorges A, Romieu C, Pellerin P (2001). Involvement of pectin methylesterase during the ripening of berrier: partial cDNA isolation, transcript expression and changes in the degree of methyl esterification of cell wall pectins. Phytochemistry, 58: 693-701.

Benhura MAN, Katayi-Chidewe C (2000). Preliminary study of the gelling properties of a polysaccharide preparation that is isolated from the fruit of *Cordia abyssinica*. Gums and Stabilizers for the Food Industry, 10: 65-75.

Benhura MAN, Chidewe C (2002). Some properties of a polysaccharide that is isolated from the fruit of Cordia abyssinica. Food Chem., 76: 343-347.

Benhura MAN, Chidewe C (2004). Emulsifying properties of a polysaccharide preparation isolated from the fruit of *Cordia abyssinica*. Int. J. Food Sci. Technol., 39: 579-583.

Chidewe C (2004). Characterisation of the polysaccharide material that isolated from the fruit of *Cordia abyssinica*. PhD thesis, University of Zimbabwe, Zimbabwe, p. 95.

Chaplin MF, Kennedy JF (1986). Carbohydrate Analysis-A practical approach. Washington, DC: IRL Press, pp. 81-87.

Churms SC, Stephen AM (1984). Structural studies of an arabinogalactan-protein from the gum exudate of *Acacia robusta*. Carbohydr. Res., 133: 105-123.

Euston SR, Hirst RL (2000). The emulsifying properties of commercial milk protein products in simple oil in water emulsions and in a model food system. J. Food Sci., 65(6): 934-940.

James CR (1995). Analytical Chemistry of Foods. London: Blackie Academic and Professional, pp. 37: 53-63.

Kossori RLE, Sanchez C, Bouston ESE, Maucourt MN, Sauvaire LM, Megean L, Villaume C (2000). Comparison of effects of prickley pear fruit, gum arabic, carrageenan, alginic acid, locust bean gum and citrus pectin on viscosity and *in-vitro* multi-enzyme digestibility of casein. J. Food Sci. Food Agric., 80: 359-364.

Laere KMV Hartemink R, Bolveld M, Schols HA, Voragen AGJ (2000). Fermentation of plant cell wall derived polysaccharides and their corresponding oligosaccharides by intestinal bacteria. J. Agric. Food Chem., 48: 1644-1652.

Osman ME, Williams PA, Menzies A, Phillips GO (1993). Characterisation of commercial samples of gum Arabic. J. Agric. Food Chem., 41: 71-77.

Palmer E, Pitman N (1972). Trees of Southern Africa, vol 3. Cape Town: A. A. Balkema, pp. 1934-1941.

Ridley BL, O'Neill MA, Mohnen D (2001). Pectins: Structure, biosynthesis and oligogalacturonic related signalling. Phytochemistry, 57: 929-967.

Rodriguez GO, De Ferrier BS, Ferrier A, Rodriguez B (2004). Characterisation of honey produced in Venezuela. Food Chem., 4: 499-502.

Rojas RM, Valverde MAD, Arroyo BM, Gonzalez TJ, Capote CJB (2004). Mineral content of *Gurumelo* (*Amanita ponderosa*). Food Chem., 85(3): 325-330.

Saka HJDK, Msonthi JD (1994). Nutritional value of edible fruits of indigenous wild trees in Malawi. Fore. Ecol. Mgt., 64: 245-248.

Sims IM, Furneaux RH (2003). Structure of the exudate gum from *Meryta sinclairii*. Carbohydr. Polymers, 52(4): 423-431.

Strahm A, Amando R, Neukom H (1981). Hydroxyproline-galactoside as a protein-polysaccharide linkage in water soluble arabinogalactan peptide from wheat endospernm. Phytochemistry, 20: 1061-1063.

Structure of pectin showing uronic acid residues and methylated groups (http://www.kjemi.uio.no/Polymerkjemi/Res./pectin.jpeg Accessed 08-02-2011).

Thomas M, Guillemin F, Guillon F, Thibault JF (2003). Pectins in the fruits of Japanese quince. Carbohy. Polymers, 53 (4): 361-372.

Van Wyk B, Van Wyk P (1998). Field Guide to Trees of Southern Africa. Cape Town: Struik Publishers (Pty) Ltd., pp. 136-137.

Vidal S, Williams P, Doco T, Moutounet M, Pellerin P (2003). The polysaccharides of red wine: Total fractionation and characterisation. Carbohydr. Polymers, 54(4): 439-447.

26

Evaluation of nutritional and antinutritional characteristics of Obeche (*Triplochition scleroxylon scleroxylon*) and several Mulberry (*Morus alba*) leaves

S. A. Adeduntan[1] and A. S. Oyerinde[2*]

[1]Department of Forestry and Wood Technology, Federal University of Technology, P. M. B. 704, Akure, Nigeria.
[2]Department of Agricultural Engineering, Federal University of Technology, P. M. B. 704, Akure, Nigeria.

The biochemical composition of some selected varieties of White Mulberry and Obeche leaves were investigated to ascertain their nutritional and antinutritional values. Three varieties of white Mulberry leaves (S_{36}, S_{54} and K_2) were harvested from Ondo State sericulture centre while Obeche leaves were harvested from Aponmu Forest Reserve located in Ondo State, Southwestern Nigeria, for analysis. Proximate chemical composition, minerals and anti-nutritional contents were determined on dry matter basis for the samples. The percentage of crude protein in all the samples was significantly higher ($P \leq 0.05$) with 34.31, 21.66, 21.585 and 21.24% in Obeche, S_{36}, S_{54} and K_2, respectively. Similarly, crude fibre follow the same trend with 20.753, 13.70, 10.81, 10.81, 13.70 and 8.74%, respectively, while the percentage water content were 73.70, 71.35, 72.16 and 76.00%, in Obeche , S_{36}, S_{54} and K_2, respectively. The results further show that the samples contain zinc in the range of 34.4 - 57.5, sodium 1069 – 1526, manganese, 14.83 – 24.37, calcium 944 – 1467, potassium 1684 – 2170, iron, 129.70 – 238.00 and magnesium, 1450 – 2196 (mg/kg). The mineral composition was generally comparable with what is obtained with other leafy vegetables. Phytate was significantly higher ($P \leq 0.05$) in Obeche. Likewise, cyanide and tannin were significantly higher ($P \leq 0.05$) in S_{36} than other treatments (Obeche, K_2 and S_{34}). However, these antinutrients (phytate, cyanide and tannin) are much lower than the pemitted values in fruits and any other food items. The result of the chemical analysis showed that all the selected Mulberry varieties and Obeche leaves contained adequate level of food nutrients required for normal body functioning.

Key words: Mulberry leaves, nutritional characteristics, Obeche leaves, proximate composition.

INTRODUCTION

The Mulberry tree is a perennial woody plant which belongs to the family Moraceae, Genus *morus Morus* and species *alba*. It is a deep-rooted perennial plant, capable of thriving under a variety of conditions ranging from temperate to tropical region. Several varieties of the tree are under cultivation in Ondo State, Nigeria, where the study was carried out. Mulberry tree is recognized as food plant for silkworm as well as an economic tree (Kasiviswanathan et al., 1988; Jaiyeola and Adeduntan, 2002). Its leaves contains high protein content and is also used in cattle feed for milk production (Kasiviswanathan et al., 1988). The timber is used for furniture, tool handle and the fruits are used for making wine, while the seeds are used for making jam (Datta and Ravikumar, 1988). The Mulberry can be grown as low bush, high trunk or deep-rooted tree and as such, could be utilized in afforestation of land and anti-erosion programmes (Datta and Ravikumar, 1988). Powder of *morus Morus alba* leaves have has been used to prepare drink by some people as a healthy diet in Japan, but its chemical composition was not known (Shimizu et al., 1992).

Obeche *Trilochition* (*Triplochiton scleroxylon*) is indigenous to the humid tropical forests of Central and West Africa. It is a commercial and important timber specie

*Corresponding author. E-mail: asoyerinde@yahoo.com.

in its natural habitat, the timber is used as veneer and for light construction. The species has shown considerable promise as plantation species in tropical areas of Africa and Pacific (particularly in the Solomon Island). It has an excellent form, self-pruning and grows very fast. The main drawback with the species is its short-lived seed that has stimulated considerable research into vegetative propagation techniques. Juvenile leaf cutting have been propagated successfully in conditions of high humidity in West Africa. Obeche has being used as hedges and for environmental stability. Its wood is used as fuel wood, while the sawn timber is used in building and for light construction and the leaves are used as food.

Some plant species with edible fruits in lowland rainforest ecosystem of Southwestern Nigeria is noted to contain large quantities of protein and vitamins especially vitamins A, B and C (Okafor, 1979; Akachuku, 1997). Their consumption therefore is able to augment the diet of people thereby preventing kwashiorkor and malnutrition especially in children. Some literature knowledge about the anti nutritional characteristics of these leaves should be stated. This study examined the nutrients and anti-nutrient potentials of Obeche leaves and three varieties of Mulberry leaves as a step towards establishing a wider and more purposeful utilization of these indigenous and exotic plant species.

MATERIALS AND METHODS

Obeche leaves which was collected at different parts of the branches (that is, top, middle and base) was collected were obtained from Akure Forest Reserve, Aponmu, while the varieties of the Mulberry leaves (S_{36}, S_{54} and K_2,) were collected from Ondo State Ministry of Agriculture Sericulture centre, Akure. The Mulberry varieties were given equal silvicultural treatments, while the analysis of the samples was carried out in triplicate using the standard procedures (AOAC, 1990). Each of the leaf samples of Obeche and the three Mulberry varieties were was oven dried at 60°C, pulverized and sieved through a 2 mm mesh screen and further dried at 60°C to constant weight, labeled and stored in an air-tight plastic jar at 4°C until required for analysis.

Proximate chemical compositions of various samples were analysed and crude protein, crude fibre, crude fat and crude ash were determined by using the methods of Association of Official Analytical Chemist (AOAC, 1990). The samples were dissolved in 10% HCL, filtered and diluted to 100 ml before estimation of their heavy metal contents. The nitrogen free extract (carbohydrate) was estimated by subtracting the sum of weights of crude protein, crude fibre, crude fat and crude ash from the total dry matter. Phosphorus was determined by the Phosphovanado molybdate method of (Ranjhan and Krisha, 1980) while the other minerals were determined after wet digestion with a mixture of nitric, sulphuric and perchloric acids using an Atomic Absorption Spectrometer (AAS: Model SP 9). A corning flame photometer - 410 was used for the determination of Na and K.

Extraction and precipitation of phytate were done by the method of Wheeler and Ferrel (1971) as used by Aletor (1995); Enujuigha and Ayodele (2003). Iron in the precipitate was determined by the method of Makower (1970.) Tannin values were obtained by adopting the method of (Markar and Goodchild, 1996), while hydrogen cyanide in the samples was obtained determined by AOAC (1990) method.

Data generated obtained from proximate, minerals and anti-nutrition properties were subjected to one-way analysis of variance (ANOVA), Steel and Torrie (1960), where significant differences were observed and mean separation was done by Duncan Multiple Range Test (DMRT) (Duncan, 1955).

RESULT AND DISCUSSION

Proximate chemical composition of the leaf samples is presented in Table 1. Here, a wide variation was noticed in the samples and the result of chemical analysis of Obeche and Mulberry leaves differ significantly. The crude protein ranged from 21.24 to 21.66% in Mulberry leaves while Obeche leaves contain 34.31%. This shows higher protein content compared with some major vegetable leaves such as (FAO, 1990). Table 1 revealed that Obeche has the highest protein content followed by S_{36}, S_{54} and K_2 with the least value. The high protein values observed is in agreement with Kasiviswanathan et al. (1988) and it is an indication that both Mulberry and Obeche can be of food value in man, silkworm and animal. The limitation to the full utilization of Obeche leaves could be due to high concentration of anti-nutritional factors mainly phytate and cyanide. This result equally indicated higher levels of crude protein compared to the commonly cultivated legumes such as cowpea; pigeon pea and lima beans as it was reported by Aletor and Adeogun (1995). Thus all species of Mulberry and Obeche leaves tested in this work could serve as substitute for existing plant protein and since the biomass yield of Mulberry plant is very high at very shortest time (Adeduntan, 2003), coupled with lower cost of production, it could be highly recommended as food for man and animal (Silkworm) man, animal and Silkworm. The level of carbohydrate in Mulberry varieties and Obeche were 56.442, 49.04, 47.227 and 30.04%, respectively with K_2 having the highest values and Obeche having the least value. There is corresponding relationship between the carbohydrate content of the leaves and their protein values. As protein values increases in some variety their corresponding carbohydrate value decreases. Thus, the higher the protein, the lower their corresponding carbohydrate, a situation that is very advantageous for rural sector of the economy that is facing food crisis today.

The fat content of the leaves were as shown in Table 1. S_{36} has the highest fat level of 8.02% followed by S_{54} with 6.05% while Obeche and K_2 are 5.46 and 5.31% respectively. These results signify that there are significant differences among the samples in fat levels and it agreed with the result obtained in the nutrient composition in Mulberry leaves carried out in Asia, UN (1993). The ash content was the lowest in K_2 (8.219%) and the highest in S_{36} (12.63%); Obeche and S_{54} has 9.22 and 9.65% respectively. High ash content was considered to be a good source of mineral food (Enujiugha and Agbede, 2000). The fibre content was (Enujiugha and Agbede, 2000). The fibre content was lowest in K_2 (8.274%) with

Table 1. Proximate composition of Mulberryleaves and Obeche leaves.

Samples	Moisture content (%)	Crude protein (%)	Crude fat (%)	Ash (%)	Crude fibre (%)	Total carbohydrate
S_{36}	79.35 ± 0.69^a	21.66 ± 0.0^b	8.02 ± 0.30^a	12.63 ± 0.20^a	10.8 ± 0.80^c	47.27 ± 0.41^c
Obeche	73.70 ± 0.13^c	34.31 ± 0.0^a	5.46 ± 0.00^c	9.22 ± 0.04^c	20.73 ± 0.15^a	30.04 ± 0.17^d
S_{54}	72.16 ± 0.41^d	21.55 ± 0.0^c	6.05 ± 0.00^b	9.65 ± 0.16^b	13.70 ± 0.18^b	49.04 ± 0.23^b
K_2	76.00 ± 0.44^b	21.24 ± 0.0^d	5.31 ± 0.03^c	8.19 ± 1.70^d	8.74 ± 0.15^d	56.42 ± 0.53^a

Values in the same column followed by the same alphabet are not significantly different ($P < 0.05$).

Table 2. Anti-nutritional composition of the samples.

Samples	Phytate (mg/kg)	Cyanide (mg/kg)	Tannin (mg /kg)
S_{36}	488.90 ± 32.47^b	2.14 ± 0.20^a	5.32 ± 0.13^a
Obeche	997.80 ± 130.27^a	1.12 ± 0.20^b	3.54 ± 0.04^c
K_2	451.30 ± 0.00^b	1.01 ± 0.00^b	3.65 ± 0.01^c
S_{54}	456.80 ± 9.45^b	1.24 ± 0.20^{ab}	3.78 ± 0.00^b

Values in the same column followed by the same alphabet are not significantly different ($P < 0.05$).

Obeche having the highest of 20.73%. The lowest level of fibre in K_2 may be attributed to its high carbohydrate content of 56.41% while Obeche that has the highest fibre resulted in the lowest carbohydrate 30.04%. Generally, the result of the proximate analysis of the samples (S_{36}, S_{54}, K_2 and Obeche) shows significant differences for crude protein, ash, crude fibre and carbohydrate but there were no significant differences in crude fat of Obeche and K_2.

Table 2 shows the level of anti-nutrients in various samples (S_{36}, S_{54}, K_2 and Obeche), which hinder the utilizable nutrient in them. The LSD (Least significant differences) shows that there are significant differences in the phytate levels of the samples. The phytate content ranged from 451.3 mg/kg in K_3, 456.8 mg/kg in S_{54}, 488.9 mg/kg in S_{36} to 9797 mg/kg in Obeche. Phytin-p is known to be the primary storage form of phosphorus in mature legume seed (Enujiugha and Agbede, 2000). The high phytin content in Obeche has nutritional significance as it does not only make phytin-p unavailable to humans and monogastric but it also lowers the availability of many other essential minerals as reported by Aletor (1995).

The phytate contents of these leaves are less than what is obtained in some fruits such as guava (327 mg/100 g), Plantain (553.08 mg/100 g), and banana (847.53 mg/100 g) whereas the lethal standard value for phytate is 2500 mg/100 g (FAO, 1990).

The cyanide content of the leaves is 1.10 01 mg/kg in K_2, 1.12 mg/kg in Obeche, 1.264 mg/kg in S_{54} and 2.14 mg/kg in S_{36} which indicated the highest level. This result is far below the standard permitted value for cyanide in fruits and in any other food value. The minimum standard cyanide value in leaves is 30 mg/kg FAO (1990). These levels of cyanide may not have much effect on man and silkworm.

Table 2 also shows the tannin levels in the samples.

S_{36} has the highest tannin content of 3.54%, 5.32 mg/kg followed by S_{54} (3.78% mg/kg), K_2 (3.65% mg/kg) and Obeche (5.32%, 3.54 mg/kg). These values obtained are considered to be lower when compared with the standard value of 37%, mg/kg (FAO, 1990). Thus these leaves are safe for consumption by man and livestocks. Goldstein and Swain (1963) described Tannin as phenolic com-pounds with degree of hydrozylethyn with molecular size that is sufficient to form complexes with proteins. Rubino and Davidoff (1979) reported that cyanide of any part of plants causes accident and often total cyanide poisoning.

There are significant differences in the zinc content between Obeche and the rest samples (S_{54} and K_2). S_{36} has the highest zinc content with 57.50 mg/kg followed by Obeche 48.60 mg/kg while both S_{54} and K_2 has 34.40 mg/kg. Zinc, which has been found to be an essential component of enzymes that plays critical role in protein and carbohydrate synthesis. The deficiency of zinc can cause break down in immense function of host defensive mechanism. S_{36} have the highest magnesium content followed by Obeche with 17.00 mg/kg, both K_2 and S_{54} has 14.50 mg/kg.

Calcium in Obeche was significantly lower than any other samples with 944.7 mg/kg while K_2 and S_{54} ranked second with value of 13.75 1375 mg/kg and S_{36} ranked the highest with 1467 mg/kg as shown in Table 2. The values of calcium detected are in order of which is noted to be good for bone formation and osmo-regulation. Potassium level in S_{36} is significantly higher than all other treatments with 2170 mg/kg while the level in Obeche, which is 1703 mg/kg, is significantly higher than those of K_2 and S_{54} with value of 1684 mg/kg each as shown in Table 3. In man, potassium is good for nerves and muscle functions. The values of iron in K2 and S54 are 141.75 and 141.57 mg/kg, respectively. The value in

Table 3. Mineral elements of Mulberry leaves and Obeche.

Samples	Zinc (mg/kg)	Sodium (mg/kg)	Manganese (mg/kg)	Calcium (mg/kg)	Potassium (mg/kg)	Iron (mg/kg)	Magnesium (mg/kg)	Co	Cd	Lead	Cu
S_{36}	57.50 ± 0.45[a]	1526 ± 0.8[a]	24.37 ± 0.47[a]	1467 ± 1.05[a]	2170 ± 0.35[a]	238 ± 0.66[d]	2196.4 ± 1.23[a]	ND	ND	ND	ND
Obeche	48.60 ± 0.46[b]	1069 ± 0.31[c]	14.83 ± 0.25[c]	944 ± 0.47[c]	1703 ± 0.44[b]	129.7 ± 0.46[c]	1700 ± 0.55[b]	ND	ND	ND	ND
S_{54}	34.40 ± 0.56[c]	1081 ± 0.32[b]	18.27 ± 0.21[b]	1375 ± 0.7[b]	1684 ± 0.53	141.7 ± 0.45[b]	1450 ± 0.36[c]	ND	ND	ND	ND
K_2	34.40 ± 0.46[c]	1081 ± 0.55[b]	18.23 ± 0.25[b]	1375 ± 0.43[b]	1684 ± 0.29	141.5 ± 0.49[b]	1450 ± 0.35[c]	ND	ND	ND	ND

Values in the same column followed by the same alphabet are not significantly different (P < 0.05). ND: Not detected.

Obeche is the least (129.7 mg/kg) while S_{36} has the highest value. The mineral element in Table 3 however indicated further those elements of copper, cadmium and lead that were not detected but their absence is not a nutritional disadvantage.

CONCLUSION AND RECOMMENDATION

Obeche (*Triplochiton scleroxylon*), contained very high protein and crude fibre that is very essential for man, livestock and silkworm's growth. Its leaves have been eaten by some communities in Ondo State, Nigeria, but its consumption has not been widespread. It is therefore suggested that the leaves be subjected to further treatments or processing to reduce the toxic level before processing for consumption and K_2 S_{54} varieties of Mulberry leaves should be encouraged by local farmers to feed their silk worms and livestock because they contain lower phytate, cyanide and tannin content, and are rich in essential minerals which are made available for utilization. This work thus, supports the consumption of Mulberry and Obeche leaves by man. As it has been reported by Datta (1992) that Mulberry leaves serves as a source of delicious vegetable, which is very rich in protein.

REFERENCES

Adeduntan SA (2003). Effect of fertilizer application on the growth and folia composition of Mulberryleaf. Appl. Trop. Agric. Res., 11: 18-22.

Aletor VA (1995). Compositional Studies on Edible Tropical Species of Mushrooms. Food Chem., 54: 265-268.

Aletor VA, Adeogun OAOA (1995). Nutrient and anti-nutrient composition of some tropical leafy vegetables. Food Chem., 53: 375-379.

Akachuku IO (1997). Status of Forest Food Plant Species and Environmental Management in southeast Nigeria. In Oduwaye, E. A; Obiaga, P. C. and Abu, J. E. (Editors), Proceedings of 24th Annual Conference of Forestry Association of Nigeria, Ibadan, Nig., pp. 21-29.

AOAC (1990). Official method of Analysis. Association of Analytical Chemists, Washington D.C, 15th ed.

Datta RK, Ravikumar C (1988). Sericulture and Rural Development International Congress on Tropical Sericulture practice. Lead paper 1, Central Silk Board, India, pp. 4-5.

Datta RK (1992). Guidelines for bivoltine rearing central silk board. Ministry of Textile, Government of India, 39, Mahatma Gandhi Board, Banglare.

Duncan B (1955). Reddy. N.R. Multiple Range f-tests Biometrics, 11: 1-41.

Enujiugha UM, Agbede JO (2000). Nutritional and Anti-nutritional Characteristics of African oil bean *pentadethra marophylle Benth* seed. Appl. Trop. Agric., 5: 1-4.

Enujiugha UM, Ayodele OM (2003). Nutritional status of lesser- known oil seeds. Int. J. Food Sci. Tech., 38: 1-4.

FAO (1990). World meat situation and outlook, commodities and trade division Food and Agriculture Organisation Italy, Rome.

Goldstein JL, Swain T (1963). Change of ripening fruits. Phytochem., 2: 271-283.

Jaiyeola V, Adeduntan SA (2002). Sericulture in Ondo state: A means of alleviating rural poverty. In Forestry and Challenges of sustainable livelihood Proceedings of the 28th Annual Conference of the Forestry Association of Nigeria held in Akure, Ondo state, pp. 202-207.

Kasiviswanathan K, Parankumar T, Chowdhary PC, Somashakara KS (1988). Moricultural practical International congress on Tropical Sericulture practices, Lead paper 2. Central silkboard Bangalore, Indian, p. 11.

Makower RV (1970). Extraction and Determination of phytic acid in beans (*phaseolus lunatus*). Cereal Chem., 47: 28-292.

Markar AOS, Goodchild AV (1996). Qualifications of tannins. A laboratory manual. ICARDA, Akeppo, Syria, 4: 25.

Okafor JC (1979). Edible Woody Plants in Rural Economy of Nigeria Forest Zone. In Okali, D. U. U. (Editor). Proceeding of a MAB state of Knowledge workshop on the Nigeria Rainforest Ecosystem, Nigeria, pp. 262-301.

Ranjhan SR, Krishna G (1980). Laboratory Manual for Nutrition Reseaarch, eds. Ranjhan, SR. and Krishna, G. Vicas publ. Co. New Delhi, India, p. 450.

Shimizu P, Yazawa M, Takede N (1992). Aromatic acid in leaves of *morus alba* and their possible medical value, Sericologia, 32(4): 633- 636.

Steel RCD, Torrie, JH (1960). Principles and procedures of statistics 1st ed. McGraw-Hill, New York.

United Nations (UN) (1993). Techniques of silkworms rearing in the tropics table 1. Nutritive compositions in Mulberryleaf page24.Campion, D. G. 1972. Some observations on the use of pheromone traps as a survey tool for Spodoptera littoralis. Cent. Overseas Pest Res. Rep., 4: 10.

Wheeler VE, Ferrel RE (1971). A method for phytic acid determination in wheat fractions. Cereal Chem., 48: 312-316.

27

Effects of soaking and boiling and autoclaving on the nutritional quality of *Mucuna flagellipes* ("ukpo")

E. A. Udensi[1], N. U. Arisa[2*] and E. Ikpa[1]

[1]Department of Food Technology, Bells University of Technology, Ota, Ogun State, Nigeria.
[2]Department of Food Science and Technology Abia State University Uturu, Abia State, Nigeria.

The effect of soaking and boiling, and autoclaving on nutritional factors (proximate and mineral) and anti-nutritional qualities of the legume, *Mucuna flagellipes* were studied. Batches of seeds were soaked for 6, 12, 18 and 24 h in distilled water at room temperature, then boiled in water for 30, 45, 60 or 90 min respectively. Another batch of *M. flagellipes* was autoclaved for different duration of 30, 45, 90 and 120 min respectively. Results showed that soaking followed by boiling produced products with lower crude fibre content (10% for soaking for 24 h followed by boiling for 90 min). However it increased the carbohydrates. Autoclaving resulted in products with lower mineral contents (1042.5 mg/100 g phosphorous for autoclaving for 120 min.

Key words: Legumes, *Mucuna flagellipes*, nutritional factors, soaking, boiling, autoclaving.

INTRODUCTION

Legumes are good sources of cheap and widely available proteins for human consumption. They are staple foods for many people in different parts of the world (Youseff et al., 1989). Legume seeds have an average of twice as much protein as cereals and the nutritive value of the proteins are usually high (Vijaykumarri et al., 1997). They range between the highly utilized legumes such as soybeans, cowpeas to the lesser known ones like African yam beans (*Sphenostylis stenoscarpa), Mucuna conchinchinesis* and *Mucuna flagellipes* ("ukpo"). Studies have shown that the lesser known legumes together with other conventional legumes can be used for combating protein malnutrition prevalant in the third world. This can be achived by the consuption of the legumes whole and in various processed forms (condiments) (Arisa and Aworh, 2007).

M. flagellipes ("ukpo") contains high percentage of proteins (20.4%), carbohydrates (61%) and fat (9.6%) on fresh weight (Enwere, 1998). The excellent nutritional value of the legume in terms of proximate, mineral composition makes it necessary for it be used as compliments in African dietss which are mainly roots and tuber based (Pirie, 1975). Due to the fact that the seedss are rich in protein and carbohydrates and the protein content content range from 11.82 ± 0.25 g/100 g to 24.94 ± 0.18 g/100 g dry matter basis, they compare favourably with high protein animal sources such as oyster, beef, pork and marine fishes (Ajayi et al., 2005).

The cotyledons of seeds are widely used as soup thickener when they have been broken, boiled sufficiently to soften them and milled into powder in the Eastern part of Nigeria. They are sometimes broken, roasted with hot charcaol and ash, milled and used soup thickners. It has been reported that nutritive value of some seeds; rape seed, *M. conchinchinensis*, *Mucuna utilis* have been improved by heat treatment due the reeduction of their antinutritional factors content by them (Manssour et al., 1993; Ukachuku and Oioha, 2000; Udensi et al., 2004).

This study was undertaken to evaluate the effect soaking and boiling or autoclaving will have on the nutrients present in the seeds with a view of recommending a process that will produce flour with high nutritional content.

MATERIALS AND METHODS

M. flagellipes seeds obtained from Umuahia main market, Abia State Nigeria were cleaned and broken to remove the seed coat. They were divided into three portions (each weighing 300 g). First

---

*Corresponding author. E-mail: chizaramekpere2006@yahoo.co.uk.

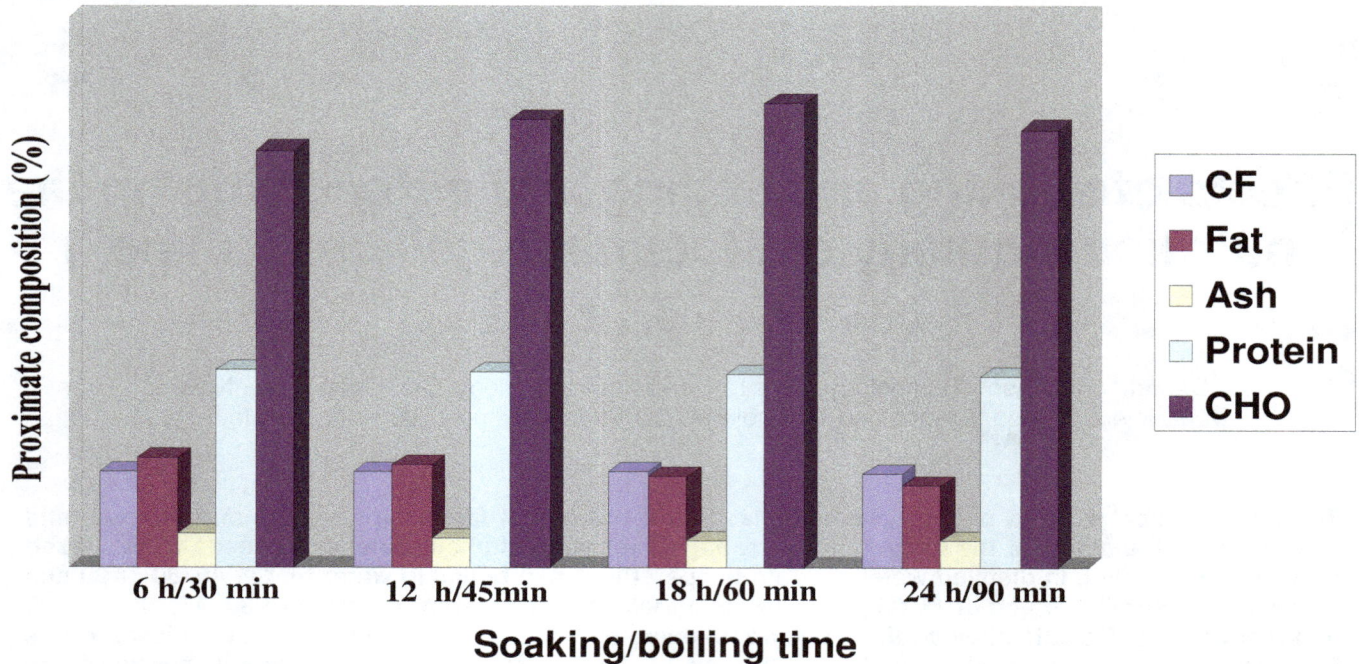

Figure 1. Effect of soaking/boiling on the proximate composition of *Mucuna flagellipes*. CF: Crude fibre, CHO: Carbohydrates, h: Hours and min: Minutes.

portion (300 g) of the dehulled cotyledons were soaked for 6, 12, 18 or 24 h in water at room temperature respectively. They were subsequently boiled in water for 30, 45, 60 and 90 min respectively, dried in a forced draught oven at 70°C and milled into powder using a hammer mill (100 μ mesh size). The other portion (300 g) of dehulled cotyledons was autoclaved at 15 lb pressure (121°C) for a period of 30, 60, 90 or 120 min respectively. Then dried in a forced draught oven at 70°C and milled into powder using a hammer mill 100 μ mesh size). The third portion (300 g) was processed as is (raw) control. The untreated and processed cotyledons and the flours from the different treatments were analyzed in trplicates for nutritional factors, antinutritional and some mineral content.

Analysis of nutritional factors

Moisture was determined by drying to constant weight at 105°C in a forced draught oven. Crude protein content of the samples was determined using the micro Kjedhal digestion method described by AOAC (1984). The method described by Kirk and Sawyer (1991) was used to determine the crude fibre content of the samples. The protocol for crude fibre content was given below. Defatted (2 g) sample was boiled in 200 cm^3 of 0.1275 M sulphuric acid solution for 30 min with constant agitation. The boiling mixture was poured into a buckner funnel and washed with boiling water twice. Then, the residue was boiled in a 0.313 M sodium hydroxide solution for 30 min with constant stirring. The residue was then washed twice with boiling water followed by 1% hydrochloric acid, then washed with boiling water until free from acid. It was then dried in an oven to a constant weight.

The fat content of the samples were determined using the procedure described by Pearson (1976). Total ash content was determined using the method of Kirk and Sawyer (1991). The total ash present in 5 g of the samples was determined by incinerating the sample in a muffle furnace at 550°C 3 h. The carbohydrate

content was determined by the method described by Pearson (1976).

Analysis of the mineral content

The mineral content of the various samples were determined using the procedures described by Kirk and Sawyer (1991).

RESULTS AND DISCUSSION

Soaking for a period of time followed by boiling (Figure 1) resulted in decrease in the crude fibre content of the *M. flagellipes* flour with soaking for 24 h followed by boiling for 90 min giving product with the least crude fibre content (10.10%). This is in line with Ukachukuwu and Obioha (2000) who reported a decrease in crude fibre content of *M. cochinchinensis* as boiling time was increased. The same trend of reduction was observed in the other proximate compositions (fat, ash and protein) as the time of soaking/boiling increased. The reduction in the nutrients content could have been as aresult of the leaching out of the some water soluble nutrients into the the soak water and boiling water. This trend of decrease is similar to the observation made by Obasi and Wogu (2008), who reorted decrease in protein content of soaked yellow maize during soaking. The reduction in the protein content may also be attributed to the progressive solubilization and leaching out of the nitrogenous substances during soaking and boiling of the legume

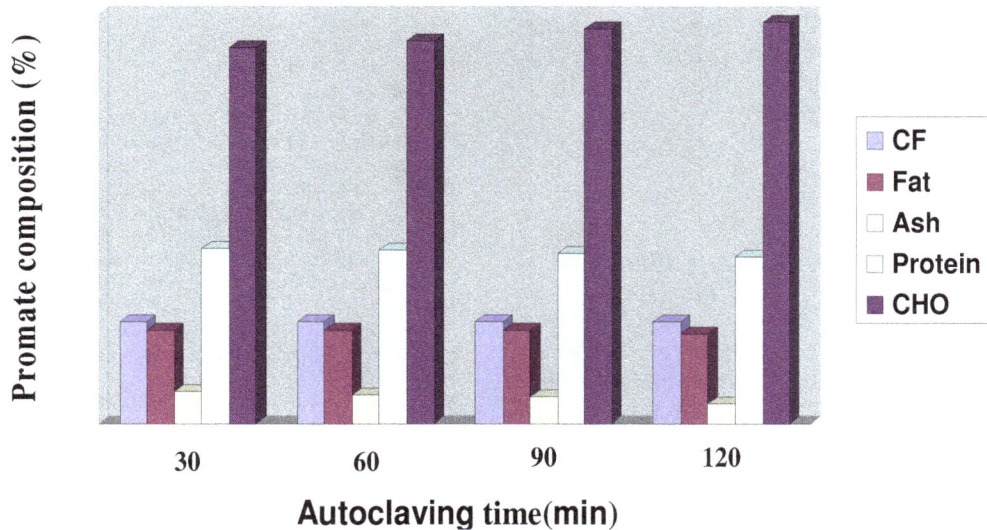

Figure 2. Effect of Autoclaving on the Proximate Composition of *Mucuna Flagellipes*. CF: Crude fibre, CHO; Carbohydrates, min; Minutes.

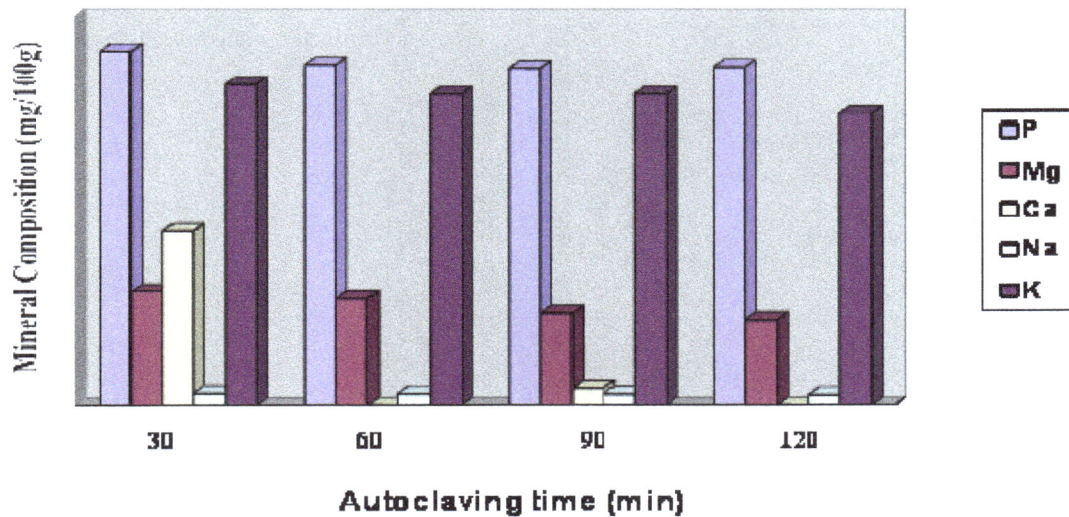

Figure 3. Effect of autoclaving on the mineral composition of *Mucuna Flagellipes*. Min; minutes.

(Ukachukwu and Obioha, 2000).

However the carbohydrate content seemed to increase (50.66% for soaking followed by boiling for period of 18 h and 60 min respectively), this could have been due to the break down of complex carbohydrates which were otherwise bound in the raw sample by boiling. Autoclaving (Figure 2) of the seeds yielded products which had lower protein and ash contents than the raw seeds. Increase in time of autoclaving progressively gave products with lower ash content (2.40% for autoclaving for 120 min). It gave products with higher fat content than the raw. Results have shown that processing (boiling,

soaking, autoclaving and roasting) leads to products with reduced anti nutritional content and consequently this increase in some nutrients content (Udensi et al., 2008). This could have been due to the fact that the bound fat in the cells of the seeds may have been released as a result of the treatment. The carbohydrate content was also increased as the time of autoclaving increased (49.90%) for autoclaving 120 min being the highest.

Soaking followed by boiling gave products with reduced mineral contents (Figure 3) generally but soaking for 24 h followed by boiling for 90 min resulted in products with lowest mineral contents (1055.00 mg P, 374.40 mg Mg,

408.82 mg Ca, 31.00 mg Na, 945.00 mg K) when compared with the raw sample except for the values of P, Mg and Ca which were the same as those of soaking for 18 h and boiling for 60 min respectively.

Conclusion

Soaking followed by boiling resulted in products with reduced fibre, fat, ash and protein contents. The use of autoclaving in the treatment of the seeds yielded products with increased carbohydrates and fat contents. This could have been as a result liberation of the cell bound nutrients which otherwise would not have been available as pure nutrients. It is therefore suggested that any of the processes can be used for processing the mucuna, however due to the fact that autoclaves may not be readily available for the locals soaking followed by boiling can be done.

REFERENCES

Ajayi IA, Oderinde RA, Kajogbola DO, Uponi JI (2005). Oil content and fatty acid composition of some underutilized legumes from Nig. Food Chem. 99(1): 115-120.

AOAC (1984).Official method of analysis (14th Edn). Ed. S. Williams. Arlington,V.A. USA.

Arisa NU, Aworh OC (2007). Production, quality assessment and acceptability of African yam bean Sphenostylis stenoscarpa sauce. J. Food Proc. Preserv. 31: 771-778.

Balagopelan CG, Padmaj SI, Nanda C, Moorthy SN (1988). Cassava food, feed and industry. CRS Press Inc. Florida, USA pp. 187-189.

Enwere JN (1998). Foods of plant origin. Afro. orbis Publ. Nsukka Nigeria pp. 64-124.

Kirk RS, Sawyer R (1991). Fats and Oils. In: Pearson's Composition and Analysis of Foods, 9th edn. Longman Group Limited, UK p. 641. Mansour EH, Dworschsk E, Lugasi A, Barna E, Gergely A (1993). Effect of processing on the antinutrional factors and nutritional value of rape seed products, Food Chem. 47: 247-252.

Obasi NE, Wogu CO (2008). Effect of soaking time on the proximate and mineral compositions and antinutritional factors of yellow maize (Zea mays). Nig. Food J. 26(2): 69-77.

Pearson D (1970). The chemical Analysis of Foods, 6th edn. Churchill Livingstone, Edinburgh pp. 487-451.

Pirie NW (1975). Food protein sources. Cambidge University Press, London p. 230.

Udensi EA, Onwuka GI,Okoli EG (2004). Effect of processing on the levels of some antinutritional factors of Mucuna utilis Plant Prod. J. 8(1): 1-6.

Udensi EA, Arisa NU, Maduka M (2008). Effect of processing method on the level of antinutritional factors in Mucuna flagellipes Nig. Food J. 26(2): 53-59.

Ukachukwu SN, Obioha FC (2000). Effect of time duration of thermal treatments on the nutritive value of Mucuna cochinchinensis. Global J. Pure Appl. Sci. 9: 11-15.

Vijayakumari K, Siddhuraji P,Janardhanan K (1997). Effect of domestic processing the levels of certain antinutrients in Prosopis chilensis (Molina) Stunz. Seeds. Food Chem. 59(3): 367-371.

Youseff MM, Abdal MA, Shekibs LAE, Ziena HM (1989). Effects of dehulling, soaking and germination of chemical composition, mineral elements and protein patterns of feba beans (Vicia feba 1) Food Chem. 23: 129 – 136.

28

Iodine and inorganic mineral contents of some vegetables, spices and grains consumed in Southeastern Nigeria

Ujowundu C. O.[1*], Kalu F. N.[2], Nwosunjoku E. C.[3], Nwaoguikpe R. N.[1], Okechukwu R. I.[4] and Igwe K. O.[1]

[1]Department of Biochemistry, Federal University of Technology Owerri, Nigeria.
[2]Department of Biochemistry, University of Nigeria, Nsukka, Nigeria.
[3]Environmental Health Department, Imo State College of Health Science and Technology, Amigbo, Imo State Nigeria.
[4]Department of Biotechnology, Federal University of Technology Owerri, Nigeria.

Selected vegetables, spices and grains which serve as staples in Southeastern Nigeria were analyzed for their iodine and some essential minerals. Our results showed that the mean iodine concentration of the vegetables were of 94.61 ± 15.28 µg/100 g, with Uha leaves (*Pterocarpus* spp.) having the highest with 117.66 ± 3.00 µg/100 g iodine, cowpea (*Vigna sinensis*) had the highest mean iodine concentration of 69.36 ± 7.94 µg/100 g in grains. The concentration of iodine in the spices ranged from 21.56 ± 3.6 µg/100 g in garlic (*Allium sativum*) to 95.66 ± 1.73 µg/100 g in Uziza (*Piper guineense*). Iodine was not detected in some of the plants sampled. The concentrations of the mineral elements in the vegetables, spices and grains were presented. The mineral contents indicated that, they are valuable and positive contributors to the overall iodine nutrition in the diets of the rural and urban people of Southeastern Nigeria.

Key words: Iodine, vegetable, spices, grains, goitrogen, iodine deficiency.

INTRODUCTION

No micro-ingredient is more critical to metabolism and overall health than iodine. Iodine is unique among the required trace elements because it is the constituent of the thyroid hormones. Iodine deficiency is a common cause of preventable mental defects (Hetzel, 1989; Hetzel and Wellby, 1997). Iodine is irregularly distributed over the earth's crust and in some areas the surface soil becomes progressively poorer in iodine through accelerated deforestation, soil erosion and leaching processes (EGVM, 2002; Singh, 2004). Foods grown in iodine deficient soil cannot provide enough iodine for the people and livestock living there (Koutras et al., 1985; WHO, 1996; Souci et al., 2000). Humans naturally receive their iodine by consuming plant and animal products. The intake of iodine generally corresponds to the amount entering the local food chain from geochemical environment and it is normally low from natural foods (US-FNB, 2001). For the proper utilization of iodine for thyroid hormone synthesis some mineral are required at the right concentration and proportion.

The mineral nutrients are interrelated and balanced against each other in human physiology. They cannot be considered as a single element with circumscribed functions. For instance, sodium, calcium, magnesium and phosphorus serve individual and collective purposes in the body fluid regulation. Inadequate mineral intake generally produces deficiency symptoms which include anaemia, impaired healing of wounds, delayed blood clotting severe diarrhea and chronic renal failure. Selenium and iodine ingestion have to be regulated as deficiency can lead to extreme fatigue, endemic goiter, cretinism and recurrent miscarriages (Vanderpas et al., 1990). Some inorganic mineral nutrients have been reported to be antagonistic and interfere with iodine metabolism (Underwood and Suttle, 1999; Wash, 2003). Iron (Fe) deficiency lowers the thyroid peroxidase activity - a heme-containing enzyme that catalyzes the initial steps in thyroid hormone synthesis. High calcium (Ca) diets or hard water high in Ca, may increase the need for additional iodine (Jooste et al., 1999). Mineral nutrient deficiencies such as zinc, copper, iron, also contribute to inability to use iodine well and this may lead to the development of Goiter (Osman and Fatah, 1981).

Mineral malnutrition can have a negative impact on thyroid function but in the presence of adequate iodine supplies, it is less common for such factors to cause problems (Gartan, 1988). High levels of minerals above the recommended daily allowance (RDA) have also been shown to be goitrogenic (Osman and Fatah 1981). This study seeks to determine the concentration of iodine and other inorganic minerals in some commonly consumed plant foods that have formed staples in the study area. The data generated may be used to work out daily intake of iodine and other inorganic minerals in the study area, thereby serving as a tool for proper nutritional planning.

MATERIALS AND METHODS

Sample collection

The samples were collected from; old market-Douglas Road, New market- Douglas Road Owerri, Umuokochi market-Nekede, Ihiagwa market and Umuokoto market-Nekede all in Imo State Nigeria. The grains sold in these markets come in from the northern region of the country while the vegetables come in from the nearby villages in Imo State. The samples were; Grains- millet (*Panicum miliaceum*) cowpea (*Vigna sinensis*) beans (*Phaseolus vulgaris*), wheat (*Triticum aeastivum*), Okpa (*Vadzeia subterranean*), groundnut (*Arachis hypogaea*), maize (*Zea may*), soybeans (*Glycine max*), rice (*Oryza sativa*) and sorghum (*Sorghum bicolor*). Spices- curry (*Muraya koenigii*), Uziza (*Piper guineense*), Nchaunwu (*Ocimium gratissium*), ginger (*Zingiber officinalis*), thyme (*Thymus vulgaris*), nutmeg (*Myristic fragrans*), rosemary (*Rosmarinus officinalis*), Uda (*Xylopia aethiopica*), bay leaves (*Laurus nobilis*), garlic (*Allium sativum*), and Utazi (*Gongronema latifolium*). Vegetables- tomatoes (*Lycopersicon esculentum*), Uha (*Pterocarpus spp*), Ugu (*Telfairia occidentalis*), Ukazi (*Gnetum africanum*) lettuce (*Lactuca sativa*), Green (*Amaranthus caudatus*), waterleaf (*Talinum triangulare*), cabbage (*Brassica oleracea*), okro (*Abelmoschus esculenta*), and garden egg leaves (*Solanum melongena*).

Sample preparation

The samples collected were oven dried at 40 °C, ground to powder using warring blender, packaged in an air-tight glass jar and stored at room temperature until analysis was carried out.

Determination of iodine in plant samples

The alkaline dry ash method as described by Fisher et al. (1986) was used for this determination. This was done by adding 0.5 g of each sample into nickel crucibles. Then 1 ml of a mixture of 0.5 M sodium hydroxide and 0.1 M potassium

nitrate was added to the samples, mixed and allowed to dry. The containers were then covered with aluminum foil and placed in a muffle furnace. The samples were heated to 250 °C, held for 15 min, heated further to 480 °C, again held for 15 min, and finally brought to 580 °C. They were maintained at this temperature for 3 h, after which they were allowed to cool to room temperature. The resultant ash was extracted with three successive, 2 ml portions of a 1.0 mM sodium hydroxide solution, made up in double-distilled water.

The solution was centrifuged at 2500 g for 20 min using polypropylene centrifuge tubes and the supernatant solution collected for iodine determination. (The heat destroyed the organic matrix. The sodium hydroxide was used to keep the iodine in a nonvolatile form, while the potassium nitrate was used to increase the oxidation of the organic matter). Iodine was determined by adding 1 ml of sample solution to a cuvette at 35 °C and 1 ml of arsenic reagent was added. The reaction was started by the addition of 1 ml of Ceric reagent. The initial reaction rate was calculated from the change in absorbance at 420 nm. The iodine concentrations of the samples were determined from a standard curve.

Mineral composition

2 g of each sample was transferred into a 75 ml digestion tube. 5 ml of each digestion mixture was added to each, swirled and placed in a fume cupboard. Digestion was allowed for 2 h at 150 °C. These were removed from the digester, cooled for 10 min, and 3 ml of 6 N HCl was added to each tube. These mixtures were digested for another 1.5 h. The set up was removed from the digester, cooled and made up to 50 ml with distilled water. Each tube was stirred vigorously using the vortex mixer. The resulting digest was used for the determination of copper, zinc, selenium, iron, calcium and magnesium with an atomic absorption spectrophotometer (Perkin Elmner, USA).

Statistical analysis

Data obtained were expressed as means ± standard deviation and presented in bar charts and tables.

RESULTS

The iodine content of grains shown in Figure 1, ranged from 0.760 ± 0.17 µg/100 g in millet to 69.36 ± 7.94 µg/100 g in cowpea. Iodine was not detected in maize (yellow), soybeans, rice (foreign) and sorghum samples. While the distribution of iodine in spices/seasonings shown in Figure 2, ranged from 21.56 ± 3.6 µg/100 g in garlic to 108.96 ± 3.00 µg/100 g in Maggi cube. Uziza had highest iodine content of 95.66 ± 1.73 µg/100 g among the spices. Iodine was not detected in nutmeg. Figure 3 showed the iodine concentration of vegetables. Iodine concentration of vegetables ranged from 66.06 ± 5.20 µg/100 g in tomato to 117.66 ± 3.00 µg/100 g in Uha leaves. The mean iodine concentration of the vegetables gave a value of 94.61 ± 15.28 µg/100 g and most of the vegetables have iodine concentration close to the upper range.

Table 1 shows the results of concentration of copper,

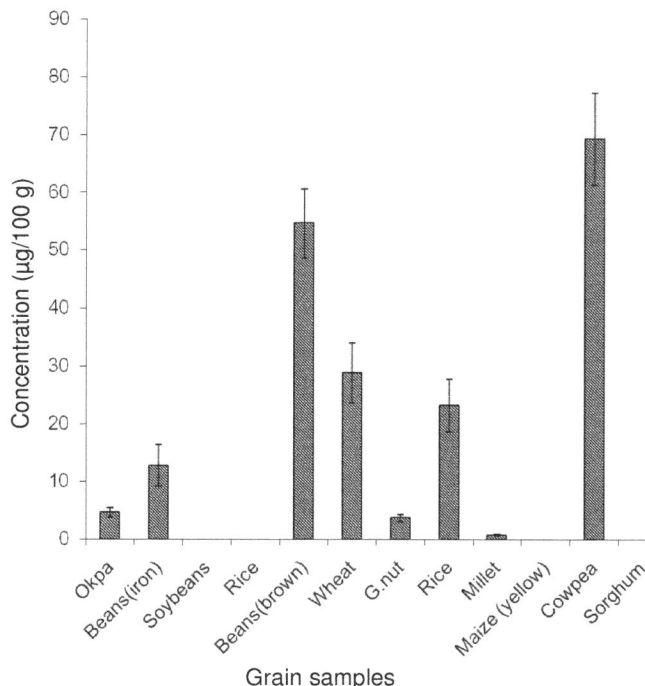

Figure 1. Profile of iodine concentration in grains.

zinc, selenium, iron, calcium and magnesium in selected spices/seasonings. The result showed that the concentration of copper ranged from 1.93 mg/kg in Uziza to 9.84 mg/kg in rosemary leaves. Zinc concentration was highest in Knor but was not detected in Uziza, garlic and Nchanwu. Selenium was not detected in Utazi, thyme, onion and rosemary leaves but showed highest concentration of 1.90 mg/kg in curry. Iron ranged from 4.80 mg/kg in Utazi to 192.70 mg/kg in Uda. Calcium concentration ranged from 1.73 mg/kg in Utazi to 77.25 mg/kg in Uziza. Magnesium ranged from 19.17 mg/kg in Onion to 261 mg/kg in Uziza. Table 2 shows the result of concentration of copper, zinc, selenium, iron, calcium and magnesium in selected vegetables. The result showed that copper concentration ranged from 3.07 mg/kg in Okro to 28.96 mg/kg in garden egg leaves. Zinc was not detected in garden egg leaf, Green, Uha leaf, water leaf and Okro but showed highest concentration of 0.25 mg/kg in lettuce. Selenium concentration was highest in Uha leaves but was not detected in Ugu. Iron ranged from 176.60 mg/kg in Uha leaf to 2004.23 mg/kg in Green. Magnesium concentration in the vegetables ranged from 135.36mg/Kg in Cabbage to 1275.70mg/Kg in Water-leaf with a mean value of 506.30mg/Kg

Table 3 shows the results of concentration of copper, zinc, selenium, iron, calcium and magnesium in selected grains. It showed that copper concentration ranged from 1.87 mg/kg in maize (yellow) to 17.29 mg/kg in groundnut. Zinc concentration of 0.22 mg/kg in maize (yellow) was the highest in the grains but was not detected in soybeans, cowpea, groundnut, and beans

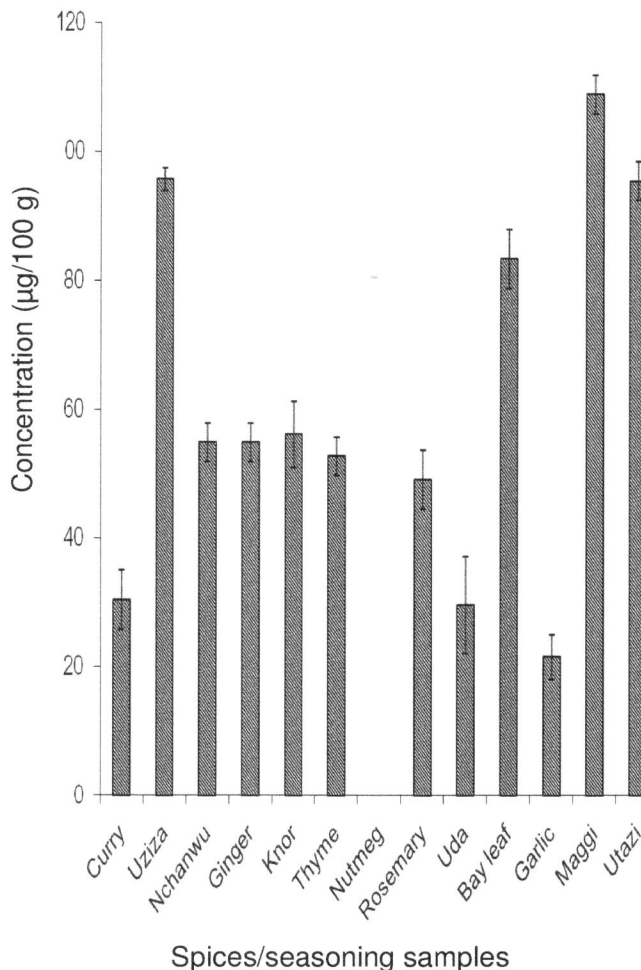

Figure 2. Profile of iodine concentration in spices/seasonings.

(brown). Selenium was not detected in wheat, maize (white), groundnut, rice (foreign), maize (yellow) and sorghum. Iron concentration ranged from 6.70 mg/kg in maize (white) to 91.90 mg/kg in wheat. Calcium concentration ranged from 0.77 mg/kg in sorghum to 22.80 mg/kg in soybeans. Also, the concentration of magnesium ranged from 39.71 mg/kg in rice (foreign) to 139.54 mg/kg in soybeans.

DISCUSSION

Our results suggest that the plants studied have varying ability to absorb iodine in its tissues, which partly supports the assumption that concentration of iodine in plant is the reflection of iodine in the soil in which it is grown (Aston and Brazier, 1979; Fuge, 2005). The result of the iodine concentration in the vegetables showed that, most of them have iodine value around the upper range of 117.66 ± 3.00 μg/100 g in Oha leaves. These vegetables iodine range are also within the upper range

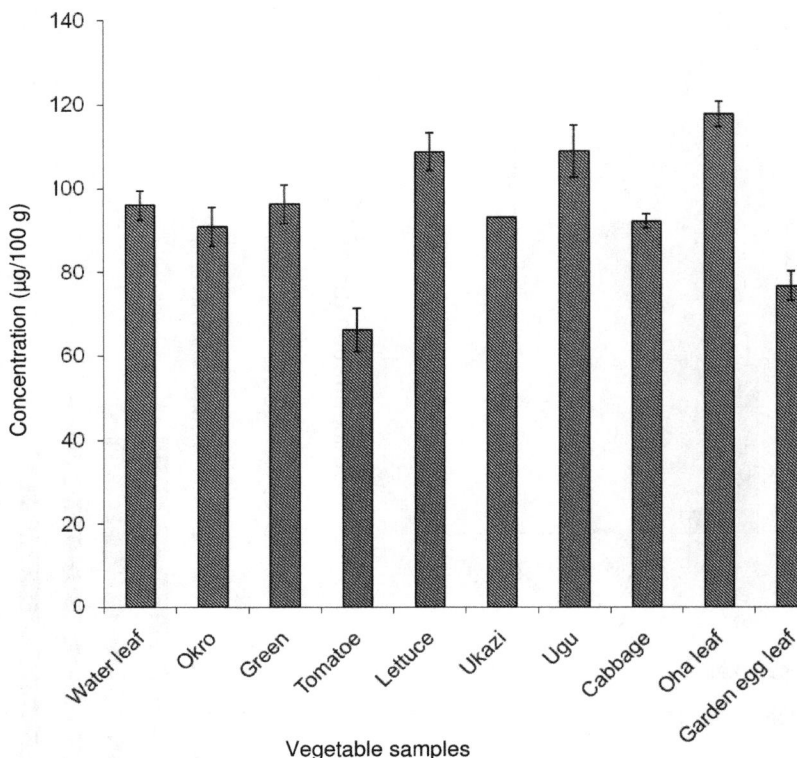

Figure 3. Profile of iodine concentration in vegetables.

of iodine level in vegetables reported by Koutras et al. (1985). The mean iodine concentration of 94.61 ± 15.28 µg/100 g of vegetables, suggests that, these vegetables studied will contribute significantly to the iodine needs of the consumers.

The results of iodine concentration of grains consumed in the south east showed that iodine concentration ranged from 0.66 ± 1.73 µg/100 g in millet to 69.36 ± 7.94 µg/100 g in cowpea and the mean was 16.49 ± 23.56 µg/100 g. These observations are quite different from the result obtained by Koutras (1985) but within the range as observed by Souci et al. (2000). This disparity can be explained by the statement given by Koutras et al. (1985) that most iodine content of food varies with geographic location because there are large variations in the iodine content of the inorganic world. These plant foods (vegetables, cereals and legumes) are staple foods of many communities in Nigeria.

The result showed that some have greater ability to concentrate iodine in their tissue than others. This is in line with the report of Howarth (1999). The knowledge of such plants would enable careful selection for consumption and iodine biofortification, where there is problem of selection. The screening of these plants grown and consumed in this geographical region (rain-forest) for their iodine content is of nutritional importance. Careful selection and combination of these plants with high iodine concentration may be of nutritional benefit for

a population whose salt intake is low, either as a result of health such as hypertension or as dietary habit (Stephen and Hoption, 2006). The plants with low iodine content are not without benefit. In iodine deficient areas, where introduction of iodine salt may trigger hyperthyroidism (Barbara, 1994), careful selection of plants low in iodine concentration may be beneficial.

The results of iodine concentration of the food seasoning showed that, in the natural food seasoning (Uziza, Nchanwu, ginger, Uda, garlic and Utazi) studied, iodine concentration was highest in Uziza. The mean iodine concentration was 61.40 ± 30.58 µg/100 g. The iodine concentration in the synthetic food seasoning ranged from 30.46 ± 4.58 µg/100 g in curry to 108.96 ± 4.58 µg/100 g in maggi. The mean iodine concentration was 62.36 ± 33.19 µg/100 g. The iodine concentration in Maggi suggests a good level of iodine, because the concentration is around the upper range given for vegetables by Koutras et al. (1985) and WHO (1996). These results clearly suggest that, vegetables have the highest mean copper concentration, followed by grains and spices. These results indicate that the concentrations of copper in the plants studied cannot pose any danger of toxicity or inhibition of thyroid function with respect to the RDA (FNB-Institute of Medicine, 2001). Similarly, our bodies can regulate copper storage through excretion via the bile (Wardlaw et al., 2004; Raju and Madala, 2005). The

Table 1. Mineral concentration in selected spices/seasonings (mg/kg).

Spices	Copper (Cu)	Zinc (Zn)	Iron (Fe)	Calcium (Ca)	Selenium (Se)	Magnesium (Mg)
Piper guineense	1.93 ±0.13	ND	188.7 ±1.70	77.25 ±4.05	1.00 ±0.01	261.12 ±0.50
Gongronema latifolium	2.22 ±0.59	0.06 ±0.00	4.8 ±0.1	1.73 ±0.09	ND	41.8 ±0.38
Thymus vulgaris	6.82 ±0.08	0.08 ±0.00	60.5 ±1.20	67.99 ±0.85	ND	130.69 ±0.68
Xylopia aethiopica	2.40 ±0.82	0.07 ±0.00	192.7 ±1.70	21.78 ±0.76	0.80 ±0.01	179.49 ±2.36
Allium sativum	3.73 ±0.02	ND	34.8 ±0.10	19.04 ±2.84	0.2 ±0.01	43.34 ±1.51
Laurus nobilis	9.19 ±0.17	0.15 ±0.01	86.2 ±3.00	54.69 ±3.24	0.3 ±0.01	72.85 ±0.08
Zingiber officinalis	3.52 ±0.14	0.02 ±0.00	138.3 ±1.90	5.64 ±0.34	0.20 ±0.01	74.65 ±0.10
Allium cepa	4.58 ±0.43	0.01 ±0.00	10.4 ±2.50	7.48 ±0.89	ND	19.17 ±0.57
Knor	3.80 ±0.12	0.63 ±0.01	5.20 ±0.30	5.90 ±0.49	0.90 ±0.01	35.76 ±1.75
Muraya koenigii	2.20 ±0.06	0.16 ±0.01	160.6 ±2.00	48.63 ±0.73	1.9 ±0.01	220.65 ±2.94
Rosmarinus officinalis	9.84 ±1.06	0.02 ±0.00	12.6 ±0.40	34.31 ±1.54	ND	64.79 ±1.11
Ocimium gratissium	6.26 ±0.20	ND	101.1 ±5.40	3.77 ±0.72	0.6 ±0.01	240.26 ±0.64

Values are mean of three determinations ± SD. ND = Not Detected

very low concentration of zinc in most of the plants studied, indicates a very serious health and nutritional challenge. This is important because zinc is involved in some catalytic reaction and can stabilize the structure of some enzymes (Wardlaw et al., 2004). Also studies have shown that low zinc intake exacerbates the effect of low iodine intake (Walsh, 2003). Although, the population of the study area have been shown to consume iodized salt, which puts median urinary iodine excretion of their pregnant women within the normal range (Ujowundu et al., 2010). Zinc intake from other food sources or supplementation should be encouraged. Zinc

supplementation has been reported to have favorable effect on thyroid hormone levels, particularly total T_3, and resting metabolic rate (Maxwell and Volpe, 2007).

Selenium was not detected in some of the plants studied and in those present, the concentration was very low. Selenium is an important component of iodoperoxidase, an enzyme that is involved in the production of the thyroid hormone and this suggests that, selenium deficiency contributes to the development of goiter even in the presence of adequate iodine intake (Sunde, 2001; Wardlaw et al., 2004). Selenium is a component of the enzymes deiodinase

Table 2. Mineral concentration in selected vegetables (mg/kg).

Vegetables	Copper (Cu)	Zinc (Zn)	Iron (Fe)	Calcium (Ca)	Selenium (Se)	Magnesium (Mg)
Telfairia occidentalis	8.46 ±0.82	0.12 ±0.06	248.83 ±0.06	9.09 ±0.38	ND	247.45 ±2.72
Lactuca sativa	8.64 ±0.04	0.25 ±0.08	416.9 ±7.60	49.02 ±2.11	0.2 ±0.00	268.3 ±1.63
Solanum melongena	28.96 ±2.26	ND	182.7 ±9.60	154.97 ±1.06	1 ±0.00	514.47 ±9.38
Amaranthus caudatus	10.21 ±0.45	ND	2004.2 ±0.10	127.8 ±6.23	0.6 ±0.03	1222.4 ±18.92
Pterocarpus spp	10.5 ±0.14	ND	176.6 ±9.60	59.91 ±1.65	1.6 ±0.05	260.71 ±0.03
Gnetum africanum	17.46 ±2.44	0.21 ±0.07	1723 ±0.03	68.18 ±2.25	0.1 ±0.00	215.94 ±0.29
Talinum triangulare	7.22 ±0.15	ND	510.56 ±1.10	68.54 ±1.95	0.2 ±0.03	1275.7 ±30.56
Abelmoschus esculenta	3.07 ±0.58	ND	250.47 ±0.1	99.87 ±2.92	1.3 ±0.02	416.38 ±2.99
Brassica oleracea	7.56 ±0.70	0.14 ±0.05	210.33 ±3.40	37.28 ±1.45	0.3 ±0.05	135.36 ±1.03

Values are mean of three determinations ± SD. ND = Not Detected

and thyroperoxidase, vital for the synthesis of thyroid hormones – T_3 and T_4 (EGVM, 2002; NNT, 2002). Selenium content of foods depends on the level of selenium in the soil (Wardlaw et al., 2004; Raju and Madala, 2005). The low concentration and total absence of selenium in some of the plants studied suggests that, dietary intake from natural sources may not supply the adequate amount required to carry out metabolic processes in which selenium are needed. Therefore, to meet the RDA of 55 µg/day selenium fortification may be necessary (FNB-Institute of Medicine, 2001). This indicates that the consumption of food seasonings rich in selenium such as curry powder and Uziza leaves as shown in this study should be encouraged.

Iron plays an important role in many parts of the body, including immune function, cognitive development, temperature regulation, energy metabolism, and work performance (FNB, 2001). Some iron rich foods are poor sources of the mineral because other compounds render it non-absorbable. Most contain considerable oxalate, which chelates iron and renders it non-absorbable (Lehninger, 1982; Raju

and Madala, 2005). This statement is in line with the results obtained which showed that, vegetables had the highest iron concentration (176.60 to 2004.23 mg/kg). The bioavailability of the iron present in a meal depends on its form and the presence or absence of factors that influence absorption (Wardlaw et al., 2004; Yip, 2001).

Calcium ions are involved in blood clotting, nerve impulse transmission, muscle contraction (Vander, 2001). Dairy products are very good sources of calcium, green vegetables are good sources also but the presence of oxalic acid for example in spinach, renders or makes it non-absorbable (Wardlaw et al., 2004). The calcium in cereals is not also readily absorbable because it is tightly bound to inositol hexaphosphate, called phytate (Lehninger, 1982). The effects of plant calcium on thyroid function have attracted less study. However, a high calcium level in drinking water may slow iodine absorption resulting in goiter, particularly if the iodine level is borderline in meeting body needs. Magnesium is found in chlorophyll, therefore green leafy vegetables are rich sources (Wardlaw et al., 2004). This was evident in the results

Table 3. Mineral concentration in selected grains (mg/kg).

Grains	Copper (Cu)	Zinc (Cu)	Iron (Fe)	Calcium (Ca)	Selenium (Se)	Magnesium (Mg)
Phaseolus vulgaris (iron)	±0.05	0.07 ±0.01	40.1 ±5.40	13.98 ±1.04	1.6 ±0.00	130.35 ±1.57
Triticum aeastivum	2.77 ±0.13	0.07 ±0.01	91.9 ±13.1	5.91 ±0.14	ND	109.03 ±0.78
Glycine max	3.45 ±0.30	ND	43.3 ±10.1	22.8 ±0.37	0.4 ±0.00	139.54 ±1.26
Vadzeia subterranean	6.46 ±0.26	0.01 ±0.00	23.1 ±4.80	1.86 ±0.69	1.8 ±0.03	137.65 ±0.40
Zea may (white)	2.00 ±0.8	0.02 ±0.00	6.7 ±0.7	3.24 ±0.84	ND	54.47 ±0.31
Vigna sinensis (akidi)	8.96 ±0.88	ND	24.6 ±0.71	5.16 ±0.30	1.8 ±0.07	138.68 ±1.15
Panicum miliaceum	3.51 ±0.04	0.1 ±0.00	29.8 ±2.60	5.61 ±0.61	3.5 ±0.60	100.98 ±40
Arachis hypogaea	17.29 ±3.16	ND	12.1 ±0.20	5.35 ±0.73	ND	105.99 ±0.59
Oryza sativa (foreign)	8.54 ±0.77	0.02 ±0.00	15.9 ±0.50	4.71 ±1.22	ND	39.71 ±2.05
Phaseolus vulgaris (brown)	4.38 ±0.19	ND	15.3 ±0.80	7.41 ±0.25	1.4 ±0.04	130.33 ±1.00
Zea may (yellow)	1.87 ±0.25	0.22 ±0.05	30 ±5.00	3.77 ±81	ND	94.9 ±0.36
Sorghum bicolor	4.25 ±0.17	0.01 ±0.00	25 ±2.30	0.77 ±0.07	ND	100.91 ±0.29

Values are mean of three determinations ± SD. ND = Not Detected

obtained, in which vegetables showed the highest mean magnesium concentration of 506.30 mg/kg. Magnesium decreases blood pressure by dilating arteries, and preventing heart rhythm abnormalities (FNB, 1997).

Conclusion

Our results suggest that, iodine concentration in the soil and water media contains iodine at a concentration that can only supply moderate iodine to the plants grown in the environment and to the population that consume the plants. The iodine concentration in the plants studied showed variation in the iodine concentration. These iodine contents could only provide moderate iodine which is not sufficient for optimal thyroid function. This suggests that the combination of these plants could provide enough to meet RDA of iodine. Also, the concentrations of the inorganic minerals were not at levels that may cause negative impact on thyroid function.

REFERENCES

Aston SR, Brazier PH (1979). Endemic goiter, the factors controlling iodine deficiency in soils. Sci. Total Environ., 11: 99-104.
Expert Group on Vitamins and Minerals (EGVM) (2002). Draft report on

"Safe upper levels for vitamins and minerals", London. pp. 205-214.

Fisher PWF, L'Abbe MR, Giroux A (1986). Colorimetric determination of total iodine in foods by iodide-catalyzed reduction of Ce^{+4}. J. Assoc. Official Anal. Chem., 69: 687-689.

Food and Nutrition Board (FNB)-Institute of Medicine (1997). Dietary Reference intakes for calcium, phosphorus, magnesium, vitamin D, and fluorine. Washington DC: Standing Committee on the Scientific Evaluation of Dietary Reference Intakes National academy Press.

Food and Nutrition Board (FNB)-Institute of Medicine (2001). Dietary Reference intakes for Vitamin A, K, arsenic, boron, chromium, copper iodine, iron, manganese, molybdenum, nickel, silicon, vanadium and zinc. Washington D.C.

Fuge R (2005). Soils and Iodine Deficiency. Essentials of Medical Geology: Impacts of the Natural Environment on Public Health. Ed. Selinus, Alloway, Centeno, Finkelman, Fuge, Lindh, and Smedley. Singapore: Elsevier Academic Press, pp. 418.

Gartan E (1988). Goitrogens. Bailliers Clin. Endocrinol. Metab., 2: 683 – 702 (review).

Hetzel BS, Wellby ML (1997). In "Handbook of Nutritionally Essential Minerals" (B.L.O'Dell and R.A. Sunde, eds.). Marecel Dekker, Inc., New York. p. 557.

Hetzel BS (1989). The Story of Iodine Deficiency, 1st edition. Oxford Medical Publications, Great Britain. p. 3.

Howarth B (1999). Breeding staple food crop that are more nutritious. International Food Policy Research Institute (IFPRI), Res. Perspect., 21: 3.

Jooste PL, Weight MJ, Krick JA, Louw AJ (1999). Endemic goitre in the absence of iodine deficiency in schoolchildren of the Northern Cape Province of South Africa. Eur. J. Clin. Nutr., 53(1): 8-12.

Koutras DA, Matovinovic J, Vought R (1985). The ecology of iodine. In: Stanbury J.B., Hetzel B.S. (eds). Endemic goitre and cretinism, iodine nutrition in health and disease. Wiley Eastern Limited New York, p. 185-95.

Lehninger AL (1982). Principles Biochemistry. 2 nd edn Worth Publisher. New York.

Maxwell C, Volpe SL (2007). Effect of Zinc Supplementation on Thyroid Hormone Function; A Case Study of Two College Females. Ann Nutr Metab., 51(2): 188-194.

Nordic Working Group on Food Toxicology and Risk Evaluation (NNT) (2002). Mechanisms of toxicity of chemicals affecting the function of the thyroid gland. Berthelsew P and Thorup I (eds.). Nordic Council of Ministers, Tema Nord.

Osman AK, Fatah AA (1981). Factors other than Iodine deficiency contributing to the endemicity of goiter in Darfur province (sudan). J. Hum. Nutr., 35: 302-9.

Raju SM, Madala B (2005). Mineral Metabolism; in Illustrated Medical Biochemistry. Medical Publishers, New Delhi. pp. 98-109.

Singh SP (2004). A Textbook of Biochemistry 3rd ed. Satish Kumar Jian Publisher India. pp. 426-435.

Souci FW, Fachmann W, Kraut H (2000). Food Consumption and Nutrition Tables, 6th rev. edn. Medpharm, Scientific Publishers, Stuttgart, Germany.

Stephen A, Hoption C (2006). Hypothesis: Dietary iodine intake in the etiology of cardiovascular disease. J. Am. Coll. Nutr., 25: 1-11.

Sunde RA (2001). Selenium. In Bowman B.A. and Rusell R.M., (eds): Present Knowledge in nutrition. 8th ed. Washington, D.C. International Life Sciences Institute.

Ujowundu CO, Kalu FN, Nwosunjoku EC (2010). Thyroid Function and Urine Iodine of Pregnant Women in Owerri, Nigeria. Nig. J. Biochem. Mol. Biol., 25(2): 91 – 97.

Underwood EJ, Suttle NF (1999). In "The Mineral Nutrition of Livestock" 3rd Ed. CABI Publishing, UK.

Vander A (2001). Human physiology. 8th ed. Boston: McGraw-Hill.

Vanderpas JB, Contempré B, Duale NL, Goossens W, Bebe N, Thorpe R, Ntambue K, Dumont J, Thilly CH, Diplock AT (1990). Iodine and selenium deficiency associated with cretinism in Northern Zaire. Am. J. Clin. Nutr., 52: 1087-1093.

Walsh S (2003). Iodine Nutrition, Vegan Society-Hosting by Spirit Quest, www.vegansociety.com/food/nutrition/iodine.

Wardlaw GM, Hampl JS, Disilvestro RA (2004). Perspectives in Nutrition. pp. 413 – 445. McGrawhill-USA.

World Health Organisation (WHO) (1996). Trace elements in human nutrition and health.WHO, Geneva.

Yip R (2001). Iron; In: Bowman B.A. and Russell R.M. (eds): Present knowledge in nutrition. 8th ed. Washington DC: ILSI Press, pp. 311-328.

29

Changes in the amino acids contents of selected leafy vegetables subjected to different processing treatments

Chinyere, G. C.* and Obasi, N. A.

Department of Biochemistry, Abia State University, Uturu, Nigeria.

Changes in the amino acids contents of *Veronica amygdaline, Gnetum africana, Gongronema latifolium* and *Ocimum gratissimum* subjected to different processing treatments were investigated. The processing treatments employed include fresh milling, sun drying, oven drying, steaming and a combination of these while the amino acids evaluated include isoleucine, leucine, lysine, methionine, cysteine, phenylanine, tyrosine, threonine, valine, histidine, alanine arginine, aspartic acids, glutamic acid and glycine. Fresh milling was the most effective method of retaining all the amino acid followed by sun drying method while steaming combined with oven drying caused the highest losses in the amino acids contents of the studied vegetables. Other processing treatment led to varying degrees of significant reduction in the levels of the amino acids content of the test vegetable at p = 0.05. These vegetables were rich sources of essential amino acids with lysine and sulphur containing amino acids, the most limiting amino acids.

Key words: *Veronica amygdaline, Gnetum africana, Gongronema latifolium, Ocimum gratissimum*, processing, amino acids.

INTRODUCTION

Vegetable are common dietary sources of amino acids (FNB, 2002). In Nigeria, as in other tropical countries where the daily diet is dominated by starchy staple foods, vegetables are the cheapest and most readily available sources of important proteins, vitamins, minerals and essential amino acids (Okafor, 1983). Amino acids play vital roles in human nutrition and health (Harper, 1999; Wardlaw and Kessel, 2002).

Among the leafy vegetables widely used in Nigeria for a variety of dietetic and medicinal purpose include *Veronica amygdaline, Gnetum africana, Gongronema latifolium* and *Ocmum gratissimum*. Their leaves are used for soup and porridge preparations, salad preparation or eaten as spinach in various areas where they are believed to serve various dietetic and medicinal purposes (Aregheore et al., 1998; Ajayi et al., 2006; Matasyoh et al., 2007). Kubmarawa et al. (2008) reported the amino acid profile of two non-conventional leafy vegetables, *sesamum indium* and *Balanites aegyptiaca* while Lohlum et al. (2010) published that of *Lophira lanceolata* seeds. Both

leaves and the seed were shown to be of high protein quality.

Leafy vegetables are seasonal, highly perishable and sometimes unpalatable when taken raw. They are subjected to various processing treatments which extend their shelf life and also improve the bioavailability of their constituent nutrients. In some cases, these processing method help reduce antinutritional factors in them. However reports (Ramberg and Mc Analley, 2002; Morris et al., 2004) have shown that processing causes changes in the physical and chemical nature of products. As such, it is imperative to evaluate the effects of processing treatments on food products to ensure their safety and wholesomeness. This underscores the aim of this study which is mainly on the amino acids contents of treated leafy vegetables.

MATERIALS AND METHODS

Sample collection

The leaves of the four vegetable plants analyzed: *V. amygdalina, G. Africana, G. latifolium,* and *O. gratissimum* were harvested from cultivated farmlands near river banks located at Afikpo,

---

*Corresponding author. E-mail: gcchinyere@yahoo.com

Afikpo North L.G.A Ebonyi State, in South-Eastern Nigeria. About 5 kg materials collected from each of the plants were thoroughly mixed, had their stalks removed, rinsed with de-ionized water and the residual moistures evaporated at room temperature before subjecting them to the different processing methods.

Processing methods

About 600 g each of the *v. amygdalina, G. Africana, G. latifolium,* and *O. gratissimum* leaves collected were subjected to the following processing techniques:

(i) Fresh milling (FRM) of the leaves using a chopping knife to cut the fresh leaves and then a Thomas- Willey milling machine to blend the leaves into fine pieces.
(ii) Sun-drying (SND) on clean papers for 2 to 3 days with constant turning over to avert fungal growth.
(iii) Oven drying (OVD) on aluminum trays at 80 to 100°C for 24 h.
(iv) Steaming (STM) of the leaf samples over wire gauze placed on top of a boiling water for 30 min.
(v) Fresh milling followed by sun-drying (FRM+SND) in which the Thomas-Willey blended sample were further sun-dried for 2 to 3 days with adequate turning over to avert fungal growth.
(vi) Fresh milling followed by oven-drying (FRM+OVD) in which the Thomas-Willey blended sample were now oven-dried on aluminum trays at 80 to 100°C for 24 h.
(vii) Steaming followed by sun-drying (STM+SND) in which the steamed leaf samples were further sun-dried for 2 to 3 days with adequate turning over to avert fungal growth.
(viii) Steaming followed by oven-drying (STM+OVD) in which the steamed leaf samples were further oven-dried on aluminum trays at 80 to 100°C for 24 h.

All the processed samples were packed into tightly sealed nylon bags and analyses commenced immediately with minimum delay to forestall further changes in the quality of the samples. Where the analysis could not be completed on the same day, samples were kept frozen at temperature of -4°C.

Amino acid determination was carried out using ion-exchange chromatography with Technicon Sequential Multisample Amino Acid Analyzer, TSM (Technicon Instruments Corporation, Dubin, Ireland), as outlined in Adeyeye and Afolabi (2004). 2.0 g of each of the processed samples were defatted with petroleum ether using soxhlet extraction methods. The defatted samples were re-dried and milled into fine powder using porcelain pestle and mortar. 30.0 mg sample in triplicate were weighed into a glass ampoules to which 5.0 cm^3 of 6 M HCl and 5.0 µm norleucine were added. The ampoules were evacuated by passing nitrogen gas (to remove oxygen so as to avoid possible oxidation of some amino acid during hydrolysis), sealed with Bunsen burner flame and hydrolyzed in an oven at 110°C for 24 h. The ampoules were cooled, broken at the tip and the content filtered. The filtrates were evaporated to dryness at 40°C under vacuum in a rotary evaporator. The residues were dissolved to 5.0 µl (for acid and neutral amino acids) or 10.0 µl (for basic amino acids) with buffer; pH 2.2 and the solutions were dispensed into the cartridge of TSM. The chromatograms (amino acids peaks) obtained from automatic pen recorder correspond to the quantity of each amino acid present. Quantification was performed by comparing the peak of each amino acid in the sample to the area of the corresponding amino acid standard of the protein hydrolysis.

All values are means of three (3) replicate determinations. Data for all determinations were subjected to analysis of variance (ANOVA) using Duncan multiple range test (DMRT) as outlined by Wahua (1999). Fisher's least significant difference (LSD) test was used to identify significant differences among treatment means at P = 0.05.

RESULTS AND DISCUSSION

The changes in the amino acids contents of selected leafy vegetables were subjected to different processing treatments (Table 1 to 4). The non-essential amino acids (alanine, arginine, aspartic acid, glutamic acid, histidine, proline and serine) were generally higher in concentration (58 to 60%) compared to essential amino acids (isoleucine, leucine, lysine, methionine, cystenine, phenylalanine, tyrosine, threonine and valine) which constituted 40 to 42% of the total amino acids in all the leaves analyzed. Among the essential amino acids, leucine and the aromatic amino acids (phenylalanine and tyrosine) were the predominant amino acids while glutamic acid was found to be the major non-essential amino acid in all the tested vegetables. Olaofe and Akintayo (2000) and Hassan and Umar (2006) reported comparatively similar results for legumes and leaves of other vegetables, respectively.

The percentage (%) score of the essential amino acids in the tested vegetables was compared with those of the reference standard amino acids profile for foods established for pre-school children and adults by WHO/FAO/UNU (1985) (Table 5). The results indicated that most of the essential amino acids in all the tested vegetables scored above 100% of the reference standard value for preschool children and adults established by WHO/FAO/UNU (1985). This shows that the tested leafy vegetables were rich sources of essential amino acids. Aremu et al. (2006) and Hassan and Umar (2006) reported comparatively similar results for some Nigerian under-utilized legume flours and *Momordica balsamine* L. leaves, respectively. However relative to preschool children reference standard values, Lysine scored below 100% in all the tested vegetable. Sulphur containing amino acids scored below 100% in all except in *V. amygdaline* while threonine and histamine scored below 100% in *G. Africana* and *G. latifolium,* respectively. On the other hand, relative to adults reference standard values, sulphur containing amino acids scored below 100% in all the tested vegetables with exception in *V. amygdaline* while histidine scored below 100% in *G. latifolium.* These results indicated lysine and sulphur containing amino acids as the limiting amino acids in the tested vegetables.

The amino acids indicated that the different processing treatments caused significant losses in the seventeen amino acids contents of the selected leafy vegetable at p = 0.05 (Tables 1 to 4). This study supports earlier literature by Lund (1975) and Ramberg and Mc Annalley (2002). Also, comparative significant losses were reported for the amino acids contents of fruits and vegetable following various processing treatment (Christian, 2000; Souci et al., 2000). These results indicated that fresh milling (FRM) was the most effectives method of retaining the amino acids in the tested vegetable followed by sun drying method while steaming combined with oven drying (STM+OVD) elicited the

Table 1. Amino acid composition in differently processed *V. amygdalina* leaves.

Process	Amino acid concentration (g/ 100 g protein)									
	Isoleucine	Leucine	Lysine	Methionine	Cysteine	Total sulphur EAAS	Phenylalanine	Tyrosine	Total aromatic EAAS	Threonine
FRM	3.99[a]±0.02	9.52[a]±0.01	5.20[a]±0.04	1.95[a]±0.01	1.41[a]±0.01	3.36±0.03	5.03[a]±0.03	3.38[a]±0.03	8.41±0.03	3.96[a]±0.03
SND	3.83[b]±0.03	9.00[b]±0.03	4.95[b]±0.03	1.87[b]±0.03	1.36[0]±0.03	3.23±0.04	4.85[b]±0.01	3.28[b]±0.01	8.13±0.04	3.87[b]±0.03
OVD	3.75[d]±0.01	8.64[e]±0.02	4.81[f]±0.03	1.78[e]±0.04	1.28[e]±0.05	3.06±0.03	4.77[c]±0.01	3.19[d]±0.01	7.96±0.04	3.72[e]±0.04
STM	3.78[d]±0.04	8.82[C]±0.01	4.85[d]±0,01	1.81[d]±0.03	1.30[d]±0.03	3.11±0.03	4.78[c]±0.03	3.21[c]±0.03	7.99±0.03	3.80[c]±0.03
FRM±SND	3.75[d]±0.03	8.85[c]±0.02	4.88[d]±0.03	184[c]±0.03	1.32[b]±0.01	3.16±0.03	4,76[b]±0.02	3.22[c]±0.05	7.98±0.02	3.81[c]±0.05
FRM±OVD	3.68[e]±0.02	8.60[e]±0.06	472[g]±0.03	1.68[f]±0.01	1.19±0.02	2.89±0.04	4.68[d]±0.02	3.10[e]±0.06	7.78±0.03	3.67[f]±0.04
STM±SND	3.74[d]±0.03	8.75[d]±0.01	4.84[e]±0.01	1.78[e]±0.03	1.28[e]±0.05	3.06±0.04	4.76[b]±0.01	3.20[d]±0.01	7.76±0.04	3.78[d]±0.03
STM±OVD	3.65[f]±0.04	8.53[f]±0.03	4.70[g]±0.03	1.65[g]±0.03	1.16[g]±0.04	2.81±0.03	4.65[e]±0.04	3.08[e]±0.01	7.73±0.03	3.62[g]±0.01

Process	Valine	Histidine	Alanine	Arginine	Asparatic acid	Glutamic acid	Glycine	Proline	Serine
FRM	5.20[a]±0.03	3.25[a]±0.03	4.98[a]±0.03	5.90[a]±0.03	9.11[a]±0.01	14.32[a]±0.01	5.63[a]±0.01	4.17[a]±0.03	4.89[a]±0.04
SND	5.09[a]±0.04	3.16[b]±0.01	4.88[b]±0.06	5.79[b]±0.04	9.02[b]±0.03	14.20[b]±0.01	5.52[b]±0.01	4.08[b]±0.02	4.74[b]±0.04
OVD	4.95[e]±0.05	3.04[e]±0.06	4.76[f]±0.03	5.68[d]±0.03	8.87[d]±0.01	14.01[e]±0.01	5.40[e]±0.01	4.97[e]±0.03	4.62[e]±0.04
STM	5.00[c]±0.04	3.10[c]±0.01	4.86[c]±0.03	5.73[c]±0.01	8.91[c]±0.03	14.09[c]±0.02	5.46[c]±0.02	4.01[c]±0.03	4.68[c]±0.01
FRM±SND	5.01[c]±0.03	3.08[d]±0.05	4.84[d]±0.04	5.72[c]±0.01	8.92[b]±0.04	14.10[c]±0.02	5.48[c]±0.01	4.02[c]±0.03	4.68[c]±0.03
FRM±OVD	4.88[f]±0.03	2.95[f]±0.01	4.70[g]±0.04	5.60[e]±0.03	8.76[e]±0.04	13.97[f]±0.01	5.32[e]±0.01	3.89[e]±0.01	4.54[f]±0.02
STM±SND	4..98[d]±0.04	3.07[d]±0.03	4.82[e]±0.03	5.69[d]±0.03	8.88[d]±0.01	14.05[d]±0.03	5.44[d]±0.04	3.97[d]±0.01	4.65[b]±0.01
STM±OVD	4.84[g]±0.03	2.92[g]±0.03	4.68[h]±0.01	5.60[e]±0.08	8.75[g]±0.01	13.94[g]±0.01	5.31[f]±0.02	3.86[f]±0.01	4.50[f]±0.06

Each data is mean of three replicates. Figures followed by the same alphabets along the column are not significantly different at P= 0.05 using Duncan multiple range test (DMRT). FRM = fresh milling; SND = sun drying; OVD = oven drying; STM = steaming; EAAS = essential amino acids.

Table 2. Amino acid composition in differently processed *G. africana* leaves.

Process	Amino acid concentration (g/ 100 g protein)									
	Isoleucine	Leucine	Lysine	Methionine	Cysteine	Total sulphur	Phenylalanine	Tyrosine	Total aromatic	Threonine
FRM	3.04[a]±0.01	8.64[a]±0.01	3.96±0.06	1.05[a]±0.01	0.76[a]±0.02	1.8±0.02	4.07[a]±0.01	2.90[a]±0.04	6.97±0.01	3.38[a]±0.06
SND	2.98[b]±0.02	8.43[b]±0.01	3.88[c]±0.03	1.02[b]±0.03	0.73[b]±0.03	1.75±0.02	4.01[b]± 0.01	2.84[b]±0.04	6.85±0.03	3.32[b]±0.03
OVD	2.90[b]±0.01	8.07[e]±0.01	8.79[b]±0.03	0.92[b]±0.01	0.65[f]±0.01	1.57±03	3.94[c]±0.03	2.75[d]±0.03	6.69±0.02	3.27[c]±0.03
STM	2.92[c]±0.06	8.31[c]±0.02	3.81[c]±0.03	0.97[c]±0.01	0.68[c]±0.02	1.65±0.02	3.93[c]±0.03	2.79[c]±0.04	6.70±0.01	3.27[c]±0.01
FRM±SND	2.92[c]±0.03	8.25[e]±0.03	3.81[c]±0.04	0.95[c]±0.03	0.67[c]±0.01	1.62±0.05	3.95[c]±0.03	2.78[c]±0.01	6.73±0.01	3.26[c]±0.04
FRM±OVD	2.88[d]±0.01	7.98[f]±0.02	3.72[e]±0.01	0.87[e]±0.02	0.60[e]±0.03	1.47±0.06	3.88[d]±0.05	2.70[e]±0.03	6.58±0.01	3.21[d]±0.01
STM±SND	2.89[d]±0.03	8.12[d]±0.03	3.77[b]±0.01	0.93[d]±0.03	0.64[d]±0.01	1.57±0.01	3.91[d]±0.04	2.76[d]±0.02	6.67±0.03	3.22[d]±0.03
STM±OVD	2.82[e]±0.03	7.66[g]±0.03	3.70[e]±0.02	0.80[f]±0.08	0.56[f]±0.06	1.36±0.03	3.87[f]±0.02	2.66[f]±0.01	6.53±0.06	3.18[g]±0.01

Table 2. Contd.

	Valine	Histidine	Alanine	Arginine	Asparatic acid	Glutamic acid	Glycine	Proline	Serine
FRM	4.53a±0.03	2.65a±0.01	4.78a±0.06	5.17a±0.01	8.78a±0.05	13.25a±0.01	4.94a±0.05	3.67a±0.01	5.04a±0.04
SND	4.49b±0.03	2.61b±0.01	4.73b±0.01	5.10b±0.03	8.61b±0.06	13.01b±0.01	4.90b±0.06	3.25b±0.05	4.95b±0.03
OVD	4.40d±0.03	2.54d±0.01	4.63d±0.01	5.02d±0.06	8.52c±0.06	12.91c±0.03	4.80e±0.05	3.44d±0.01	4.89d±0.02
STM	4.43c±0.03	2.56c±0.03	4.67c±0.01	5.05c±0.03	8.53c±0.05	12.90c±0.06	4.85c±0.05	3.48c±0.03	4.94c±0.03
FRM±SND	4.42c±0.04	2.55c±0.01	4.65c±0.02	5.04c±0.03	8.48c±0.06	12.86d±0.08	4.83d±0.01	3.47e±0.07	4.94c±0.03
FRM±OVD	4.31e±0.03	2.47f±0.04	4.58f±0.03	5.98e±0.04	8.45c±0.03	12.80e±0.05	4.74f±0.01	3.40e±0.01	4.80e±0.03
STM±SND	4.39d±0.01	2.51e±0.04	4.62e±0.01	5.01d±0.02	8.43f±0.01	12.82e±0.01	4.80e±0.03	3.43d±0.04	4.89d±0.02
STM±OVD	4.32e±0.03	2.44f±0.03	4.53g±0.02	5.82f±0.01	8.40g±0.03	12.08f±0.04	4.72g±0.05	3.55f±0.01	4.94c±0.03

Each data is mean of three replicates. Figures followed by the same alphabets along the column are not significantly different at P= 0.05 using Duncan mutiple range test (DMRT). FRM = fresh milling; SND = sun drying; OVD =oven drying; STM = steaming; EAAS = essential amino acids.

Table 3. Amino acid composition in differently processed *Gongronema latifolium* leaves.

Process	Amino acid concentration (g/100 g protein)									
	Isoleucine	Leucine	Lysine	Methionine	Cysteine	Total sulphur	Phenylalanine	Tyrosine	Total aromatic	Threonine
FRM	4.71a±0.01	9.21a±0.01	5.78a±0.01	0.76a±0.02	0.43a±0.02	1.19a±0.01	6.38a±0.03	3.39a±0.02	9.77±0.04	3.83a±0.03
SND	4.62±0.02	8.99b±0.02	5.71b±0.01	0.71b±0.02	0.40b±0.03	1.14b±0.01	6.30b±0.04	3.31b±0.03	9.61±0.01	3.79b±0.06
OVD	4.50e±0.03	8.80d±0.01	5.63d±0.03	0.65d±0.01	0.30d±0.03	0.95±0.02	6.22d±0.03	3.21d±0.04	9.43±0.05	3.69d±0.04
STM	4.56c±0.03	8.81d±0.01	5.65d±0.03	0.68d±0.01	0.33d±0.03	1.01±0.03	6.24d±0.04	3.25d±0.05	9.49±0.01	3.72c±0.04
FRM±SND	4.55d±0.01	8.78d±0.01	5.62d±0.01	0.64d±0.04	0.31d±0.06	0.95±0.04	6.22d±0.03	3.24d±0.06	9.46±0.07	3.72c±0.05
FRM±OVD	4.48f±0.01	8.56f±0.02	5.60e±0.01	0.61e±0.03	0.27e±0.04	0.88±0.03	6.17f±0.01	3.18f±0.03	9.35±0.06	3.62c±0.03
STM±SND	4.51e±0.03	8.73d±0.02	5.61e±0.01	0.64d±0.03	0.30d±0.03	0.94±0.03	6.20d±0.03	3.21d±0.03	9.41±0.03	3.70d±0.04
STM±OVD	4.45e±0.04	8.44g±0.01	5.56e±0.01	0.60e±0.01	0.27e±0.02	0.87±0.02	6.15g±0.01	3.17e±0.01	9.32±0.02	3.60f±0.01

	Valine	Histidine	Alanine	Arginine	Asparatic acid	Glutamic acid	Glycine	Proline	Serine
FRM	7.70b±0.05	1.45b±0.01	7.60b±0.01	7.71b±0.03	13.79b±0.03	11.86b±0.05	10.20b±0.01	1.19b±0.08	6.22b±0.01
SND	7.60b±0.04	1.36d±0.03	7.50d±0.03	7.60d±0.01	13.72c±0.06	11.80c±0.06	10.11c±0.01	1.10d±0.02	6.13b±0.03
OVD	7.64c±0.03	1.38d±0.05	7.53d±0.06	7.64c±0.02	13.66e±0.03	11.73e±0.03	10.11c±0.03	1.14c±0.01	6.15c±0.02
STM	7.63d±0.06	1.37c±0.04	7.50d±0.05	7.62d±0.03	13.67e±0.05	11.75d±0.06	10.12c±0.02	1.13e±0.03	6.14c±0.01
FRM±SND	7.56e±0.03	1.30e±0.02	7.42e±0.03	7.56e±0.01	13.69g±0.06	11.75d±0.03	10.08d±0.02	1.07e±0.07	6.08e±0.04
FRM±OVD	7.60d±0.05	1.36d±0.06	7.51d±0.05	7.58e±0.03	13.58f±0.03	11.65f±0.01	10.11c±0.03	1.08e±0.05	6.10d±0.05
STM±SND	7.54f±0.06	1.29d±0.03	7.40f±0.03	7.52f±0.03	13.50g±0.04	11.60g±0.01	10.01d±0.04	1.01f±0.03	6.04c±0.01

Each data is mean of three replicates. Figures followed by the same alphabets along the column are not significantly different at P= 0.05 using Duncan mutiple range test (DMRT).FRM = fresh milling; SND = sun drying; OVD =oven drying; STM = steaming; EAAS = essential amino acids.

Table 4. Amino acid composition in differently processed *Ocimum gratissimum* leaves.

Amino acid concentration (g/100 g protein)

	Isoleucine	Leucine	Lysine	Methionine	Cysteine	Total sulphur	Phenylalanine	Tyrosine	Total aromatic	Threonine
FRM	3.87a±0.03	8.88a±0.03	4.86a±0.01	0.85a±0.04	0.52a±0.01	1.37±0.01	5.21a±0.01	3.02±0.01	8.23±0.01	3.48a±0.01
SND	3.80b±0.05	8.70b±0.04	4.81b±0.03	0.80b±0.02	0.48b±0.02	1.28±0.02	5.12b±0.03	2.98±0.02	8.10±0.02	3.43b±0.02
OVD	3.69f±0.05	8.41d±0.04	4.70e±0.03	0.68e±0.01	0.37e±0.01	1.05±0.03	5.00e±0.05	2.89d±0.05	7.89±0.04	3.32f±0.01
STM	3.75d±0.03	8.54c±0.03	4.78c±0.04	0.74c±0.02	0.42c±0.01	1.16±0.06	5.08c±0.04	2.93d±0.03	8.01±0.01	3.37c±0.08
FRM±SND	3.77c±0.05	8.52c±0.04	4.75d±0.01	0.74c±0.01	0.41c±0.03	1.15±0.07	5.05d±0.02	2.94c±0.01	7.99±0.01	3.35d±0.03
FRM±OVD	3.66g±0.01	8.24g±0.04	4.68e±0.03	0.65g±0.03	0.35e±0.02	1.00±0.05	4.98e±0.03	2.84d±0.04	7.84±0.03	3.30e±0.04
STM±SND	3.71e±0.05	8.39d±0.01	4.75d±0.01	0.71d±0.02	0.39d±0.01	1.10±0.01	5.04d±0.01	2.91d±0.03	7.95±0.03	3.30e±0.07
STM±OVD	3.60h±0.01	8.13f±0.01	4.63f±0.01	0.61g±0.03	0.29f±0.03	0.90±0.03	4.90±0.01	2.82f±0.03	7.72±0.04	3.25f±0.05

	Valine	Histidine	Alanine	Arginine	Asparatic acid	Glutamic acid	Glycine	Proline	Serine
FRM	5.90b±0.01	1.99b±0.03	5.89b±0.05	6.29b±0.04	11.01b±0.03	12.15b±0.02	7.44b±0.04	2.21b±0.03	5.08b±0.03
SND	5.83d±0.02	1.90d±0.03	5.78d±0.04	6.20e±0.03	10.91d±0.05	12.04d±0.04	7.35d±0.03	2.12e±0.06	5.02d±0.03
OVD	5.87c±0.03	1.94c±0.07	5.83c±0.06	6.26d±0.07	10.93d±0.06	12.07d±0.06	7.41c±0.03	2.18d±0.05	5.06d±0.02
STM	5.84d±0.04	1.90d±0.06	5.84c±0.03	6.22d±0.04	10.90d±0.02	12.09c±0.01	7.34d±0.04	2.15d±0.01	5.06d±0.05
FRM±SND	5.80e±0.02	1.88e±0.05	5.75f±0.06	6.17f±0.03	10.86e±0.01	12.01d±0.03	7.32e±0.06	2.10d±0.06	5.01d±0.07
FRM±OVD	5.80d±0.01	1.91d±0.03	5.81d±0.02	6.22d±0.03	10.89d±0.01	12.00f±0.04	7.36d±0.08	2.13e±0.06	4.98g±0.05
STM±OVD	5.77f±0.02	1.82f±0.04	5.71g±0.04	6.10g±0.06	10.81f±0.08	12.97g±0.03	7.27f±0.01	2.04g±0.02	4.93f±0.01

Each data is mean of three replicates. Figures followed by the same alphabets along the column are not significantly different at P= 0.05 using Duncan multiple range test (DMRT). FRM = fresh milling; SND = sun drying; OVD =oven drying; STM = steaming; EAAS = essential amino acids.

Table 5. Amino acid composition in differently processed leafy vegetables.

Essential amino acids	WHO ideal protein		Vernonia amygdalina			Gnetum Africana			Congronema latifolium			Ocimum gratissimum		
	A	B	Amino acids concentration (g/100g protein)	Percentage amino acid Score Children	Adult	Amino acids concentration (g/100 g protein)	Percentage amino acid score Children	Adult	Amino Acids concentration (g/100 g protein)	Percentage amino acid score Children	Adult	Amino acids concentration (g/100 g protein)	Percentage amino acid score Children	Adult
Isoleucine	2.8	1.3	3.65-3.99	130-143	281-307	2.82-3.04	101-109	217-234	4.45-4.71	159-168	342-362	3.60-3.87	129-138	277-298
Leucine	6.6	1.9	8.53-9.52	129-144	449-501	7.66-8.64	116-131	403-455	8.44-9.21	128-140	444-485	8.13-8.88	123-135	428-467
Lysine	5.8	1.6	4.70-5.20	81-90	294-325	3.70-3.96	64-68	231-248	5.56-5.78	96-100	348-361	4.63-4.86	80-84	289-304
Total sulphur amino acid	2.5	1.7	2.81-3.36	112-134	165-198	1.36-1.81	54-72	80-106	0.87-1.18	35-47	51-69	0.90-1.37	36-55	53-81
Total aromatic amino acid	6.3	1.9	7.73-8.41	123-133	407-443	6.58-6.97	104-111	344-367	9.23-9.77	148-155	49-514	7.72-8.23	123-131	406-433
-Threonine	3.4	0.9	3.63-3.96	107-116	403-440	3.18-3.38	94-99	3553-376	3.60-3.83	106-113	400-426	3.25-3.48	96-102	361-387

Table 5. Contd.

valine	3.5	1.3	4.84-5.20	138-149	372-400	123-129	332-348	4.32-4.53	7.54-7.80	215-223	580-600	5.77-595	165-170	444-465
Histidine ±	1.9	1.6	2.92-3.25	154-171	183-203	128-139	153-166	2.44-2.65	1.29-148	68-79	81-93	1.82-2.03	96-107	114-127

A = WHO/FAO/UNU ideal protein for pre-school children aged 2 to 5 years; B = WHO/FAO/UNU idea protein for adult; ± = essential for children. Percentage (%) amino acid score = [(amino acid in sample/ideal)] × 100. Source: WHO/FAO/UNU (1985).

highest losses of amino tested vegetables. The percentage loss however differs from one amino acid to another and among the test vegetables.

In this study, tryptophan was not quantified while glutamine and asparagines were quantified as glutamic and aspartic acids, respectively. This is because during the analytical quantification process, glutamine and asparagines were converted to glutamic and aspartic acid respectively (Salo-Vananen and Koivistoinen, 1996). The inclusion of these test vegetable in diet is vital for the supplementation of essential amino acids. This study has shown that fresh milling followed by sun drying methods reflected the desired effects in the treatment of vegetable required for amino acids supplements.

Conclusion

V. amygdaline, G. Africana G, latifolium and O. gratissimum are rich sources of amino acids supplying essential amino acids in diets except in cases where lysine and sulphur containing amino acids are limiting. Fresh milling followed by sun drying was the most effective method for retaining all the amino acids while steaming combined with oven drying method elicited the most significant loss of all the amino acids in all the test vegetable. The different processing treatments caused varying degrees of significant losses in the amino acids contents of all the test vegetables. This study indicated fresh milling followed by sun drying treatment as most desirable when vegetables are to be used for amino acids supplementation. However, the use of this method should be effectively monitored to avoid fungal growth and contamination of processed vegetables.

REFERENCES

Adeyeye EI, Afolabi EO (2004). Amino acids composition of three different types of land snails consumed in Nigeria. Food Chem., 85: 535-539.

Ajayi IA, Oderinde RA, Kajogbola DO, Ukponi JU (2006). Oil content and fatty acids composition of some underutilized legumes from Nigeria. Food Chem., 99(1): 115-120.

Aregheore EMK, Makkar HPS, Becker K (1998). Feed value of some browse plant from the central zone of delta state. Nig. Trop. Sci., 38(2):97-104.

Aremu MO, Olafe O, Akintayo TE (2006). A comparative study on the chemical and amino acids composition of some Nigerian under-utilized legume flours. Pak. J. Nutr. 5(1): 34-38.

Christian JC (2000). Nutrient changes in vegetables and fruits, 1951 to 1999. CTV, Ca http://www.ctv.ca \ servlets / airtcles news /story /CTVnews 20020705/ favaro nutrients- chat-020705/ Health/ story. (Retrieved May 13, 2009)

FNB (2002). Food and Nutrition Board, Institute of Medicine. National Academy of Sciences. Dietary reference intake for energy, carbohydrate, fibre, fat, fattyacids, cholesterol, protein and amino acid (micro nutrients). www.nap.edu (Retrieved on 14\5\2010).

Harper A (1999). Defining the essentiality of nutrients. In Modern Nutrition in health and Diseases (9th ed.). Shills, M.E (ed). Williams and Wilkins, Baltimore, MD. pp. 201-216.

Hassan LG, Umar KJ (2006). Nutritional value of Balsam apple (Momordica balsamina L.) leaves. Pak. J. Nutr., 5(6): 522-529.

Kubmarawa D, Andenyang FH, Magomya AM (2008). Amino

acid profile of two non-conventional leafy vegetables, Sesamum indicum and Balanites aegptiaca. Afr. J. Biotechnol. 7(19): 3502-3504.

Lohlum SA, Maikidi GH, Solomon M (2010). Proximate composition of amino acid profile and phytochemical screening of Lophira lanceolata seeds. Afr. J. Food, Agric. Nutr. Develop., 10(1): 2012-2023.

Lund DB (1975). Effect of heat processing on nutrients. In: Nutritional evaluation of food processing. Harris, R.S. and Karmas, E: (eds) AVI Publishing Co. Inc., Westport, CT., pp. 205-230.

Matasyoh IG, Matasyoh JC, Mukiama TK (2007). Chemical composition and antimicrobaial activity of the essential oil of Ocimum gratissimum L. growing in Eastern Kenya. Afr. J. Biotechnol., 6(6): 760-765.

Morris A, Barnett A, Burrows O (2004). Effects of processing on nutrient content of foods. Articles, 37: 160-164

Okafor JC (1983). Horticultural promising indigenous wild plant species of the Nigeria forest zone. ACTA Hort., 123: 165-176.

Olaofe O, Akintayo ET (2000). Production of isoelectric points of legume and oil seed proteins from their amino acid composition. J. Technol. Sci., 4: 49-53.

Ramberg J, Mc Annalley B (2002). From the farm to the kitchen tables: A review of the nutrient losses in foods. Glycosci. Nutrition, 3(5): 1-12.

Salo-Vaavanen PP, Koivistoinen PE (1996). Determination of protein in foods: Comparison of net protein and crude protein (N × 6.25). Values Food Chem., 57: 27-31.

Souci SW, Fachmann W, Kraut H (2000). Food composition and nutrition tables (6th ed.) CRC Press, Boca Raton, Florida, pp. 62-158.

Wahua TAT (1999). Applied Statistics for Scientific Studies. African Link Press. Aba, Nigeria. pp. 129-155.

Wardlaw GM, Kessel M (2002). Perspective in Nutrition (5th ed). McGraw Hill, Boston. pp. 418-461.

WHO/FAO/UNU (1985). Report: Energy and protein requirements. WHO technical report series No. 724. WHO, Geneva, pp. 220-223.

30

The effect of allelic variation on forage quality of brown midrib sorghum mutants with reduced caffeic acid O-methyl transferase activity

R. K. Vogler[1], T. T. Tesso[2], K. D. Johnson[1] and G. Ejeta[1]*

[1]Department of Agronomy, Purdue University, West Lafayette, IN 47907-2054, USA.
[2]Department of Agronomy, Kansas State University, Manhattan, KS 66506, USA.

Sorghum brown midrib (bmr) mutants have reddish-brown vascular tissues in their leaves and stems as a result of changes in lignin content and subunit composition. Past research at Purdue University has generated a set of bmr sorghum mutants via chemical mutagenesis and established some to be allelic to each other. More recently, we identified additional spontaneous mutants in true breeding lines with marked phenotype and a range of agronomic characteristics. One such mutant, bmr-26, is of particular interest because it arose in a drought-tolerant sorghum line. Analysis of testcross hybrids between this spontaneous bmr mutant and the chemically induced mutants, bmr-6 and bmr-12, showed that the bmr-26 allele was allelic to bmr-12 and not to bmr-6. Both the bmr-12 and the bmr-26 mutations significantly reduced lignin content in leaf, blade, sheath, stem, and panicle tissue. The effect of the mutation was relatively more severe in bmr-12 than in bmr-26. The impact of the two mutations on cell wall composition in different tissues varied. The biggest effect of the bmr-12 mutation was in reduction of lignin in the sheath, whereas lignin content in panicles was more affected by the bmr-26 mutation. This suggested an allele-specific effect in tissue lignin reduction of these mutants. Cellulose and hemicellulose concentrations were also significantly higher in certain tissue types for both the induced and spontaneous mutants. Forage quality traits including percent NDF and ADF were significantly increased by both mutations. Improvement in in vitro dry matter digestibility as a result of the bmr-26 mutation was relatively small and was not proportional to the reduction in the lignin content.

Key words: Acid detergent fiber (ADF), bmr, brown midrib, in vitro dry matter digestibility (IVDMD), lignin, neutral detergent fiber (NDF), Sorghum bicolor.

INTRODUCTION

Grain sorghum (Sorghum bicolor (L.) Moench) is a major food crop in semi-arid regions of Africa and Asia and the second most important feed grain in the United States. Although the grain is often used as a major ingredient in cattle feed, sorghum and sorghum x sudangrass (Sorghum sudanense (Piper) Stapf.) hybrids are widely used as forage, especially in the US dairy industry (Rook et al., 1977; Stalling et al., 1982; Oba and Allen, 1999; Cox and Cherney, 2001).

The value of a crop plant as forage is determined primarily by the degradability of the vegetative tissue, which in turn is affected by the property of its cell wall structure

structure (Åman, 1993). Cellulose and hemicellulose in the cell wall provide a major energy source for ruminant animals, when they can be degraded into oligo- and mono-saccharides (Moore and Hatfield et al., 1994). The plant cell wall is a complex matrix in which cellulose microfibrils are embedded in a matrix of hemicellulose, pectin, proteins, and aromatic compounds such as lignin and hydroxycinnamic acids (Carpita and Gibeaut, 1993). The availability of cellulose and hemicellulose as a source of energy, however, depends on the overall structural properties of the cell wall, which often varies between species, genotypes, and tissue, and the interaction between these three factors. For example, the presence of the cell wall polymer lignin has been thought to impede access of hydrolytic enzymes to the cell wall polysaccharides. Cell wall digestibility of maize (Zea mays L.)

*Corresponding author. E-mail: gejeta@purdue.edu.

has been shown to be affected by lignin content (Wolf et al., 1993; Lundvall et al., 1994), and lignin composition (Oba and Allen, 1999; Fontaine et al., 2003), although developmental changes as a result of changes in lignin composition can account, in part, for improved digestibility (Grabber et al., 1997).

Following the discovery that the brown midrib (bm) mutants of maize had altered lignin composition (Kuc and Nelson, 1964; Gee et al., 1968) and that some of them were more digestible (Barrière and Argillier, 1993), chemical mutagenesis experiments with sorghum resulted in several brown midrib (bmr) mutants (Porter et al., 1978) that resembled those in maize. Additional spontaneous bmr mutants of sorghum were identified later in breeding populations (Vogler et al., 1994). Brown midrib mutations, both in sorghum and maize, are phenotypically characterized by the presence of brown vascular tissues in the leaf blade and sheath, as well as in the stem. The bmr phenotype becomes apparent once plants have reached the four-leaf stage and tends to begin to fade as the plants approach physiological maturity (Porter et al., 1978).

Brown midrib mutants significantly reduce the level of enzyme-resistant lignin in plants and increase their palatability and digestibility (Rook et al., 1977; Cherney et al., 1991). Brown midrib silage with and without protein supplements significantly increased milk yield of lactating cows (Frenchik et al., 1976; Keith et al., 1979; Stalling et al., 1982; Cherney et al., 1991; Oba and Allen, 1999). Similarly, the rate of in vitro dry matter digestibility (IVDMD) and cell wall degradation by rumen bacterium of leaf blades from bmr-12 sorghum was shown to be significantly higher than those from their respective wild-type isolines (Akin et al., 1986a, b).

Allelism tests on the sorghum bmr mutants derived through chemical mutagenesis showed that several of the mutations are allelic, and that the total number of independent bmr loci was smaller than the number of mutant lines assembled (Bittinger et al., 1981). The molecular genetic basis of many of the bmr mutations has not been clearly elucidated. The first sorghum bmr gene was recently cloned. The mutant alleles bmr-12, bmr-18 and bmr-26 were shown to contain premature stop codons in the lignin biosynthetic enzyme caffeic acid O-methyltransferase (COMT; Bout and Vermerris, 2003). These mutations ultimately result in reduced synthesis of the monolignol sinapyl alcohol. The lignin of the bmr-6 mutant was shown to contain more cinnamaldehydes (Bucholtz et al., 1980) and displayed reduced activity of the lignin biosynthetic enzyme cinnamyl alcohol dehydrogenase (Pillonel et al., 1991).

The effect of the bmr mutations on forage quality varies depending on the genetic background of the line in which the mutation is introduced (Cherney et al., 1991; Peder-sen et al., 2005). This suggests the need to either identify a suitable genetic background that allows for optimal impact of the mutation. To do this effectively, the effect of each mutation on forage quality and agronomic characteristics needs to be determined. The impact of the bmr-26 mutation on forage quality has not been studied yet. The objective of this study was to establish the allelic relationship of the spontaneous mutant bmr-26 with other known bmr mutants, and provide an assessment of the effect of the bmr-26 mutation on forage quality.

MATERIALS AND METHODS

Plant materials

The mutants, bmr-6, bmr-12 and bmr-18 were identified in a segregating population of sorghum that was obtained from seeds mutagenized with diethyl sulfate (DES) (Porter et al., 1978). A spontaneous mutant bmr-26 was later identified in a drought-tolerant stay-green sorghum inbred line, P898012 (Vogler et al., 1994). For ease of description, the parental genotype P898012 will be referred to as N-26 and similarly the wild-type versions of bmr-6, bmr-12 and bmr-18 will be referred to as N-6, N-12 and N-18, respectively. Seeds from the bmr-6, bmr-12, bmr-18 and bmr-26 lines and their respective wild-type lines were planted in the greenhouse. Allelism tests were performed by fertilizing emasculated florets of bmr-6, bmr-12 and bmr-18 with pollens collected from panicles of bmr-26 to produce F_1 hybrids. Unwanted florets on a panicle were removed using clippers leaving only emasculated florets covered with pollination bag to prevent contamination by unwanted pollen. When stigma in the emasculated florets were receptive, usually two days after emasculation, each of the emasculated panicles were dusted with pollens collected from bmr-26, and covered with a pollination bag to avoid contamination. At physiological maturity, the F_1 seeds were collected from each of the pollinated plant panicle into a labeled bag and stored separately for planting the next season. The F_1 seeds along with parental bmr lines and their wild-type counterparts were planted in separate rows at the Purdue University Agronomy Center for Research and Education near West Lafayette, IN, USA. Prior to flowering, each of the hybrids were scored for the bmr trait in the leaves (present or absent). F_2 seeds were harvested from selfed panicles of the hybrids and the F_2 populations were planted the next season and scored for segregation of the brown midrib phenotype.

Determination of cell-wall components and in vitro dry matter digestibility (IVDMD)

Plant cell wall composition was analyzed by determining the neutral detergent fiber (NDF), acid detergent fiber (ADF), lignin, cellulose and hemicellulose. Tissue samples from bmr-12 and bmr-26, their normal counterparts, and F_1 hybrids were collected when the plants were at the half-bloom stage. Analyses were performed on whole plants (excluding roots) and on individual tissues: stem, leaf sheath, leaf blade and panicle. For individual tissue analysis, a pool representing four plants (three for the F_1 hybrid) was analyzed, whereas for the whole-plant analysis three plants were pooled. The tissues were dried at 100°C for 48 h, and ground with a UDY cyclone mill to pass a 1-mm screen. Tissues were analyzed sequentially for NDF, ADF, and permanganate lignin concentration as described by Van Soest (1991). Hemicellulose was calculated as the loss of NDF residue upon subsequent extraction with acid detergent. Cellulose was determined as the loss of ADF residue upon extraction with sulfuric acid (72% H_2SO_4). In vitro dry matter disappearance (IVDMD) was determined from the ground tissue samples following the procedure described by Tilley and Terry (1963) and modified by Barnes et al. (1971). Samples were collected and analyzed in two replicates for each of the laboratory procedures.

Table 1. Midrib phenotypes of parental, F_1 and F_2 genotypes derived from crosses of bmr-6 with bmr-12 and bmr-18.

Genotype	Midrib phenotype	F₂ segregation				χ^2	
		Observed		Expected*			
		brown	white	brown	white		
bmr-26	brown						
N-26	white						
bmr-12	brown						
N-12	white						
bmr-6	brown						
N-6	white						
(bmr-6 × bmr-26)F₁	white						
(bmr-12 × bmr-26)F₁	brown						
(bmr-18 × bmr-26)F₁	brown						
(bmr-6 × bmr-26)F₂	white	72	74	64	82	1.84[a]	3.84[b]
(bmr-12 × bmr-26)F₂	brown	297	0	297	0	0.00	1.0
(bmr-18 × bmr-26)F₂	brown	165	0	165	0	0.00	1.0

[a]calculated χ^2 statistic; [b]value from χ^2 table at 1 degree of freedom.

Statistical analysis

Data were analyzed using a two-factor factorial model in a randomized complete block design. Factor I consisted of five genotypes (bmr-12, bmr-26, N-12, N-26 and bmr-12 x bmr-26) and factor II consisted of five tissues (whole plant, stem, sheath, blade and panicles). The general linear model (PROC GLM) of the statistical analysis systems (SAS, 2002) was used to analyze the data. The analysis was performed independently for different plant tissues. The main effects were tested using appropriate mean squares and a contrast statement was drawn to determine significant differences between bmr and normal genotypes.

RESULTS

The bmr-26 mutation is allelic to bmr-12 and bmr-18

The midrib phenotypes of the F_1 progenies resulting from crosses of bmr-26 with bmr-6, bmr-12 and bmr-18 along those of the parental lines are presented in Table 1. Hybrids of bmr-12 x bmr-26 and bmr-18 x bmr-26 expressed the mutant phenotype indicating that the bmr-26 allele is allelic to bmr-12 and bmr-18. On the other hand, crosses of bmr-6 x bmr-26 produced hybrids with a wild-type phenotype, which indicates bmr-26 is non-allelic to bmr-6. The F_2 progeny of the cross bmr-6 x bmr-26 was therefore expected to segregate for the bmr phenotype in a 9:7 ratio. Indeed, evaluation of 146 F_2 individuals resulted in 74 plants with a wild-type phenotype and 72 bmr mutants. A chi-squared goodness-of-fit test (df = 1) resulted in a value of 1.84 (p = 0.175), which is consistent with the 9:7 segregation ratio (Table 1).

Variation in NDF and ADF

Significant variation in cell wall composition was observed among the genotypes in all tissue samples examined (Tables 2 and 3). Percent NDF and ADF across genotypes were significantly higher in panicle tissues followed by the leaf sheath (Table 2). This was consistent in both the bmr mutants and normal genotypes. The lowest NDF and ADF values were obtained for leaf blade tissues (Table 2). The NDF of bmr stems was significantly higher than the wild-type stems, but the differences in NDF values of the other tissues were not statistically significant (Table 2). Differences between individual bmr mutants and their respective wild-type counterparts, however, were in most cases significant (Table 3). The bmr-12 mutant had significantly higher NDF compared to its wild-type isoline in all tissues except the panicle. The ADF values showed a similar pattern to NDF, except that on a whole-plant basis, the difference was not significant. The impact of the bmr-26 mutation on a whole-plant basis was considerably less than that of the bmr-12 mutation. No significant differences were observed for NDF values, and ADF values were only different in the panicle. The F_1 bmr hybrids generally had high NDF and ADF values in all tissues except when the whole plant was evaluated.

The bmr-26 and bmr-12 mutations have different impact on permanganate lignin content

Mean permanganate lignin content was lowest in leaf blades and highest in panicle tissues across genotypes (Table 2). Permanganate lignin content of bmr mutants averaged across tissue and genotypes was 21% lower than normal genotypes. The largest difference was observed in stem and leaf sheath tissues, where the mutants had a 30 and 27% lower lignin content, respec-

Table 2. Percent cell wall and *in vitro* dry matter digestibility (IVDMD) of normal and brown midrib sorghum genotypes.

Tissue	NDF	ADF	Lignin	Hemicellulose	Cellulose	IVDMD
Across genotypes						
Whole plant	58.1 c	31.8 c	4.8 c	26.2 c	27.4 c	59.2 b
Stem	55.0 d	31.8 c	4.9 c	23.1 d	27.6 c	59.8 ab
Sheath	64.5 b	36.3 b	5.4 b	28.2 b	31.0 b	56.2 c
Blade	53.9 e	27.6 d	3.9 d	26.3 c	23.9 d	61.3 a
Panicle	75.7 a	37.3 a	5.8 a	38.3 a	31.8 a	51.9 d
Bmr genotypes						
Whole plant	58.0 c	31.4 d	4.3 c	26.6 c	27.7 d	59.5 ab
Stem	57.5 c	32.8 c	4.2 c	24.7 d	29.1 c	56.7 bc
Sheath	65.5 b	35.8 b	4.7 b	29.7 b	31.3 b	57.0 bc
Blade	55.0 d	28.3 e	3.8 d	26.7 c	24.8 e	61.9 a
Panicle	75.9 a	36.9 a	5.4 a	39.0 a	31.8 a	54.2 c
Normal genotypes						
Whole plant	58.2 c	32.4 c	5.6 b	25.7 b	27.0 c	58.6 b
Stem	51.1 e	30.3 d	6.0 ab	20.7 c	25.4 d	64.4 a
Sheath	63.1 b	37.0 b	6.5 a	26.0 b	30.6 b	55.0 c
Blade	52.3 d	26.5 e	4.0 c	25.7 b	22.4 e	60.4 b
Panicle	75.2 a	38.0 a	6.5 a	37.2 a	31.7 a	48.5 d

Means followed by the same letter in a column within a genotype are not significantly different. NDF - neutral detergent fiber; ADF - acid detergent fiber; IVDMD *in vitro* dry matter digestibility.

pectively, than the normal genotypes. In contrast, the difference in permanganate lignin content in leaf blades was only 5%, whereas whole plant and panicle tissues showed differences of 23 and 18%, respectively.

Both bmr-12 and bmr-26 displayed significant reductions in permanganate lignin content. The impact of the bmr-26 mutation on permanganate lignin content, however, is different compared to the bmr-12 mutation (Table 3). On a whole-plant basis bmr-12 and bmr-26 contained 22 and 16% less lignin, respectively, than their normal counterparts. Lignin reduction among stem tissues was 22 and 34%, while among leaf sheaths 29 and 23%, and among panicles 5 and 25%, respectively, in bmr-12 and bmr-26. Among leaf blades only bmr-12 results in a significant reduction in lignin content (14%). The lignin content in the bmr hybrid was, depending on the tissue, more similar to bmr-12 or bmr-26.

Variation in cellulose and hemicellulose content

Hemicellulose and cellulose contents across genotypes were significantly higher in panicle tissues both in the bmr mutants and the normal genotypes (Table 2). The lowest cellulose and hemicellulose levels were detected in stem and leaf blade tissues, respectively, in both bmr and normal genotypes, but the levels were generally higher in bmr lines than in normal genotypes. The bmr-26 mutant had higher hemicellulose content in the sheath, and slightly less in the leaf blade. Cellulose content was higher in the bmr-26 stem and blade. In contrast, the

bmr-12 mutant had significantly higher hemicellulose and cellulose content than its normal counterpart in all tissues except the panicle (Table 3).

In vitro dry matter digestibility

There were significant differences in IVDMD among different tissues in both bmr and normal genotypes (Table 2). Mean IVDMD was remarkably higher in leaf blade tissues for the bmr mutants and in the stem tissues for normal genotypes, but it was lowest in the panicle tissues of both bmr and normal genotypes. Among genotypes, bmr-12 and bmr-12 x bmr-26 hybrid had the highest IVDMD for the whole plant tissue, while bmr-26 had the lowest IVDMD though not significantly different from its normal isoline (Table 3). The IVDMD of the stem their normal counterparts by 10 and 12%, respectively.

The bmr hybrid also had a low IVDMD score. In other tissues IVDMD was not significantly affected as a result of the mutations, except in the panicle tissues, where the bmr mutants had better digestibility scores than their normal counterparts.

DISCUSSION

28 bmr mutants of sorghum have been identified to date, most of which were developed via chemical mutagenesis. The number of independent loci appears to be smaller than the number of mutants, based on allelism tests per-

Table 3. Tissue cell wall and *in vitro* dry matter digestibility (IVDMD) of normal and brown midrib sorghum genotypes.

Genotype	NDF	ADF	Lignin	Hemicellulose	Cellulose	IVDMD
Whole plant						
bmr-26	60.2 a	33.1 a	4.7 b	27.1 a	29.1 a	56.1 c
N-26	60.4 a	33.7 a	5.6 a	26.6 a	28.4 a	58.3 bc
bmr-12	58.3 b	31.2 b	4.3 b	27.1 a	27.3 b	61.7 a
N-12	56.0 c	31.1 b	5.6 a	24.9 b	25.7 c	58.9 abc
F1 (bmr)	55.5 c	29.9 b	4.0 b	25.6 ab	26.7 b	60.8 ab
Stem						
bmr-26	57.4 a	33.4 ab	4.4 c	23.9 ab	29.8 a	59.5 b
N-26	56.5 a	34.3 a	6.7 a	22.1 bc	28.4 c	65.8 a
bmr-12	58.8 a	32.6 b	4.1 c	26.2 a	29.0 b	55.3 c
N-12	45.7 b	26.4 c	5.2 b	19.3 c	22.5 d	63.1 ab
F1 (bmr)	56.4 a	32.4 b	4.1 c	24.1 ab	28.5 bc	55.3 c
Sheath						
bmr-26	63.2 c	36.3 b	4.7 d	26.8 b	31.8 a	58.9 a
N-26	63.5 c	38.0 a	6.1 b	25.4 c	31.8 a	57.2 ab
bmr-12	66.1 b	34.8 c	4.8 c	31.3 a	30.3 b	56.5 ab
N-12	62.6 c	36.0 b	6.9 a	26.6 c	29.4 c	52.8 b
F1 (bmr)	67.3 a	36.3 b	4.7 d	30.9 a	31.8 a	55.7 ab
Blade						
bmr-26	53.2 c	28.5 ab	4.0a	24.6 c	24.7 ab	61.5 ab
N-26	53.0 cd	27.4 bc	3.9 a	25.5 b	23.4 c	56.9 b
bmr-12	54.9 b	27.2 c	3.5 a	27.6 a	24.1 bc	62.7 a
N-12	51.6 d	25.6 d	4.0 a	25.9 b	21.5 d	63.8 a
F1 (bmr)	57.0 a	29.1 a	3.9 a	27.9 a	25.6 a	61.4 ab
Panicle						
bmr-26	72.1 c	35.9 c	5.8 b	36.1 bc	30.5 c	54.6 b
N-26	74.2 bc	38.8 a	7.7 a	35.3 c	31.4 bc	46.7 a
bmr-12	77.7 a	37.0 bc	5.0 b	40.7 a	32.4 ab	58.1 a
N-12	76.3 ab	37.1 bc	5.3 b	39.2 ab	31.9 ab	50.3 a
F1 (bmr)	78.1 a	37.9 ab	5.3 b	40.1 a	32.6 a	49.9 a

Means followed by the same letter in a column within tissue type are not significantly different. NDF - neutral detergent fiber; ADF - acid detergent fiber; IVDMD *in vitro* dry matter digestibility.

formed by Bittinger et al. (1981). The results from the tissue in bmr-26 and bmr-12 was significantly lower than allelism tests reported here are consistent with this finding where bmr-12 and bmr-18 are reported to be allelic and different from bmr-6, as well as with recent sequence and expression data reported by Bout and Vermerris (2003).

The most dramatic effect of the bmr-12 and bmr-26 mutations appears to be on the lignin content. Lignin concentration measured as permanganate lignin was consistently reduced in bmr mutations compared to the wild-type iso-lines. However, the degree to which the lignin content was reduced varied among tissues. The effect of the bmr-26 mutation was most severe in the panicle tissue and the least severe in the leaf blade, whereas the impact of the bmr-12 mutation was most severe in the sheath and least severe in the panicle. On

whole plant basis, lignin content appeared to have been severely reduced in bmr-12 than bmr-26. These results are in agreement with previous reports from studies conducted on same mutations (Oliver et al., 2004, 2005). Such differences may indicate the possibility of multiple mutant alleles occurring at the same locus. In this respect, it is of interest to note that the bmr-12 and bmr-26 mutations are in the first and second exon of the COMT gene, respectively (Bout and Vermerris, 2003), perhaps resulting in differential expression of the mutations. Further analyses of the effect of these two mutations in the same genetic background may produce additional information on background effect on the degree of expression of mutations. Such information may be used to selectively alter lignin content and subunit composition in specific parts of the plant and also provide means for addressing the potential negative agronomic attributes of the mutations such as lodging and susceptibility to pests and diseases supposedly associated with modified lignin content or composition. It is also interesting to note that there is a differential effect on cellulose and hemicellulose contents of the mutants. In the bmr-12 mutant, the degree of reduction in lignin content is paralleled with an increase in cellulose and hemicellulose contents. The reduction in lignin content is likely to have an effect on the overall structural integrity of the cell wall. The increase in cellulose and hemicellulose contents may reflect the existence of a mechanism that compensates for the reduction in lignin content. This has been shown to occur in aspen (Hu et al., 1999) and rice (Li et al., 2003). Alternatively, the reduction in lignin content may make the polysaccharide fractions more susceptible to chemical or enzymatic degradation, which could account for an increased release of sugars.

The lack of significant improvement in IVDMD as a result of the bmr mutations are somewhat unexpected. The maize bm3 mutant, which also has a mutation in the COMT gene (Vignols et al., 1995), was shown to result in significantly better intake and digestibility in animal feeding studies (Oba and Allen, 1999), and similar results were expected for bmr-12 and bmr-26 mutants. One explanation for improved intake would be increased palatability, which may primarily reflect taste and physical attributes such as breaking strength of the tissue. Digestibility, however, has been shown to depend on chemical composition of the cell wall (Jung and Buxton, 1994; Lundvall et al., 1994; Fontaine et al., 2003), although the exact relationship is quite complex. The IVDMD analysis is performed *in vitro*, and may therefore not be an accurate reflection of *in-vivo* digestibility in this particular case. It may be worth performing animal feeding studies with these two specific mutants to develop a better understanding on their effect on dry matter digestibility. However, feeding studies conducted on dairy cows have shown that bmr sorghum silage has significantly improved dry matter digestibility and resulted in higher milk yield compared to conventional sorghum silage (Oliver et al., 2005; Dann et al., 2008). The bmr mutants had 4% higher NDF than the normal lines, though this varied according to tissue type and the genetic background in which the mutations are constituted. The mutant bmr-12 had consistently higher NDF in all tissue types than N-12, while bmr-26 was exceeded by its normal counterpart in the whole plant and leaf sheath tissues. There was no noticeable difference for mean ADF between bmr mutants and the normal lines. Both N-12 and N-26 had slightly higher ADF than their mutant counterparts in all tissues except the stem and leaf blade for bmr-12 and the leaf blade tissue for bmr-26. This result is in agreement with previous work by Cox and Cherney (2001) who reported significantly higher NDF in maize brown midrib hybrids than normal hybrids of different backgrounds. Porter et al. (1978) also reported significantly higher percent ADF in normal genotypes than their mutant counterparts, but the difference in the current study was comparatively small. Variation in cellulose and hemicellulose contents in both backgrounds followed a similar pattern. The bmr mutants generally tended to have higher cellulose and hemicellulose contents though there were few exceptions for certain tissue types in the bmr-26 background. This again agrees with the report by Porter et al. (1978) who observed consistently higher cellulose and hemicellulose concentrations in bmr-12 than its normal counterpart. But this is not always the case for all genotypes. Similar to the bmr-26 in this study, many other mutants have been shown to have lower concentrations of cellulose, hemicellulose or both than their normal counterparts. However, the result for lignin content was very different. Almost all previous reports revealed significantly lower concentration of lignin in mutants compared to their normal sister lines in both sorghum (Porter et al., 1978) and corn (Cardinal et al., 2003). The results of our current study also corroborates these findings as we found the average lignin content combined across tissue types of bmr mutants to be 21% lower than the normal lines. This varied for different tissue types with the range spanning from only 6% difference in the leaf blade tissue to 30% in the stem. This variation, however, was not reflected in the difference in IVDMD between the mutants and the normal lines in this study. The mutation that reduced the lignin content of the genotypes may have altered the composition of lignin that renders them undigestable in certain tissues. Previous studies have shown that in the cell walls of bmr-6, the concentration of certain phenolic compounds, such as aldehydes and ferulates, were increased while the content of p-coumarate decreased (Akin et al., 1986b; Pillonel et al., 1991). In bmr-12 and bmr-18, the concentration of syringyl moieties and p-coumarate were reduced (Akin et al., 1986) and in a related cereal, pearl millet (*Pennisetum glaucum* (L.) R. Br.), p-coumarate was again reduced but the concentration of the guaiacyl units were increased (Cherney et al., 1991). This decrease in one and increase in another component of cell wall, though ultimately resulting in lower lignin content, might have altered its composition such that in certain tissues and at a certain growth stage, it becomes more resistant to degradation. However, previous feeding studies have shown that brown midrib mutants in both corn and sorghum have resulted in improved milk yields in dairy cows and growth in sheep (Frenchik et al., 1976; Keith et al., 1979; Stalling et al., 1982; Aydin et al., 1999; Grant et al., 1995; Oba et al., 1999). This could be due to the combined effects of improved digestibility of the cell wall associated with reduced lignin content and increased intake as a result of improved palatability of the brown midrib genotype.

ACKNOWLEDGEMENT

This research was funded by the United States Agency for International Development (USAID) Grant # Dan 254-

G-00-002 through INTSORMIL, the International Sorghum and Millet Collaborative Research Support Program. This is article number 2005-17753 in the Purdue Agricultural Experiment Station series.

REFERENCES

Akin AE, Hanna WW, Rigsby LL (1986). Normal-12 and Brown Midrib-12 sorghum. I. Variations in Tissue Digestibility. Agron. J., 78:827-832.

Akin AE, Hanna WW, Snook ME, Himmelsbach DS, Barton FE II, Windham WR (1986), Normal-12 and Brown Midrib-12 sorghum. II. Chemical variations and digestibility. Agron. J. 78:832-837.

Åman P, Composition and structure of cellwall polyscaharides in forages. In: Jung HG, Buxton DR, Hatfield RD, Ralph L (1993) (eds) Forage cell wall structure and digestibility.ASA-CSSA-SSSA, Madison., 183-196.

Aydin g, Grabt RJ, O'Rear J (1999). Brown midrib sorghum in diets for lactating dairy cows. J. Dairy. Sci. 82:2127-2135.

Barnes RF, Muller LO Bauman LF, Colenbrander VF (1971) In vitro dry matter disappearance of brown midrib mutants of maize (Zea mays L.). J. Anim. Sci. 33:881-884.

Barriere Y, Argillier O (1993). Brown midrib genes of maize: A review. Agronomie 13:865-876.

Bittinger TS, Cantrel RP, Axtell JD (1981). Allelism tests of the brown midrib mutants of sorghum. J. Hered 72:147-148.

Boudet AM (2000) Lignin and lignification: Selected issues. Plant .Physiol. Biochem. 38:81-96.

Bout S, Vermerris W (2003) A candidate gene approach to clone the sorghum Brown midrib gene encoding caffeic acid O-methyl transferase. Mol. Gen. Genomics. 269:205-214

Bucholtz DL, Cantrell RP, Axtell JD, Lechtenberg VL (1980) Lignin biochemistry of normal and brown midrib mutant sorghum. J. Agric. Food Chem. 28:1239-1245.

Campbell MM, Sederoff RR (1996). Variation in lignin content and composition. Plant Physiol. 110:3-13.

Cardinal AJ, Lee M, Moore KJ (2003). Genetic mapping and analysis of quantitative trait loci affecting fiber and lignin content in maize. Theor. Appl. Genet. 106:866-874.

Carpita NC, Gibeaut DM (1993) Structural models of the primary cell walls in flowering plants: consistency of molecular structure with the physical properties of the wall during growth. Plant J. 3:1-30.

Cherney JH, Cherney JR, DE Akin DE, Axtell JD (1991). Potential of brown midrib, low lignin mutants for improving forage quality. Adv Agron. 46:157-198.

Cox WJ, Cherney DL (2001) Influence of brown midrib, leafy, and transgenic hybrids on corn forage production. Agron. J. 93:790-796.

Dann HM, Grant KW, Cotanch ED, Thomas CS, Ballard R (2008). Rice. Comparison of brown midrib sorghum-sudangrass with corn silage on lactational performance and nutrient digestibility in Holstein dairy cows J. Dairy Sci. 91: 663 - 672.

Fontaine AS, Bout S, Barriere Y, Vermerris W (2003). Variation in Cell Wall Composition among Forage Maize (Zea Mays L.) Inbred lines and its impact in digestibility analysis of neutral detergent fiber composition by pyrolysis-gas chromatography-mass spectrometry. J. Agric. Food Chem, 51:8080-8087.

Frenchik GE, Johnson DG, Murphy JM, Otterby DE (1976) . Brown midrib corn silage in dairy cattle rations. J. Dairy Sci. 59:2126-2129.

Gee MS, Nelson OE, Kuc J (1968). Abnormal lignin produced by brown-midrib mutants of maize. II Comparative studies on normal and brown midrib-1 dimethyl formamide lignins. Arch. Biochem. Biophys.123:403-408.

Grabber JH, Ralph J, Hatfield RD, Quideau S (1997) p-Hydroxyphenol, guaiacyl, and syringyl lignins have similar inhibitory effects on wall degradability. J. Agric. Food Chem. (45:2530-2532.

Grant JR, Haddad SG, Moore KJ, Pedersen JF (1995). Brown mid rib sorghum silage for mid lactation dairy cows. J. Dairy Sci. 78:1970-1980.

Hu WJ, Harding SA, Lung J, Popko JL, Ralph J, Stokke DD, Tsai C-J, Chiang VL (1999). Repression of lignin biosynthesis promotes

cellulose accumulation and growth in transgenic trees. Nat. Biotechnol. 17: 808-812.

Jung H-JG, Buxton DR (1994). Forage quality variation among maize inbreds: relationships of cell-wall composition and in-vitro degradability for stem internodes. J. Sci. Food Agric. 66:313-322.

Keith EA, Colenbrander VF, Lechtenberg VL, Bauman LF (1979) Nutritional value of brown midrib corn silage for lactating dairy cows. J. Dairy. Sci. 62:788-792.

Kuc J, Nelson OE (1964). The abnormal lignins produced by the brown midrib mutants of maize. I. The brown midrib-1 mutants. Arch. Biochem. Biophys., 105: 103-113.

Kuc J, Nelson OE Flanagan P(1968). Degradation of abnormal lignins in the brown midrib mutants and double mutants of maize. Phytochemistry, 7: 1435-1436.

Li Y, Qian Q, Zhou Y, Yan M, Sun L, Zhang M, Fu Z, Wang Y, Han B, Pang X, Chen M, Li J (2003). BRITTLE CULM1, which encodes a Cobra-like protein, affects the mechanical properties of rice plants. The Plant Cell, 15: 2020-2031.

Lundvall JP, Buxton DR, Hallauer AR, George JR (1994). Forage quality variation among maize inbreds: In vitro digestibility and cell wall components. Crop Sci. 34:1671-1678.

Moore KJ, Hatfield RD (1994). Carbohydrates and Forage quality. In: Fahey, G.C. Jr. (ed) Forage quality, evaluation and utilization. ASA-CSSA-SSSA, Madison., 229-280.

Oba M, Allen MS (1999). Effects of brown midrib-3 mutation in corn silage on dry matter intake and productivity of high yielding dairy cows. J. Dairy. Sci. 82:135-142.

Oliver AL, Grant RJ, Pedersen JF, O'Rear J (2004). Comparison of brown midrib 6 and 18 forage sorghum with conventional sorghum and corn silage in diets of lactating cows. J. Dairy. Sci. 87: 637-644.

Oliver AL, Pedersen J F, Grant RJ, Klopfenstein TJ (2005) Comparative Effects of the Sorghum bmr-6 and bmr-12 Genes: I. Forage Sorghum Yield and Quality. Crop Sci. 45: 2234-2239.

Oliver AL, Pedersen JF, Grant RJ, Klopfenstein TJ, Jose HD (2005). Comparative Effects of the Sorghum bmr-6 and bmr-12 Genes: II. Grain Yield, Stover Yield, and Stover Quality in Grain Sorghum. Crop Sci. 45: 2240-2245.

Pillonel C, Mulder MM, Boon JJ, Forster B, Binder A (1991) Involvement of cinnamyl-alcohol dehydrogenase in the control of lignin formation in Sorghum bicolor L. Moench. Planta, 185:538-544.

Porter KS, Axtell JD, Lechtenberg VL, Colenbrander VF(1978). Phenotype, fiber composition, and in vitro dry matter disappearance of chemically induced brown midrib (bmr) mutants of sorghum. Crop Sci. 18:205-209.

Rook JA, Muller LD, Shank DB (1977). Intake and digestibility of brown midrib corn silage by lactating dairy cows. J. Dairy Sci. 60:1894-1904.

SAS® (2002). User's Guide: Statistics, Version 6, 4th Edition. SAS Inst., Inc., Cary, NC.

Stallings CC, Donaldson BM Thomas JW, Rosman EC (1982). In vivo evaluation of brown midrib corn silage by sheep and lactating dairy cows. J. Dairy Sci. 65:1945-1949.

Tilley JM, Terry AR (1963). A two stage technique for the in vitro digestion of forage crops. J. Br. Grassl. Soc. 18: 104-111.

Van Soest PJ, Robertson JB, Lewis BA (1991). Methods for dietry fiber, neutral detergent fiber and non-starch polysaccharides in relation to animal nutrition. J. Dairy Sci.74:3583-3597.

Vignols F, Rigau J, Torres MA, Capellades M, Puigdomenech P (1995) The brown-midrib3 (bm3) mutation in maize occurs in the gene encoding caffeic acid O-methyl transferase. Plant Cell 7, 407-416.

Vogler R, Ejeta G, Johnson K, Axtell J (1994). Characterization of a new brown midrib sorghum line. Agron. Abst. p. 124.

Wolf DP, Coors JG, Albrecht KA, Undersander DJ, Carter PR (1993). Forage quality of maize genotypes selected for extreme fiber concentrations. Crop Sci. 33: 1353-1359.

31

Predicting α-amylase yield and malt quality of some sprouting cereals using 2nd order polynomial model

C. Egwim Evans[1] and O. Adenomon Monday[2]

[1]Biochemistry Department, Federal University of Technology, P. M. B. 65, Minna, Niger State, Nigeria.
[2]Department of Mathematics, Statistics and Computer Studies, Federal Polytechnic, Bida, Niger State, Nigeria.

Alpha amylase yield in sprouting Maize, Acha, Rice and Sorghum were studied for 180 h. The result was analyzed using 2nd order polynomial model. The result showed that the rate of α- amylase secretion with growth period is significantly high ($p < 0.05$) and the R^2 for each ranged from 67 - 90%, while the R^2 for sprouting vigour ranged within 99% for all the cereals studied. The prediction for amylase activity from sprouting vigour was significant ($p < 0.05$) for all the cereals studied, the R^2 for all the cereals ranged between 63 - 91%. The results conclude that α-amylase and malt quality can be predicted in sprouting cereals from the growth vigour.

Key words: Cereals, amylase, growth vigour, model.

INTRODUCTION

Alpha amylase is synthesized during cereal development and stored in matured endosperms (Evans, et al, 2003). Alpha-amylase, as other amylases increase markedly during germination. It has been shown that alpha amylase yield will peak within 3 - 4 days of cereal germination (Egwim and Oloyede, 2006).

George-Kraemer et al. (2001) have shown that amylase activity is a good predictor of Diastatic Power (DP) which is required in brewing processes and an important characteristic for estimating the quality of malt for beer production (Evans, et al, 1995).

Malting forms a critical stage in the production of cereal-based-beverages in which amylase and proteases inherently embedded in the cereal grain are activated for the purpose of hydrolysis of starch and protein into sugars and amino acids respectively (Okafor, 1987). The determination of alpha-amylase yield in sprouting cereals is a rigorous one involving several stages of chemical reactions and calculations. A quick method of predicting alpha-amylase yield from sprouting cereals is investigated using 2nd order polynomial model.

The goal of prediction is to determine either the value of a new observation of the response variable, or the values of a specified proportion of all future observations of the response variable (www.iti.nist.gov/div898/handbook).

In this paper, this goal will be achieved using polynomial model. Polynomials are particularly important in the experimental sciences since they often give a simple theoretical description of experimental results (Eason et al., 1989).

MATERIALS AND METHODS

Cereals (Maize, Acha, Rice and Sorghum) used were obtained from National Cereal Research Institute (NCRI) Badeggi, Niger State, Nigeria.

The cereals were sprouted for 180 h. Cereal ascrospire was measured with a meter rule while alpha amylase activity was assayed as reported earlier (Egwim and Oloyede, 2006). Briefly, an aliquot (0.1ml) of crude enzyme was pipetted into clean test tube, and 0.9ml of 2% starch solution was added and incubated in a shaking water bath at 50°C for 30 min. The reaction was stopped by adding 3 ml of DNSA reagent and boiled for 3 min for colour development. Absorbance was read at 550 nM against reagent blank. Enzyme activity was thereafter computed from a standard glucose curve (0.1 – 1 mM glucose mL^{-1}). The data mean was analyzed using the 2nd order polynomial regression model.

MODEL SPECIFICATION

The polynomial response model $y = \beta_o + \beta_1 x + \beta_2 x^2 + \beta_3 x^3 + \ldots$ is another example of linear model despite the fact that y is described by non-linear function of the explanatory variable x (Everitt and Dunn,

---
*Corresponding author. E-mail: evanschidi@gmail.com.

1991). A polynomial model may be a 0, 1, 2, 3, . . . etc order, in this present study; a 2nd order polynomial model was adopted given by

$$y = \beta_o + \beta_1 x + \beta_2 x^2 \text{ , where } \beta_2 \neq 0.$$

The method of least squares is used to estimate the model coefficients. The deviations around the regression line (ε- error term) are assumed to be normally and independently distributed with mean of 0 and a standard deviation sigma which does not depend on x (www.iti.nist.gov/div 898/handbook).
The estimate of the model coefficient is obtained from the normal equations:

$$\sum Y = n\hat{\beta}_o + \hat{\beta}_1 \sum X + \hat{\beta}_2 \sum X^2 \qquad \text{. . . (i)}$$

$$\sum XY = \hat{\beta}_o \sum X + \hat{\beta}_1 \sum X^2 + \hat{\beta}_2 \sum X^3 \qquad \text{. . . (ii)}$$

$$\sum X^2 Y = \hat{\beta}_0 \sum X^2 + \hat{\beta}_1 \sum X^3 + \hat{\beta}_2 \sum X^4 \qquad \text{. . . (iii)}$$

A matrix was then formed

$$\begin{bmatrix} n & \sum X & \sum X^2 \\ \sum X & \sum X^2 & \sum X^3 \\ \sum X^2 & \sum X^3 & \sum X^4 \end{bmatrix} \begin{bmatrix} \hat{\beta}_o \\ \hat{\beta}_1 \\ \hat{\beta}_2 \end{bmatrix} = \begin{bmatrix} \sum X \\ \sum XY \\ \sum X^2 Y \end{bmatrix}$$

$$let\ A = \begin{bmatrix} n & \sum X & \sum X^2 \\ \sum X & \sum X^2 & \sum X^3 \\ \sum X^2 & \sum X^3 & \sum X^4 \end{bmatrix}$$

The determinant method can be used to estimate the model parameters

$$\hat{\beta}_0 = \frac{\begin{vmatrix} \sum X & \sum X & \sum X^2 \\ \sum XY & \sum X^2 & \sum X^3 \\ \sum X^2 Y & \sum X^3 & \sum X^4 \end{vmatrix}}{|A|} , \quad \hat{\beta}_1 = \frac{\begin{vmatrix} n & \sum X & \sum X^2 \\ \sum X & \sum XY & \sum X^3 \\ \sum X^2 & \sum X^2 Y & \sum X^4 \end{vmatrix}}{|A|}$$

$$\hat{\beta}_2 = \frac{\begin{vmatrix} n & \sum X & \sum X \\ \sum X & \sum X^2 & \sum XY \\ \sum X^2 & \sum X^3 & \sum XY \end{vmatrix}}{|A|}$$

The correlation coefficient can be obtained using the formular below

$$r = \frac{n \sum XY - \sum X \sum Y}{\sqrt{\left(n \sum X^2 - (\sum X)^2\right)\left(n \sum Y^2 - (\sum Y)^2\right)}}$$

While the coefficient of determination is given as $R^2 = (r)^2$ x 100% and the

$$\text{Adjusted } \overline{R}^2 = \left[1 - \left[(1 - R^2)\left(\frac{n-1}{n-k}\right)\right]\right] \text{x } 100\%$$

The t statistic can be used to test whether or not the coefficient is significantly different from 0. The t-statistic is given as:

$$t_c = \frac{\hat{\beta}_i - \beta_i}{S\left(\hat{\beta}_i\right)}$$

The hypothesis of interest
H_o: Coefficient equal 0 verses H_1: Coefficient not equal 0.

We reject H_o if $t_c > t_{1-\alpha/2}^{n-k}$.

Lastly, the F-test is used to test how suitable is the model obtained. It is given as

$$F_c = \frac{R^2/(k-1)}{(1-R^2)/(T-k)}, \quad F_c > F_{0.05,(k-1)(T-k)} \text{ where T is}$$

the number of observation, k is the number of parameter estimated. The model obtained is suitable for forecasting if $F_c > F_{0.05,(k-1)(T-k)}$ (Omotosho, 2000). For this study SPSS software, version 10.0 was employed for polynomial regression analysis.

RESULTS AND DISCUSSION

The summary of cereal acrospire vigour in relation with time is shown in Table 1. The result showed that time was significant to the acrospire vigour; this implies that the growth of acrospire increases significantly with time. The relationship between vigour and time is linear as shown in Figure 1. The degree of relationship between acrospire vigour and time is very strong (correlation (r) is within 0.998). The R^2 for the relationship is within 99% for all the cereals studies, which further reveals that the models can be used to predict future value of acrospire vigour for any known time.
The summary of alpha amylase yield in relation with time is shown in Table 2. The result shows that time was significant to alpha amylase yield for all cereals studied, that is, as time increases, alpha-amylase yield also increase. It further shows that amylase yield can be predicted using 2nd order polynomial model since all (t * * 2) were significant in the models for all the cereals studied, this finding is further expressed in Figure 2. The

Table 1. Summary of Results on the Acrospire Vigour with Respect to Time (Polynomial Regression Model, n = 10).

Cereal Acrospire (mm)	Rate of Increase t	Rate of Change t**2	R^2	R^2(Adjusted)	r
Maize	1.062882	-0.00175	0.99701	0.99651	0.99850
Acha	0.423099	0.000495	0.99241	0.99114	0.99620
Rice	0.575556	0.000101	0.99591	0.99523	0.99795
Sorghum	1.253353	0.000140	0.98584	0.98348	0.99290

Table 2. Summary of Results on the Amylase Activity in sprouting Cereal with Respect to Time (Polynomial Regres-sion Model, n = 10).

Source of Amylase	Rate of Increase t	Rate of Change t**2	R^2	R^2 (Adjusted)	r
Maize-amy	0.002924	-1.54667×10^{-5}	0.87716	0.85668	0.93657
Acha-amy	0.004150	-2.1634×10^{-5}	0.90099	0.88449	0.94921
Rice-amy	0.001142	-7.5078×10^{-6}	0.67693	0.62309	0.82276
Sorghum-amy	0.002374	-1.386×10^{-5}	0.78354	0.74746	0.88518

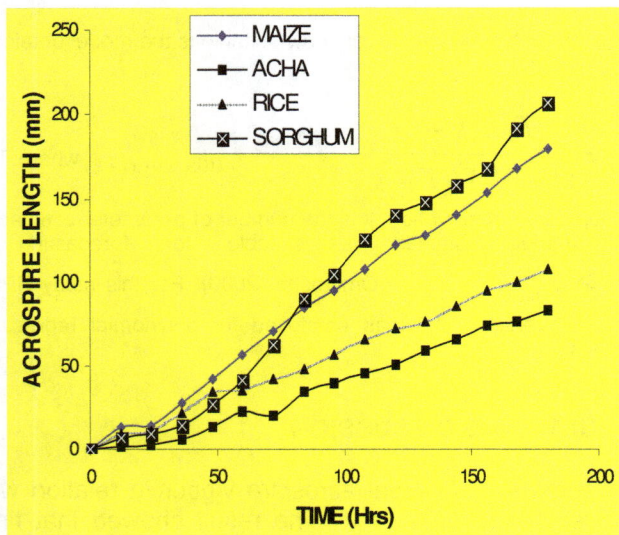

Figure 1. Acrospire vigour with time.
Model: Maize=-6.19086* + 1.0629t** - (0.00175t**2)*.
Acha= -6.802154** + 0.423099t** + (0.00495t**2)*.
Rice= 1.068879* + 0.57556t** + (0.000101t**2)*.
Sorghum= -22.37833** + 1.253353t** + (0.000140t**2)*.
Remark: ** - Significant, * - Not Significant at 5% level of Significance.

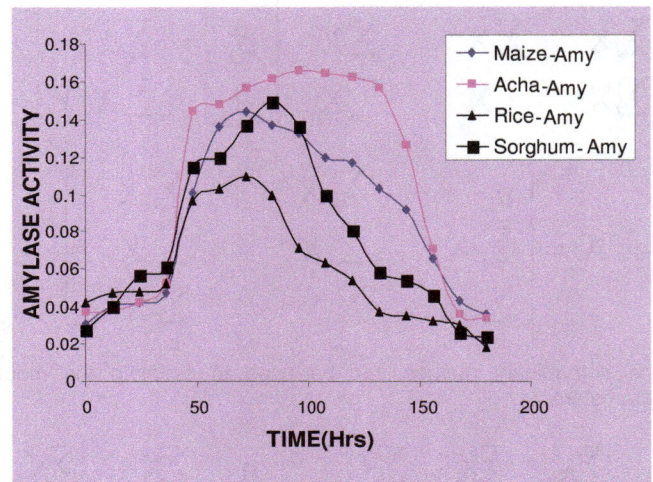

Figure 2. Amylase activity in sprouting cereals with time.
Model
Maize-amy = -0.00622* + 0.002924t** - (1.54667 x 10^{-5}t**2)**.
Acha-amy = -0.03065** + 0.004150t** - (2.1634 x 10^{-5}t**2)**.
Rice-amy = 0.0395* + 0.001142t** - (7.5078 x 10^{-6}t**2)**
Sorghum-amy = 0.017316* + 0.002374t** - (1.38608 x 10^{-5}t**2)**.
Remark: ** - Significant, * - Not Significant at 5% level of Significance.

result also reveals that the degree of relationship between amylase yield and time is very strong (correlation (r) ranges from 0.823 - 0.937). The R^2 for each cereal ranges from 67 - 90%, which further reveals that the models can be used to predict future value of amylase yield for any known time. The implication of this finding is quite interesting because it give predictive information of the yield of alpha amylase with time in sprouting cereal.

The summary of alpha amylase yield in relation with acrospire vigour is shown in Table 3. The result shows that amylase yield is significant with acrospire vigour for all the cereals studied, that is, as acrospire vigour increases, amylase yield also increase. The result further reveals that the degree of relationship between amylase

Table 3. Summary Results on the Amylase Activity in Sprouting With Respect to Acrospire Vigour of the Cereals (Polynomial Regression Model, n = 10).

Source of amylase	Rate of increase t	Rate of change t**2	R^2	R^2 (Adjusted)	r
Maize-amy	0.002751	-1.495×10^{-5}	0.90717	0.89170	0.95246
Acha-amy	0.007155	-8.8226×10^{-5}	0.92649	0.91424	0.96254
Rice-amy	0.001943	-2.1167×10^{-5}	0.68399	0.63132	0.82704
Sorghum-amy	0.001594	-9.1850×10^{-6}	0.75156	0.71016	0.86693

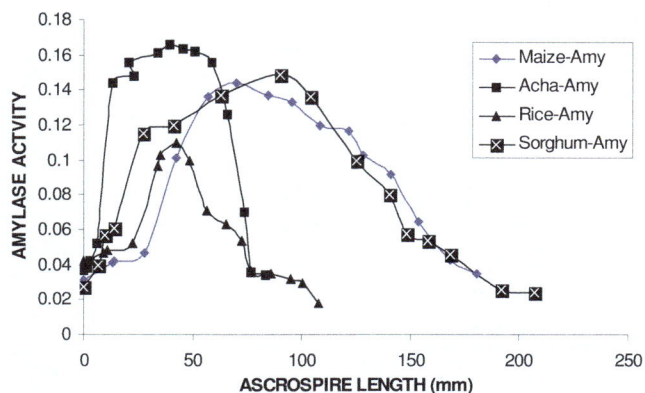

Figure 3. Amylase activity with acrospire length.
Model
Maize-amy = -0.006628* + 0.00275Maize** - (1.49508 x 10⁻⁵Miaze**2)**
Acha-amy = 0.029563** + 0.007155Acha** - (8.82226 x 10⁻⁵Acha**2)**
Rice-amy = 0.038269** + 0.001943Rice** - (2.11668 x 10⁻⁵Rice**2)**
Sorghum-amy = 0.053624** + 0.001594Sorghum** - (9.1850 x 10⁻⁶Sorghum**2)**
Remark: ** - Significant, * - Not Significant at 5% level of Significance

yield and acrospire vigour is very strong (correlation (r) ranges from 0.827 to 0.963). The predictions model for amylase yield from acrospire vigour was significant (p < 0.05) for all the cereals studied, and the result further reveals that 2nd polynomial model is very suitable for modeling amylase yield from acrospire vigour because the model relating alpha amylase yield with acrospire vigour was significant (p < 0.05) for both 1st and 2nd order polynomial models. The 2nd order polynomial model therefore would be a good predictive model. The R^2 for all the cereal studied ranges between 63 - 91% which reveal that the models obtained can be used to predict future value of amylase yield from any known acrospire vigour in the cereal studied (Figure 3).

The present finding suggests that the 2nd order polynomial model is suitable for predicting amylase yield in sprouting cereals. The implication of this finding is that the yield of alpha-amylase is a good indicator of malting quality. This finding agrees with the report of Jin-Xin et al.

(2006) who have shown a strong relationship between amylase activity and malting quality.

The present work therefore conclude that it is possible to predict alpha-amylase yield as well as malting quality of sprouting cereals by measuring the acrospire vigour using a 2nd order polynomial model.

REFERENCES

Eason G, Coles CW, Gettinby G (1989). Mathematics and Statistics for the Bio-Sci. New York: John Wiley & Sons.
Egwim EC, Oloyede OB (2006). Comparison of α-Amylase Yield in Sprouting Nigeria Cereals. Biochem. 18(1): 15-20
Evans DE, Van Weger B, Ma YF, Eghinton J (2003). The Impact of the Thermostability of α-amylase, β-amylase and limit Dextrinase on Potential work Fermentability. J. Am. Society of Brewing Chem. 61: 210-218.
Evans DE, Lance RCM, Elington JK (1995). The Influence of β-Amylase Isofirm Pattern on β-Amylase Activity in Barley and Malt. Proc. 25th Austr. Cer. Chem. Conf. Adelaid. pp 357-364.
Everitt BS, Dunn G (1991). Applied Multivariate Data Analysis. Great Britain: Edward Arnold.
George-Kraemer JE, Mundstock EC, Cawalli-Mohina S (2001). Developmental Expression of Amylase during Barley Malting. J. Cereal Sci. 33: 279-288.
Jin-Xin C, Fei D, Kang W, Guo-ping Z (2006). Relationship Between Malt Qualities and β-amylase Activities and Protein Content as Affected by Timing of Nitrogen fertilizer Application. J. Zhejiang University Sci. B. 7(1): 79-84.
Okafor N (1987). Processing of Nigerian Indigengenous Foods: A Chance of Innovation, Nig. Food J. 1: 32-34.
Omotosho MY (2000). Econometrics "A Practical Approach". Ibadan: Yosode Book publishers.

Appendix:

Amylase activity in sprouting cereals (nM glucose min^{-1})				
Time(h)	Maize-Amy	Acha-Amy	Rice-Amy	Sorghum-Amy
0	0.031	0.037	0.042	0.027
12	0.041	0.038	0.047	0.04
24	0.042	0.042	0.048	0.057
36	0.047	0.052	0.052	0.061
48	0.101	0.144	0.097	0.115
60	0.136	0.148	0.103	0.12
72	0.144	0.156	0.11	0.137
84	0.137	0.161	0.1	0.149
96	0.133	0.166	0.071	0.136
108	0.12	0.164	0.063	0.1
120	0.117	0.162	0.054	0.08
132	0.103	0.156	0.037	0.058
144	0.092	0.126	0.035	0.054
156	0.065	0.07	0.032	0.046
168	0.043	0.036	0.03	0.026
180	0.035	0.034	0.018	0.024

Acrospire vigour				
Time(hrs)	Maize	Acha	Rice	Sorghum
0	0	0	0	0
12	13.1	2	9.3	6.8
24	14.1	2.9	10.9	9.6
36	28.1	6.1	22.2	14
48	42.4	13.2	34.1	27
60	56.7	22.9	35.3	41.1
72	70.5	20.6	42.3	62.7
84	84.7	34.1	48.4	90.1
96	95.5	39.2	56.3	104.1
108	108.1	45.3	65.4	125.2
120	121.9	50.8	72.8	140.2
132	128.1	58.8	76.9	148.1
144	140.8	65.8	86.1	158.2
156	153.9	73.7	94.9	168.4
168	168.09	77.02	100.39	191.64
180	180.45	83.24	107.53	207

Permissions

List of Contributors

Victor N. Enujiugha
Department of Food Science and Technology, Federal University of Technology. P.M.B. 704, Akure, Nigeria

Bikanga Raphaël
Laboratory of Natural Substances and Organometallic synthesis, University of Sciences and Technology of Masuku, P. O. Box 901 Franceville, Gabon

Obame-Engonga Louis-Clément
Laboratory of Natural Substances and Organometallic synthesis, University of Sciences and Technology of Masuku, P. O. Box 901 Franceville, Gabon

Agnaniet Huguette
Laboratory of Natural Substances and Organometallic synthesis, University of Sciences and Technology of Masuku, P. O. Box 901 Franceville, Gabon

Makani Thomas
Laboratory of Natural Substances and Organometallic synthesis, University of Sciences and Technology of Masuku, P. O. Box 901 Franceville, Gabon

Anguile Jean Jacques
Laboratory of Natural Substances and Organometallic synthesis, University of Sciences and Technology of Masuku, P. O. Box 901 Franceville, Gabon

Lebibi Jacques
Laboratory of Natural Substances and Organometallic synthesis, University of Sciences and Technology of Masuku, P. O. Box 901 Franceville, Gabon

Menut Chantal
Team "Glyco and nanovecteurs for therapeutic targeting", Institute of biomolecules Mousseron, Faculty of Pharmacy, 15 avenue Charles Flahault, 34093 Montpellier, France

H. Belguith
Department of Biology, Faculty of Science, Bizerte, Tunisia

S. Fattouch
Biological Engineering Laboratory, INSAT, Tunis, Tunisia

T. Jridi
Department of Biology, Faculty of Science, Bizerte, Tunisia

J. Ben Hamida
Department of Biology, Faculty of Science, Bizerte, Tunisia

Valdinéia Soares Freitas
Department of Biochemistry and Molecular Biology, National Institute of Science and Technology in Salinity (INCTSal), Federal University of Ceará, Fortaleza-CE, Brazil

Nara Lídia Mendes Alencar
Department of Biochemistry and Molecular Biology, National Institute of Science and Technology in Salinity (INCTSal), Federal University of Ceará, Fortaleza-CE, Brazil

Claudivan Feitosa de Lacerda
Departament of Agronomy Engineering, Center of Agricultural Sciences, Federal University of Ceará, Fortaleza-CE, Brazil

José Tarquinio Prisco
Department of Biochemistry and Molecular Biology, National Institute of Science and Technology in Salinity (INCTSal), Federal University of Ceará, Fortaleza-CE, Brazil

Enéas Gomes-Filho
Department of Biochemistry and Molecular Biology, National Institute of Science and Technology in Salinity (INCTSal), Federal University of Ceará, Fortaleza-CE, Brazil

Amir Jalili
Department of Biology, Islamic Azad University, Urmia Branch, Urmia, Iran

Asra Sadeghzade
Department of Biology, Islamic Azad University, Urmia Branch, Urmia, Iran

Edmond Ahipo Dué
Laboratoire de Biochimie et Technologie des Aliments de l'Unité de Formation et de Recherche en Sciences et Technologie des Aliments de l'Université d'Abobo-Adjamé, 02 BP 801 Abidjan 02, Côte d'Ivoire

Hervé César B. L. Zabri
Laboratoire de Chimie Bio-organique de l'Unité de Formation et de Recherche des Sciences Fondamentales Appliquées de l'Université d'Abobo-Adjamé, 02 BP 801 Abidjan 02, Côte d'Ivoire

Jean Parfait E.N. Kouadio
Laboratoire de Biochimie et Technologie des Aliments de l'Unité de Formation et de Recherche en Sciences et Technologie des Aliments de l'Université d'Abobo-Adjamé, 02 BP 801 Abidjan 02, Côte d'Ivoire

Lucien Patrice Kouamé
Laboratoire de Biochimie et Technologie des Aliments de l'Unité de Formation et de Recherche en Sciences et Technologie des Aliments de l'Université d'Abobo-Adjamé, 02 BP 801 Abidjan 02, Côte d'Ivoire

T. Ogunmoyole
Department of Biochemistry, Federal University of Technology, P. M. B 704, Akure, Ondo State, Nigeria

I. J. Kade
Department of Biochemistry, Federal University of Technology, P. M. B 704, Akure, Ondo State, Nigeria

O. D. Johnson
Department of Biochemistry, Federal University of Technology, P. M. B 704, Akure, Ondo State, Nigeria

O. J. Makun
Department of Science Laboratory Technology, Auchi Polytechnic, Auchi, Edo State, Nigeria

A. Mahadeva
Residential Coaching Academy, Babasaheb Bhimrao Ambedkar University, Vidya Vihar, Rae Barely Road, Lucknow - 226 025, India

V. Nagaveni
Department of Crop Physiology, University of Agricultural Sciences, GKVK, Bangalore – 560 065. India

A. Olagunju
Department of Biochemistry Ahmadu Bello University, Zaria, Kaduna, Nigeria

E. Onyike
Department of Biochemistry Ahmadu Bello University, Zaria, Kaduna, Nigeria

A. Muhammad
Department of Biochemistry Ahmadu Bello University, Zaria, Kaduna, Nigeria

S. Aliyu
Department of Biochemistry Ahmadu Bello University, Zaria, Kaduna, Nigeria

A. S. Abdullahi
Department of Biochemistry Ahmadu Bello University, Zaria, Kaduna, Nigeria

L. I. Osumah
Department of Biochemistry, Faculty of Science, Delta State University, P. M. B. 1, Abraka, Delta State, Nigeria

N. J. Tonukari
Department of Biochemistry, Faculty of Science, Delta State University, P. M. B. 1, Abraka, Delta State, Nigeria

A. Laroubi
Laboratory of Animal Physiology unit of Ecophysiology, Cadi-Ayyad University, Faculty of Science Semlalia Marrakech, Morocco

L. Farouk
Laboratory of Animal Physiology unit of Ecophysiology, Cadi-Ayyad University, Faculty of Science Semlalia Marrakech, Morocco

R. Aboufatima
Laboratory of Animal Physiology unit of Ecophysiology, Cadi-Ayyad University, Faculty of Science Semlalia Marrakech, Morocco

A. Benharref
Chemistry Laboratory of the Natural Substances and the Heterocycles, Cadi-Ayyad University, Faculty of Science Semlalia Marrakech, Morocco

A. Bagri
Laboratory of Biochimestry and Neurosciences, Faculty of Sciences and Technology, University Hassan 1er, B.P.: 577, Settat 26000, Morocco

A. Chait
Laboratory of Animal Physiology unit of Ecophysiology, Cadi-Ayyad University, Faculty of Science Semlalia Marrakech, Morocco

Tewolde-Berhan Sarah
Department of Land Resources and Environmental Protection/Department of Crop and Horticultural Sciences, Mekelle University, P.O. Box 231, Mekelle, Ethiopia
Department of Chemistry, Biotechnology and Food Science/Department of Plant and Environmental Sciences, University of Life Sciences, P.O. Box 5003, NO-1432 Aas, Norway

Remberg Siv Fagertun
Department of Chemistry, Biotechnology and Food Science/Department of Plant and Environmental Sciences, University of Life Sciences, P.O. Box 5003, NO-1432 Aas, Norway

Abegaz Kebede
School of Human Nutrition, Food Science and Technology, Hawassa University, P.O. Box 5, Hawassa, Ethiopia

Narvhus Judith
Department of Chemistry, Biotechnology and Food Science/ Department of Plant and Environmental Sciences, University of Life Sciences, P.O. Box 5003, NO-1432 Aas, Norway

Abay Fetien
Department of Land Resources and Environmental Protection/Department of Crop and Horticultural Sciences, Mekelle University, P.O. Box 231, Mekelle, Ethiopia

Wicklund Trude
Department of Chemistry, Biotechnology and Food Science/Department of Plant and Environmental Sciences, University of Life Sciences, P.O. Box 5003, NO-1432 Aas, Norway

R. K. Naresh
Department of Agronomy, Sardar Vallabhbhai Patel University of Agriculture and Technology, Meerut (U. P.), India

Purushottam
Department of Plant pathology & Microbiology, Sardar Vallabhbhai Patel University of Agriculture and Technology, Meerut (U. P.), India

S. P. Singh
Department of Soil Science, Sardar Vallabhbhai Patel University of Agriculture and Technology, Meerut (U. P.), India

Ashish Dwivedi
Department of Agronomy, Sardar Vallabhbhai Patel University of Agriculture and Technology, Meerut (U. P.), India

Vineet Kumar
Department of Soil Science, Sardar Vallabhbhai Patel University of Agriculture and Technology, Meerut (U. P.), India

Awadhesh Kumar Singh
Department of Process and Food Engineering, Punjab Agricultural University, Ludhiana-141004, Punjab, India

Dattatreya Mahadev Kadam
Food Grains and Oilseeds Processing Division, Central Institute of Post-Harvest Engineering and Technology, PO: PAU, Ludhiana-141004, Punjab, India

Mili Saxena
R. P. Singh
Jalal Shakhes
Department of Wood and Paper Science and Technology, Faculty of Forest and Wood Technology, Gorgan University of Agricultural Sciences and Natural Resources, (Postal code: 15339-95911), Gorgan, Iran

Farhad Zeinaly
Department of Wood and Paper Science and Technology, Faculty of Forest and Wood Technology, Gorgan University of Agricultural Sciences and Natural Resources, (Postal code: 15339-95911), Gorgan, Iran

Morteza A. B Marandi
Department of Wood and Paper Science and Technology, Faculty of Forest and Wood Technology, Gorgan University of Agricultural Sciences and Natural Resources, (Postal code: 15339-95911), Gorgan, Iran

Tayebe Saghafi
Department of Forestry Science and Technology, Faculty of Agricultural Sciences and Natural Resources, Tehran University, Karaj, Iran

Nyerhovwo J. Tonukari
Department of Biochemistry, Faculty of Science, Delta State University, P. M. B. 1, Abraka, Delta State, Nigeria

Linda I. Osumah
Department of Biochemistry, Faculty of Science, Delta State University, P. M. B. 1, Abraka, Delta State, Nigeria

M. A. Russo
Department of Agrochemistry, University of Catania, Italy

F. Sambuco
Department of Agrochemistry, University of Catania, Italy

A. Belligno
Department of Agrochemistry, University of Catania, Italy

Gueye Papa El hadji Omar
Molecular Biology Unit, Bacteriology-Virology Laboratory at Aristide Le Dantec Hospital, 30 Pasteur Avenue, BP 7325 Dakar, Senegal

Diallo Mouhamadou
Molecular Biology Unit, Bacteriology-Virology Laboratory at Aristide Le Dantec Hospital, 30 Pasteur Avenue, BP 7325 Dakar, Senegal

Deme Awa Bineta
Molecular Biology Unit, Bacteriology-Virology Laboratory at Aristide Le Dantec Hospital, 30 Pasteur Avenue, BP 7325 Dakar, Senegal

Badiane Aida Sadikh
Molecular Biology Unit, Bacteriology-Virology Laboratory at Aristide Le Dantec Hospital, 30 Pasteur Avenue, BP 7325 Dakar, Senegal
Department of Parasitology and Mycology, University Cheikh Anta Diop, Dakar, Senegal

Dior Diop Mare
Molecular Biology Unit, Bacteriology-Virology Laboratory at Aristide Le Dantec Hospital, 30 Pasteur Avenue, BP 7325 Dakar, Senegal

Ahouidi Ambroise
Molecular Biology Unit, Bacteriology-Virology Laboratory at Aristide Le Dantec Hospital, 30 Pasteur Avenue, BP 7325 Dakar, Senegal

Abdoul Aziz Ndiaye
Department of Parasitology and Mycology, University Cheikh Anta Diop, Dakar, Senegal

Dieng Thérése
Department of Parasitology and Mycology, University Cheikh Anta Diop, Dakar, Senegal

Lutgen Pierre
Iwerliewen Fir Bedreete Volleker, Luxemburg NGO, Luxemburg

Mboup Souleymane
Molecular Biology Unit, Bacteriology-Virology Laboratory at Aristide Le Dantec Hospital, 30 Pasteur Avenue, BP 7325 Dakar, Senegal

Sarr Ousmane
Molecular Biology Unit, Bacteriology-Virology Laboratory at Aristide Le Dantec Hospital, 30 Pasteur Avenue, BP 7325 Dakar, Senegal

Mamounata Diao,
Laboratoire de Biochimie Alimentaire, Enzymologie, Biotechnologie Industrielle et Bioinformatique (BAEBIB), Département de Biochimie et Microbiologie, Université de Ouagadougou, Burkina Faso

Oumou H. Kone
Laboratoire de Biochimie Alimentaire, Enzymologie, Biotechnologie Industrielle et Bioinformatique (BAEBIB), Département de Biochimie et Microbiologie, Université de Ouagadougou, Burkina Faso

Nafissétou Ouedraogo
Laboratoire de Biochimie Alimentaire, Enzymologie, Biotechnologie Industrielle et Bioinformatique (BAEBIB), Département de Biochimie et Microbiologie, Université de Ouagadougou, Burkina Faso

Romaric G. Bayili
Laboratoire de Biochimie Alimentaire, Enzymologie, Biotechnologie Industrielle et Bioinformatique (BAEBIB), Département de Biochimie et Microbiologie, Université de Ouagadougou, Burkina Faso

Imael H. N. Bassole
Laboratoire de Biochimie Alimentaire, Enzymologie, Biotechnologie Industrielle et Bioinformatique (BAEBIB), Département de Biochimie et Microbiologie, Université de Ouagadougou, Burkina Faso

Mamoudou H. Dicko
Laboratoire de Biochimie Alimentaire, Enzymologie, Biotechnologie Industrielle et Bioinformatique (BAEBIB), Département de Biochimie et Microbiologie, Université de Ouagadougou, Burkina Faso

A. O. Ebunlomo
Department of Physiology, Nnamdi Azikiwe University, Nnewi, Anambra State, Nigeria

A. O. Odetola
Department of Physiology, Nnamdi Azikiwe University, Nnewi, Anambra State, Nigeria

O. Bamidele
Department of Physiology, Bowen University, Iwo, Osun State, Nigeria

J. N. Egwurugwu
Department of Physiology, Imo State University, Owerri, Imo State, Nigeria

S. Maduka
Department of Physiology, Nnamdi Azikiwe University, Nnewi, Anambra State, Nigeria

J. Anopue
Department of Physiology, Madonna University, Elele, Rivers State, Nigeria

Preeti Mishra
Department of Botany, Plant Pathology, Tissue Culture and Biotechnology Laboratory, University of Rajasthan, JLN Marg, Jaipur-302004, India

Vidya Patni
Department of Botany, Plant Pathology, Tissue Culture and Biotechnology Laboratory, University of Rajasthan, JLN Marg, Jaipur-302004, India

Naa Ayikailey Adamafio
Department of Biochemistry, University of Ghana, P. O. Box LG 54, Legon, Ghana

Maxwell Sakyiamah
Department of Biochemistry, University of Ghana, P. O. Box LG 54, Legon, Ghana

Josephyne Tettey
Department of Biochemistry, University of Ghana, P. O. Box LG 54, Legon, Ghana

Gloria N. Elemo
Department of Food and Analytical Services, Federal Institute of Industrial Research, Lagos, Nigeria

Babajide O. Elemo
Biochemistry Department, Lagos State University, Lagos, Nigeria

Ochuko L. Erukainure
Department of Food and Analytical Services, Federal Institute of Industrial Research, Lagos, Nigeria

Zhi-Cai Zhang
School of Food Science and Bio-Technology, Jiangsu University, Zhenjiang 212013, P. R. China

Mingxia Chen
School of Food Science and Bio-Technology, Jiangsu University, Zhenjiang 212013, P. R. China

Xin Li
School of Food Science and Bio-Technology, Jiangsu University, Zhenjiang 212013, P. R. China

Wangli Shen
School of Food Science and Bio-Technology, Jiangsu University, Zhenjiang 212013, P. R. China

M. A. N. Benhura
Department of Biochemistry, University of Zimbabwe, P.O. Box Mp 167, Mt Pleasant , Harare, Zimbabwe

C. Chidewe
Department of Biochemistry, University of Zimbabwe, P.O. Box Mp 167, Mt Pleasant , Harare, Zimbabwe

S. A. Adeduntan
Department of Forestry and Wood Technology, Federal University of Technology, P. M. B. 704, Akure, Nigeria

A. S. Oyerinde
Department of Agricultural Engineering, Federal University of Technology, P. M. B. 704, Akure, Nigeria

E. A. Udensi
Department of Food Technology, Bells University of Technology, Ota, Ogun State, Nigeria

N. U. Arisa
Department of Food Science and Technology Abia State University Uturu, Abia State, Nigeria

E. Ikpa
Department of Food Technology, Bells University of Technology, Ota, Ogun State, Nigeria

C. O. Ujowundu
Department of Biochemistry, Federal University of Technology Owerri, Nigeria

F. N. Kalu
Department of Biochemistry, University of Nigeria, Nsukka, Nigeria

E. C. Nwosunjoku
Environmental Health Department, Imo State College of Health Science and Technology, Amigbo, Imo State Nigeria

R. N. Nwaoguikpe
Department of Biochemistry, Federal University of Technology Owerri, Nigeria

R. I. Okechukwu
Department of Biotechnology, Federal University of Technology Owerri, Nigeria

K. O. Igwe
Department of Biochemistry, Federal University of Technology Owerri, Nigeria

G. C. Chinyere
Department of Biochemistry, Abia State University, Uturu, Nigeria

N. A. Obasi
Department of Biochemistry, Abia State University, Uturu, Nigeria

R. K. Vogler
Department of Agronomy, Purdue University, West Lafayette, IN 47907-2054, USA

T. T. Tesso
Department of Agronomy, Kansas State University, Manhattan, KS 66506, USA

K. D. Johnson
Department of Agronomy, Purdue University, West Lafayette, IN 47907-2054, USA

G. Ejeta
Department of Agronomy, Purdue University, West Lafayette, IN 47907-2054, USA

C. Egwim Evans
Biochemistry Department, Federal University of Technology, P. M. B. 65, Minna, Niger State, Nigeria

O. Adenomon Monday
Department of Mathematics, Statistics and Computer Studies, Federal Polytechnic, Bida, Niger State, Nigeria